高等职业教育智能制造专业系列教材
Textbook Series for Intelligent Manufacturing in Higher Vocational Education

工业机器人自动化生产线集成与运维

The Integration, Operation and Maintenance of Industrial Robot Automated Production Lines

主　编　杨　铨　黄　洁
副主编　梁倍源　辛华健　谢　雨　杨万叶
参　编　黄熙彡　张　朋　黄子钊　韦河光　李海桦　张锦成
主　审　曲宏远

Editor-in-Chief	Yang Quan　Huang Jie
Associate Editor-in-Chief	Liang Beiyuan　Xin Huajian　Xie Yu　Yang Wanye
Co-Editor	Huang Xiwen　Zhang Peng　Huang Zizhao
	Wei Heguang　Li Haihua　Zhang Jincheng
Examiner-in-Chief	Qu Hongyuan

机械工业出版社
CHINA MACHINE PRESS

本书是高等职业教育智能制造专业系列教材之一，主要内容包括：认知工业机器人自动化生产线集成工作站、典型工业机器人机床上下料工作站系统的设计及应用、典型工业机器人搬运工作站系统的设计及应用、典型工业机器人弧焊工作站系统的设计及应用、数字化生产线的构架及技术特点。

本书可作为高等职业教育智能制造、工业机器人、电气自动化等专业的教材，也可作为智能制造、工业机器人、电气自动化相关专业技术人员的参考用书。

This book is one of the series of textbooks for intelligent manufacturing in higher vocational education. The main content includes: Understanding of the Integrated Workstations of Industrial Robot Automated Production Lines, Design and Application of a Typical Machine Tool Loading and Unloading Workstation System of the Industrial Robot, Design and Application of a Typical Industrial Robot Handling Workstation System, Design and Application of a Typical Industrial Robot Arc Welding Workstation System, Architecture and Technical Characteristics of Digital Production Lines.

This book can serve as a textbook for higher vocational education majors in intelligent manufacturing, industrial robots, and electrical automation, as well as a reference book for technical personnel in intelligent manufacturing, industrial robots, and electrical automation related fields.

图书在版编目（CIP）数据

工业机器人自动化生产线集成与运维 / 杨铨，黄洁主编 . -- 北京：机械工业出版社，2024.8. --（高等职业教育智能制造专业系列教材）. -- ISBN 978-7-111-76948-4

Ⅰ . TP242.2

中国国家版本馆 CIP 数据核字第 20246N6U22 号

机械工业出版社（北京市百万庄大街 22 号　邮政编码 100037）
策划编辑：王振国　　　　　　责任编辑：王振国
责任校对：梁　园　王　延　　封面设计：陈　沛
责任印制：常天培
北京机工印刷厂有限公司印刷
2025 年 2 月第 1 版第 1 次印刷
184mm×260mm・22 印张・498 千字
标准书号：ISBN 978-7-111-76948-4
定价：69.80 元

电话服务　　　　　　　　网络服务
客服电话：010-88361066　　机　工　官　网：www.cmpbook.com
　　　　　010-88379833　　机　工　官　博：weibo.com/cmp1952
　　　　　010-68326294　　金　　书　　网：www.golden-book.com
封底无防伪标均为盗版　　机工教育服务网：www.cmpedu.com

前 言
Preface

为帮助工业机器人相关专业学习者和兴趣爱好者快速全面地掌握机器人技术技能，培养更多的从事工业机器人技术应用和开发的创新人才，普及工业机器人集成开发与应用的基础知识，我们结合广西机械工业研究院智能制造生产线系统编写了此书。本书获批立项为"十四五"首批广西壮族自治区职业教育规划教材，可以作为工业机器人集成开发与应用的入门教材。本书从项目情景引出项目任务，项目任务先是介绍需要掌握的理论内容，然后安排对应的实训案例，整书内容合理、全面，有良好的系统性和可操作性，突出素质教育，将价值塑造、知识传授和能力培养有机融合。

本书结合智能制造生产线系统实训平台重点讲解机器人集成应用的上下料、搬运、焊接等应用案例，突出工业机器人集成应用的可操作性。全书共5个项目，从整体上认知工业机器人自动化生产线集成工作站，从工艺要求分析和硬件选型、系统设计方案编写、施工图的设计和建模、工作站的系统仿真、程序的编写与调试、技术交底材料的整理与编写等方面来介绍典型工业机器人机床上下料工作站、搬运工作站、弧焊工作站系统的设计及应用，最后总结整条数字化生

We have compiled this course with reference to the intelligent manufacturing production line system of Guangxi Research Institute of Mechanical Industry Co., Ltd. for the purpose of helping learners of and those who are interested in the industrial robot related majors master the robot technology and skill quickly and comprehensively, cultivating more innovative talents engaged in the application and development of the industrial robot technology, and popularizing the basic knowledge of integrated development and application of the industrial robot. This course has been approved as one of the textbooks in the first batch for vocational education plan of Guangxi Zhuang Autonomous Region during the 14th Five-Year Plan period, and can serve as an introductory textbook for the integrated development and application of industrial robots. In this course, program scenarios lead to program tasks, for which we first introduce the theoretical knowledge needed to be mastered, and then arrange corresponding practical training cases. With the reasonable and comprehensive content, this course is systematic and operable, highlights the ideological and quality education, and integrates organically value shaping, knowledge transference, and ability cultivation.

This course focuses on the explanation of integrated application cases of robot, such as loading and unloading, handling, and welding with reference to the intelligent manufacturing production line system, highlighting the operability of integrated applications of industrial robots. In this course, we provide five programs, from which readers can understand the integrated workstations of industrial robot automated production lines from a whole picture, and introduce the design and application of typical machine tool loading

产线的构架及技术特点；同时又兼顾"1+X 工业机器人操作与运维职业技能等级证书"的知识点要求进行教学内容的安排，主要培养学生的综合素质能力。

本书的编写得到了广西玉柴机器股份有限公司和广西机械工业研究院有限责任公司的大力支持和帮助，同时在教材编写中参阅了大量的相关图书和互联网资料，在此向相关人员表示衷心的感谢。

由于作者水平有限，且技术不断发展，书中难免会有错漏和不足之处，恳请读者提出宝贵意见和建议。

编 者

and unloading workstations, handling workstations, and arc welding workstation systems of industrial robot from the following aspects: process requirement analysis and hardware selection, compilation of design schemes for systems, design and modeling of detailed drawings, workstation system simulation, and program writing and commissioning, as well as organization and compilation of technical disclosure materials. Finally, we summarize the architecture and technical characteristics of the entire digital production line. In addition, the teaching content is arranged by taking into account the knowledge requirements of the "1+X Certificate of Vocational Skill Level for Operation and Maintenance of Industrial Robot", with the main aim of cultivating the comprehensive competence of the student.

During the compilation process of this course, we have received great support and assistance from Guangxi Yuchai Machinery Co., Ltd. and Guangxi Research Institute of Mechanical Industry Co., Ltd., referred to a large number of relevant books and data on the internet, and would like to express our sincere gratitude to the relevant persons.

Due to the limited level of the author and the continuous development of technology, it is inevitable that there will be errors, omissions, and shortcomings in the book. We sincerely request readers to provide valuable opinions and suggestions.

Editors

目 录
Contents

前言
Preface

项目 1　认知工业机器人自动化生产线集成工作站 ……………………………………… 1
Program 1　Understanding of the Integrated Workstations of Industrial Robot Automated Production Lines …………………………………………… 1

　任务 1.1　认识工业机器人自动化生产线集成工作站 …………………………………… 3
　Task 1.1　Recognition of the Integrated Workstations of Industrial Robot Automated Production Lines …………………………………………………… 3
　　1.1.1　工业机器人自动化生产线集成工作站概述 …………………………………… 4
　　1.1.1　Overview of Integrated Workstations of Industrial Robot Automated Production Lines ……………………………………………………………… 4
　　1.1.2　生产线集成技术的发展现状及趋势 ……………………………………………… 7
　　1.1.2　Development Status Quo and Trends of Production Line Integration Technology ………………………………………………………………………… 7
　任务 1.2　典型的工业机器人自动化生产线集成工作站 ……………………………… 10
　Task 1.2　Typical Integrated Workstations of Industrial Robot Automated Production Lines …………………………………………………………… 10
　　1.2.1　机床上下料工作站 ……………………………………………………………… 11
　　1.2.1　Machine Tool Loading and Unloading Workstation ……………………… 11
　　1.2.2　搬运工作站 ……………………………………………………………………… 13
　　1.2.2　Handling Workstation ……………………………………………………… 13
　　1.2.3　焊接工作站 ……………………………………………………………………… 15
　　1.2.3　Welding Workstation ………………………………………………………… 15
　任务 1.3　本书的学习方法 …………………………………………………………… 18
　Task 1.3　Learning Methods for This Course …………………………………… 18

项目 2　典型工业机器人机床上下料工作站系统的设计及应用 ·············· 22
Program 2　Design and Application of a Typical Machine Tool Loading and Unloading Workstation System of the Industrial Robot ·············· 22

任务 2.1　工艺要求分析及硬件选型 ·············· 24
Task 2.1　Process Requirement Analysis and Hardware Selection ·············· 24
2.1.1　工艺要求分析 ·············· 26
2.1.1　Process Requirement Analysis ·············· 26
2.1.2　主要硬件选型 ·············· 33
2.1.2　Main Hardware Selection ·············· 33

任务 2.2　设计方案的编写 ·············· 48
Task 2.2　Compilation of Design Schemes ·············· 48
2.2.1　设计方案的结构和要素 ·············· 49
2.2.1　Structure and Elements of Design Schemes ·············· 49
2.2.2　设计方案编写示例 ·············· 49
2.2.2　An Example of Design Scheme Compilation ·············· 49

任务 2.3　施工图的设计及绘制 ·············· 52
Task 2.3　Design and Preparation of Detailed Drawings ·············· 52
2.3.1　设备布局图 ·············· 54
2.3.1　Equipment Layout Diagram ·············· 54
2.3.2　系统框图 ·············· 55
2.3.2　System Chart ·············· 55
2.3.3　电气原理图 ·············· 55
2.3.3　Electrical Schematic Diagram ·············· 55
2.3.4　非标件工程图 ·············· 60
2.3.4　Engineering Drawings of Non-standard Parts ·············· 60

任务 2.4　机器人机床上下料工作站的仿真 ·············· 65
Task 2.4　Simulation of Machine Tool Loading and Unloading Workstation of Robot ·············· 65
2.4.1　仿真环境搭建 ·············· 66
2.4.1　Establishment of Simulation Environment ·············· 66
2.4.2　工作站仿真 ·············· 67
2.4.2　Workstation Simulation ·············· 67

任务 2.5　工作站安装与调试及程序的编写 ·············· 72
Task 2.5　Workstation Installation, Commissioning and Their Programming ·············· 72
2.5.1　工作站安装 ·············· 74
2.5.1　Workstation Installation ·············· 74
2.5.2　PLC 程序编写及交互界面设计 ·············· 75

2.5.2　PLC Programming and Interactive Interface Design ························ 75
　　2.5.3　工作站调试 ·· 79
　　2.5.3　Workstation Commissioning ·· 79
任务 2.6　技术交底材料的整理和编写 ·· 82
Task 2.6　Organization and Compilation of Technical Disclosure Materials ·············· 82
　　2.6.1　主要技术资料 ·· 83
　　2.6.1　Main Technical Documents ··· 83
　　2.6.2　使用说明书的编写 ·· 84
　　2.6.2　Compilation of User Manual ··· 84

项目 3　典型工业机器人搬运工作站系统的设计及应用 ······························ 100
Program 3　Design and Application of a Typical Industrial Robot Handling Workstation System ·· 100

任务 3.1　搬运工作站系统分析及硬件选型 ·· 103
Task 3.1　Handing Workstation System Analysis and Hardware Selection ················ 103
　　3.1.1　搬运工作站系统分析 ·· 105
　　3.1.1　Handling Workstation System Analysis ·· 105
　　3.1.2　主要硬件选型 ·· 106
　　3.1.2　Main Hardware Selection ·· 106
任务 3.2　设计方案的编写 ·· 111
Task 3.2　Compilation of Design Schemes ·· 111
　　3.2.1　设计方案的结构和要素 ·· 112
　　3.2.1　Structure and Elements of Design Schemes ······································ 112
　　3.2.2　设计方案编写示例 ·· 113
　　3.2.2　An Example of Design Scheme Compilation ······································ 113
任务 3.3　施工图的设计及建模 ·· 115
Task 3.3　Design and Modeling of Detailed Drawings ······································ 115
　　3.3.1　设备布局图 ·· 116
　　3.3.1　Equipment Layout Diagram ·· 116
　　3.3.2　系统框图 ·· 117
　　3.3.2　System Chart ·· 117
　　3.3.3　电气原理图 ·· 118
　　3.3.3　Electrical Schematic Diagram ·· 118
　　3.3.4　非标件工程图 ·· 124
　　3.3.4　Engineering Drawings of Non-standard Parts ···································· 124
任务 3.4　机器人搬运工作站的仿真 ·· 128
Task 3.4　Simulation of Robot Handling Workstation ······································ 128

3.4.1	仿真环境搭建	132
3.4.1	Establishment of the Simulation Environment	132
3.4.2	工作站仿真	132
3.4.2	Workstation Simulation	132

任务 3.5　视觉系统的调试 143

Task 3.5　Commissioning of the Visual System 143

3.5.1	视觉概述	144
3.5.1	Visual Overview	144
3.5.2	机器视觉应用	146
3.5.2	Machine Vision Applications	146
3.5.3	PLC 与视觉软件通信	147
3.5.3	Communication between the PLC and Visual Software	147
3.5.4	视觉流程与通信程序设计	151
3.5.4	Visual Process and Communication Program Design	151

任务 3.6　工作站安装与调试及程序的编写 157

Task 3.6　Workstation Installation, Commissioning and Their Programming 157

3.6.1	工作站安装	158
3.6.1	Workstation Installation	158
3.6.2	PLC 程序编写	159
3.6.2	PLC Programming	159
3.6.3	工作站调试	174
3.6.3	Workstation Commissioning	174

任务 3.7　技术交底材料的整理和编写 177

Task 3.7　Organization and Compilation of Technical Disclosure Materials 177

3.7.1	主要技术资料	178
3.7.1	Main Technical Documents	178
3.7.2	操作说明书的编写	179
3.7.2	Compilation of Operating Manual	179

项目 4　典型工业机器人弧焊工作站系统的设计及应用 186

Program 4　Design and Application of a Typical Industrial Robot Arc Welding Workstation System 186

任务 4.1　工艺要求分析及硬件选型 191

Task 4.1　Process Requirement Analysis and Hardware Selection 191

4.1.1	焊接机器人基础知识	193
4.1.1	Basic Knowledge of Welding Robots	193
4.1.2	机器人焊接工艺的制订及硬件选型	198

4.1.2　Formulation of Robot Welding Process and Hardware Selection ·········· 198
　　4.1.3　弧焊机器人的示教编程 ·········· 211
　　4.1.3　Teaching Programming of Arc Welding Robots ·········· 211
任务 4.2　设计方案的编写 ·········· 213
Task 4.2　Compilation of Design Schemes ·········· 213
　　4.2.1　设计方案的结构和要素 ·········· 215
　　4.2.1　Structure and Elements of Design Schemes ·········· 215
　　4.2.2　设计方案编写示例 ·········· 216
　　4.2.2　An Example of Design Scheme Compilation ·········· 216
任务 4.3　施工图的设计及建模 ·········· 232
Task 4.3　Design and Modeling of Detailed Drawings ·········· 232
　　4.3.1　设备布局图 ·········· 233
　　4.3.1　Equipment Layout Diagram ·········· 233
　　4.3.2　系统框图 ·········· 234
　　4.3.2　System Chart ·········· 234
　　4.3.3　电气原理图 ·········· 235
　　4.3.3　Electrical Schematic Diagram ·········· 235
　　4.3.4　机械零件图 ·········· 238
　　4.3.4　Drawing of Mechanical Parts ·········· 238
任务 4.4　机器人弧焊工作站的仿真 ·········· 240
Task 4.4　Simulation of Robot Arc Welding Workstation ·········· 240
　　4.4.1　仿真环境搭建 ·········· 241
　　4.4.1　Establishment of the Simulation Environment ·········· 241
　　4.4.2　工作站仿真 ·········· 242
　　4.4.2　Workstation Simulation ·········· 242
任务 4.5　程序编写、安装与调试 ·········· 246
Task 4.5　Programming, Installation and Commissioning ·········· 246
　　4.5.1　上电开机和操作移动机器人 ·········· 247
　　4.5.1　Startup and Operating of the Robot ·········· 247
　　4.5.2　弧焊参数的选择与设定 ·········· 250
　　4.5.2　Selection and Setting of Arc Welding Parameters ·········· 250
任务 4.6　技术交底材料的整理和编写 ·········· 259
Task 4.6　Organization and Compilation of Technical Disclosure Materials ·········· 259
　　4.6.1　主要技术资料 ·········· 260
　　4.6.1　Main Technical Documents ·········· 260
　　4.6.2　操作说明书的编写 ·········· 261
　　4.6.2　Compilation of Operating Manual ·········· 261

项目 5　数字化生产线的构架及技术特点 ··· 269
Program 5　Architecture and Technical Characteristics of Digital Production Lines ··· 269

任务 5.1　数字化生产线的系统构架 ··· 271
Task 5.1　System Architecture of Digital Production Lines ··· 271

 5.1.1　功能模块 ··· 273
 5.1.1　Function Modules ··· 273
 5.1.2　关键技术与特点 ··· 274
 5.1.2　Key Technologies and Characteristics ··· 274
 5.1.3　工作站调试 ··· 277
 5.1.3　Workstation Commissioning ··· 277

任务 5.2　数字化生产线各模块的设计与仿真 ··· 278
Task 5.2　Design and Simulation of Various Modules of Digital Production Lines ··· 278

 5.2.1　基于 Process Simulation 的仿真建模 ··· 279
 5.2.1　Simulation and Modeling Based on the Process Simulation ··· 279
 5.2.2　基于 Process Simulation 的运动学创建 ··· 282
 5.2.2　Creation of Kinematics Based on the Process Simulation ··· 282
 5.2.3　基于 Process Simulation 的装配仿真 ··· 284
 5.2.3　Assembly Simulation Based on the Process Simulation ··· 284

任务 5.3　识读数字化生产线的设计图样 ··· 289
Task 5.3　Reading of the Design Drawings of Digital Production Lines ··· 289

 5.3.1　识读机械图样 ··· 290
 5.3.1　Reading of Mechanical Drawings ··· 290
 5.3.2　识读电气原理图 ··· 294
 5.3.2　Reading of Electrical Schematic Diagrams ··· 294
 5.3.3　识读其他图样 ··· 297
 5.3.3　Reading of Other Drawings ··· 297

任务 5.4　数字化生产线的操作规范及方法步骤 ··· 300
Task 5.4　Operating Specifications, Methods and Steps for Digital Production Lines ··· 300

 5.4.1　基于 PROFINET 的生产网络 ··· 301
 5.4.1　Production Network Based on the PROFINET ··· 301
 5.4.2　基于触摸屏的操作规范 ··· 302
 5.4.2　Operating Specifications Based on Touch Screens ··· 302
 5.4.3　模块化生产单元调试 ··· 308
 5.4.3　Commissioning of Modular Production Units ··· 308

任务 5.5　典型数字化生产线系统的运行和维护 ································ 324
Task 5.5　Operation and Maintenance of a Typical Digital Production Line System ······ 324
 5.5.1　MES 概述 ··· 325
 5.5.1　Overview of the MES ·· 325
 5.5.2　系统生产计划实施 ·· 327
 5.5.2　Production Plan Implementation of the System ······················ 327
 5.5.3　系统维护 ·· 328
 5.5.3　System Maintenance ··· 328

参考文献 ··· 336
References ·· 336

项目1　认知工业机器人自动化生产线集成工作站

Program 1　Understanding of the Integrated Workstations of Industrial Robot Automated Production Lines

【项目场景】

[Program Scenario]

智能制造生产线系统数字仿真与实际生产虚实结合,满足产品多品种、多规格、个性化制造。智能制造生产线系统配有物流系统、斜轨车床及自动控制同步系统,能够进行工序内容多且复杂的作业,而且能同时完成几项工作任务。智能制造生产线系统如图1-1所示。该系统使用了6台6轴FANUC M-20iA机器人,机器人如图1-2所示。在智能制造生产线系统中,机器人工作站是相对独立的,但又与外界有着密切联系,需要协同周边设备的运行。其在作业内容、周边装置、动力系统方面往往是独立的,但在控制系统、生产管理和物流系统等方面,又与其他工作站及计算机控制处理系统成为一体。

The intelligent manufacturing production line system combines digital simulation with actual production to meet the requirements of multiple varieties, specifications, and customized manufacturing of products, and is equipped with a logistics system, inclined rail lathe, and automated control synchronization system, which enable the perform of tasks with procedures that have multiple and complex contents. The system is shown as Figure 1-1, adopting six sets of 6-axis FANUC M-20iA robot as shown in Figure 1-2. Although the robot workstations are relatively independent in the system, they are closely connected with the outside of the workstations as they need to synergize with the peripheral equipment. They are usually independent in terms of operation content, peripheral devices, and power systems, but integrated with other workstations and computer control processing systems when control systems, production management, and logistics systems are involved.

图1-1　智能制造生产线系统
Figure 1-1　Intelligent Manufacturing Production Line System

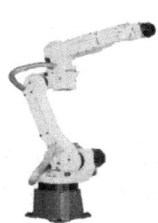

图1-2　FANUC M-20iA机器人
Figure 1-2　FANUC M-20iA Robot

【项目描述】

认知工业机器人自动化智能制造生产线集成工作站，包括机器人及其控制系统、辅助设备和其他周边设备。

【知识目标】

1. 熟悉工业机器人机床上下料、搬运、焊接自动线工作站的构成。
2. 熟悉生产线集成技术的发展现状及趋势。
3. 了解典型的工业机器人自动化生产线集成工作站。
4. 掌握工业机器人的技术参数及选择依据。
5. 熟悉工业机器人工作站外围控制系统的作用。

【技能目标】

1. 能描述智能制造生产线系统的构成。
2. 能描述工业机器人工作站的相关应用场合。
3. 能掌握工业机器人技术参数的意义。
4. 掌握课程教学方法设计。

【"工业机器人操作与运维职业技能等级标准"对中级的相关要求】

2.1.7 能根据工业机器人典型应用（搬运码垛、装配）的任务要求，编写工业机器人程序。

2.3.1 能根据工业机器人典型应用（搬运码垛、装配）的任务要求，创建相应的触摸屏工程。

[Program Description]

Recognize the integrated workstations of an industrial robot automation intelligent manufacturing production line, including robots and their control systems, auxiliary equipment, and other peripheral equipment.

[Knowledge Objectives]

1. To be familiar with the composition of the machine tool loading and unloading, handling, and automated welding line workstations of the industrial robot.
2. To be familiar with the development status quo and trends of production line integration technology.
3. To understand typical integrated workstations of an industrial robot automated production lines.
4. To master the technical parameters and selection basis of industrial robots.
5. To be familiar with the functions of the peripheral control system of industrial robot workstations.

[Skill Objectives]

1. To be able to describe the composition of the intelligent manufacturing production line system.
2. To be able to describe the relevant application scenarios of industrial robot workstations.
3. To master the significance of the technical parameters of industrial robots.
4. To master the design of teaching methods for the course.

[Relevant Requirements for Intermediate Level in the "Standard of Vocational Skill Level for Operation and Maintenance of Industrial Robot"]

2.1.7 To be able to write industrial robot programs based on the task requirements of typical applications (handling, stacking, and assembly) of industrial robots.

2.3.1 To be able to create corresponding touch screen engineering based on the task requirements of typical applications (handling, stacking, assembly) of industrial robots.

2.3.2 能完成触摸屏组态画面制作、报警信息显示、状态信息显示、变量连接、程序加密保护等。

2.3.2 To be able to complete the preparation of touchscreen configuration graphics, alarm information display, status information display, variable connection, program encryption, etc.

任务1.1　认识工业机器人自动化生产线集成工作站
Task 1.1　Recognition of the Integrated Workstations of Industrial Robot Automated Production Lines

【知识目标】

[Knowledge Objectives]

1. 了解"中国制造2025"战略。
2. 了解生产线集成技术的发展现状及趋势。
3. 了解工业机器人在国内的发展前景。
4. 了解工业机器人自动化生产线集成工作站。

1. To understand the strategy of "Made in China 2025".
2. To understand the development status quo and trends of production line integration technology.
3. To understand the development prospects of industrial robots in China.
4. To understand the integrated workstations of industrial robot automated production lines.

【技能目标】

[Skill Objectives]

1. 能描述工业机器人自动化生产线集成工作站的应用。
2. 能描述生产线集成技术的发展现状。

1. To be able to describe the application of integrated workstations of industrial robot automated production lines.
2. To be able to describe the development status quo of production line integration technology.

【素质目标】

[Competence Objectives]

1. 培养学生使其具有一定的全局观念，以及信息收集和处理能力，分析、解决问题能力，交流、合作能力。
2. 引导学生树立职业理想，增强其家国情怀。

1. To cultivate students to have a proper holistic perspective, and be able to collect and process information, analyze and solve problems, as well as communicate and cooperate with others.
2. To guide students to establish their career aspirations when enhancing their patriotism.

【任务情景】

[Task Scenario]

智能制造生产线系统是一条全自动的组装生产线，一共有6台机器人，分属四大制造环节：OP1、OP2、OP3和OP4（工件加工、加工中心、冲洗中心和组装中心）。每一个制造环节都有

The intelligent manufacturing production line system is a fully automated assembly and production line with a total of 6 robots which belong to four major manufacturing links: OP1, OP2, OP3, and OP4 (workpiece processing, machining center, flushing center, and assembly center). For each manufacturing link, they

一个巨型机械臂，能执行多种不同任务，包括机床上下料、搬运、焊接等。

【任务分析】

学习智能制造生产线的结构及功能，了解工业机器人系统集成典型工作站所需的主要设备。在智能制造生产线系统中，OP2 工位由 2 台 FANUC 机器人、2 台斜轨车床、1 套废料仓库、1 个操作台、1 个显示看板组成，该工位主要完成轴类工件的自动上下料和加工过程。OP3 工位由 6 条 8m 直线滚筒线、2 个 180°转弯滚筒线、4 套托盘阻挡定位机构、1 个显示看板组成，电动机由变频器控制。OP4 工位由 1 台 FANUC 机器人、2 台立式加工中心、1 套工件翻转机构、1 套废料仓库、1 个操作台、1 个显示看板组成。该工位主要完成法兰类工件的自动上下料和加工过程。

【知识准备】

1.1.1 工业机器人自动化生产线集成工作站概述

1. 工业机器人工作站的组成

工业机器人工作站是指使用一台或多台机器人，配有控制系统、辅助装置及周边设备，进行简单生产作业，从而达到完成特定工作任务的生产单元。工业机器人工作站一般由以下部分组成：机器人本体、机器人末端执行器、夹具和变位机、机器人架座、配套及安全装置、动力源、工件储运设备、检查监视控制系统等。OP1 主要是工件的搬运，如图 1-3 所示；OP2 主要是斜轨车床工位，如图 1-4 所示；OP3 主要是滚筒线工位，如图 1-5

have a giant robotic arm that can perform various tasks, including machine tool loading and unloading, handling, and welding.

[Task Analysis]

Learn the structures and functions of intelligent manufacturing production lines, and understand the main equipment required for typical workstations of industrial robot system integration. In the intelligent manufacturing production line system, the OP2 position is composed of 2 FANUC robots, 2 inclined rail lathes, 1 waste store, 1 operation console, and 1 display board. It is mainly for the automated loading and unloading, and processing of shaft workpieces. The OP3 position consists of 6 straight drum lines of 8 meters, 2 drum lines of 180 degree turning, 4 sets of positioning mechanisms with tray blocking, and 1 display board. The motor is controlled by a frequency converter. The OP4 position is composed of 1 FANUC robot, 2 vertical machining centers, 1 workpiece turnover mechanism, 1 waste store, 1 operation console, and 1 display board. It is mainly for the automated loading and unloading, and processing of flange workpieces.

[Assumed Knowledge]

1.1.1 Overview of Integrated Workstations of Industrial Robot Automated Production Lines

1. Composition of industrial robot workstations

Industrial robot workstations refer to production units that use one or more robots, equipped with control systems, auxiliary devices, and peripheral equipment, to carry out simple production operations and complete specific work tasks. They generally consist of the following parts: robot body, end effector of robot, clamp and positioner, robot frame, supporting and safety devices, power supply, equipment for workpiece storage and transportation, inspection and monitoring control system, etc. OP1 is mainly for the handling of workpieces, as shown in Figure 1-3; OP2 is mainly an inclined rail lathe position, as shown in Figure 1-4; OP3 is

所示；OP4主要完成法兰类工件的自动上下料和加工过程，如图1-6所示。

mainly a drum line position, as shown in Figure 1-5, when OP4 is mainly for the automated loading and unloading, and processing of flange workpieces, as shown in Figure 1-6.

图 1-3　OP1 工位

Figure 1-3　OP1 Station

图 1-4　OP2 工位

Figure 1-4　OP2 Station

图 1-5　OP3 工位

Figure 1-5　OP3 Station

图 1-6　OP4 工位

Figure 1-6　OP4 Station

2. 工业机器人工作站的发展

工业机器人是面向工业领域的多关节机械手或多自由度的机器人。工业机器人是自动执行工作的机器装置，是靠自身动力和控制能力来实现各种功能的一种机器。它可以接受人类指挥，也可以按照预先编排的程序运行，现代的工业机器人还可以根据人工智能技术制定的策略行动。

工业机器人工作站在工业生产中

2. Development of industrial robot workstations

Industrial robots are multi-joint robotic arms or robots with multi-degree of freedom oriented towards the industrial sector. As mechanical devices that automatically work, they rely on their own dynamics and control capabilities to achieve various functions. Industrial robots can either be commanded by man, or operate according to pre-programmed programs, and the modern ones can also act based on strategies formulated by artificial intelligence technology.

Industrial robot workstations can replace humans

能代替人做某些单调、频繁和重复的长时间作业，或是危险、恶劣环境下的作业，例如在冲压、压力铸造、热处理、焊接、涂装、塑料制品成形、机械加工和简单装配等工序，以及在原子能工业等部门，完成对人体有害物料的搬运或工艺操作。

工业机器人工作站配上外围辅助装置、辅助设备及输送线物流自动化系统，具有广泛的用途。智能制造自动化生产线物流系统主要由以下几个部分组成：

1) 自动化输送线将产品自动输送，并将产品工装板在各装配工位精确定位，装配完成后能使工装板自动循环；设有电动机过载保护，驱动链与输送链直接啮合，传递平稳，运行可靠。

2) 机器人系统通过机器人在特定工位上准确、快速完成部件的装配，能使生产线工业机器人工作站系统与应用达到较高的自动化程度：机器人可遵照一定的原则相互调整，满足工艺点的节拍要求，与上层管理系统的通信接口。

3) 自动化立体仓储供料系统自动规划和调度装配原料，并将原料及时向装配生产线输送，同时能够实时对库存原料进行统计和监控。

4) 全线主控制系统采用基于现场总线的控制系统，不仅有极高的实时性，更有极高的可靠性。

5) 条码数据采集系统使各种产品制造信息具有规范、准确、实时和可追溯的特点，系统采用高档文件服务器和大容量存储设备，快速采集和管理现场的生产数据。

in industrial production to perform certain monotonous, frequent, and repetitive long-term operations, or work under hazardous and harsh environments such as stamping, pressure casting, heat treatment, welding, coating, plastic product forming, and processes of mechanical processing and simple assembly, or handling or process operations of materials harmful to human health in such sectors as the atomic energy industry.

The industrial robot workstations are widely used when they are equipped with peripheral auxiliary devices and equipment, and an automation system for conveying line logistics. The logistics system of intelligent manufacturing automated production line mainly consists of the following parts:

1) An automated conveying line that automatically conveys products and accurately positions the product fixture plates at each assembly position. After assembly is completed, the fixture plates can be automatically cycled; equipped with motor overload protection device, the driving chain directly engages with the conveying chain to ensure smooth transmission and reliable operation.

2) A robot system that achieves a high degree of automation in the industrial robot workstation system and its application of the production line by accurately and quickly assembling components at specific stations: the robots can adjust to each other according to certain principles, meet the rhythm requirements of the process points, and have communication interfaces with the upper management system.

3) An automated three-dimensional storage supply system that automatically plans and schedules raw materials for assembly, and timely conveys them to the assembly production line. Meanwhile, real-time statistics and monitoring of inventory raw materials are achieved.

4) A main control system of the entire line that adopts a fieldbus based control system, which has not only extremely high real-time performance, but extremely high reliability.

5) A barcode data collection system that enables various product manufacturing information to be normative, accurate, real-time, and traceable. The system uses high-end file servers and large capacity storage devices to quickly collect and manage on-site production data.

6) 产品自动化测试系统测试最终产品性能指标,将不合格产品转入返修线。

7) 生产线监控/调度/管理系统采用管理层、监控层和设备层三级网络对整个生产线进行综合监控、调度及管理,能够接受车间生产计划,自动分配任务,完成自动化生产。

我国工业机器人产业的发展对地区的工业基础和相关科研实力有较高要求。目前,我国工业机器人产业主要集中于东北、京津冀、珠三角和长三角地区。东北地区是我国老工业基地,是最早从事工业机器人生产的地区;京津冀地区因其技术优势,工业机器人产业也有所发展,主要企业覆盖领域包括工业机器人及其自动化生产线、工业机器人集成应用、工业机器人技术咨询等产品和服务。长三角地区是我国汽车制造业和电子制造企业集中地,也是重要的机器人公司集聚地,江苏省有五座城市正在建设机器人产业园,珠三角地区工业机器人企业主要集中在深圳、佛山、东莞、广州和中山。

6) A product automation testing system that tests the performance indicators of the final product and transfers unqualified products to the rework line.

7) A production line monitoring/scheduling/management system that adopts a three-level network of management layer, monitoring layer and equipment layer to comprehensively monitor, schedule and manage the entire production line. It can accept the production plan of the plant, automatically assign tasks to have automated production.

The development of industrial robot industry in China has high requirements for the industrial foundation and related scientific research strength of the regions. At present, China's industrial robot industry is mainly concentrated in the Northeast, Beijing-Tianjin-Hebei Region, Pearl River Delta, and Yangtze River Delta regions. The Northeast region is an old industrial base in China and the earliest region engaged in the production of industrial robots; relying on its technological advantages, the Beijing-Tianjin-Hebei region also witnesses the development of the industrial robot industry, and the main areas covered by enterprises in the region include the products and services of industrial robots and their automated production lines, integrated applications of industrial robots, and industrial robot technology consulting. The Yangtze River Delta region is a concentration area for China's automobile and electronic manufacturing enterprises, as well as an important gathering place for robot companies. There are five cities in Jiangsu Province that are building robot industrial parks, while industrial robot enterprises in the Pearl River Delta region are mainly concentrated in the cities of Shenzhen, Foshan, Dongguan, Guangzhou, and Zhongshan.

1.1.2 生产线集成技术的发展现状及趋势

1. 工业机器人集成的现状

在工业机器人集成中,工业机器人是集成的核心,机器人本体的性能决定了集成的水平。

我国的工业机器人发展时间比较短,在性能、功能以及工艺上都跟国外很多工业机器人有很大的差距,但

1.1.2 Development Status Quo and Trends of Production Line Integration Technology

1. Status quo of the industrial robot integration

In the industrial robot integration, industrial robots are the core of integration, and the performance of their body determines the level of integration.

The industrial robots have been developed in China for a relatively short time, and there is a significant gap in performance, function, and technology

是我国也正在努力追赶,国产工业机器人研发水平显著提升,国家也在大力推进智能制造行业发展。

在工业机器人领域,国内80%的机器人企业都集中在系统集成领域,系统集成主要围绕工业机器人进行整线集成。汽车制造产业是机器人应用体量最大的行业,随着汽车行业增速放缓,冲压、焊装、涂装等系统集成应用逐渐普及。从相关市场数据来看,现阶段国内集成公司规模都不大,年产值不高,面临强大的竞争压力。

现阶段工业机器人集成有如下特点:

(1) 不能大批量生产

随着我国产业结构调整升级的不断深入和国际制造业中心向中国的转移,我国的机器人市场将会进一步加大,市场扩展的速度也会进一步提高。但每个企业的集成需求都是不一样的,机器人集成项目就没有统一的标准,必须根据企业的不同需求进行设计,因此难以大批量生产。

(2) 需要熟悉下游行业的工艺

由于机器人集成是次开发产品,因此从事该领域的人员需要熟悉下游行业的工艺,同时完成重新编程、布放等工作。机器人系统如果针对某一行业进行系统集成,那么集成商必须了解该行业的工艺过程,技术壁垒形成后,集成商可依靠该行业生存,但是要跨行业拓展业务,打破行业势力范围是很难的。所以现阶段国内的系统集成商多数都专注于某个行业。

compared to those of foreign countries. However, the research and development level of Chinese industrial robots has significantly improved, as China is striving to catch up with other countries and vigorously promoting the development of the intelligent manufacturing industry.

In the field of industrial robot, 80% of Chinese robot enterprises focus on the field of system integration, especially the entire line integration with industrial robots as the core. The automobile manufacturing industry demands the largest number of robot applications. As the growth rate of the automotive industry slows down, system integration applications such as stamping, welding, and painting are gradually becoming popular. From relevant market data, it can be seen that the Chinese integration companies do not have either large scales or large annual output values at present, facing huge competitive pressure.

At present, the integration of industrial robot is characterized as follows:

(1) It cannot be produced in large quantities

With the continuous deepening of China's industrial restructuring and upgrading, as well as the transfer of international manufacturing centers to China, the robot market in China will be further expanded at a faster speed. However, it is difficult to produce the industrial robots in large quantities, as the integration demands of different enterprise vary, and there is no unified standard for the items of robot integration, which have to be designed according to the different needs of the enterprises.

(2) It needs to be familiar with downstream industry processes

As the robot integration is a secondarily developed product, it is necessary for those who are engaged in this field to be familiar with the downstream industry processes and complete tasks such as reprogramming and deployment. For a robot system to be integrated for a certain industry, the integrators have to understand the technological process of the industry, and can rely on the industry for survival when technical barriers are formed. However, it is difficult to expand business across industries and break the industries' sphere of influence. Therefore, most of the system integrators in China focus on a certain industry.

2. 工业机器人集成的发展趋势

基于国内机器人行业的多样化，中国机器人产业同时也进入了高质量发展的阶段，针对工业、医疗、服务等领域，系统集成商如雨后春笋般涌现，而各集成商针对自己擅长的领域，不断开发新的先进集成系统。

(1) 系统集成产业的规模和发展

现阶段，汽车行业是国内工业机器人最大的应用市场。随着市场对机器人产品认可度的不断提高，机器人产品应用正从汽车行业向一般工业延伸。一般工业的集成研发将由难进入易的阶段，不同行业的机器人将慢慢被研发并应用到行业中去，系统集成技术将不断完善。

(2) 项目标准化程度

项目标准化程度将持续提高，所以在市场经济的调配下，机器人集成产业面临一番整合；而系统集成产业也将会打破汽车行业一家独大的格局，向其他行业延伸并细化，这就要求系统集成商能够掌握更多的行业工艺以适应行业发展的需要。如果只有机器人本体是标准的，那么整个项目标准化程度仅为30%～50%。现在很多机器人集成商在推动机器人本体加工工艺的标准化，未来机器人集成项目的标准化程度将有望达到75%。

(3) 数字化智能工厂

数字化智能工厂是现代工厂信息化发展的一个新阶段，其核心是数字化。智能化、数字化将贯穿生产的各个环节，降低从设计到生产制造之间的不确定性，缩短产品从设计到生产的转化时间，并且提高产品的可靠性与成功率。机器人集成商向数字化工厂方向发展，将来不仅能做硬件设备的集成，更多的是顶层架构设计和软

2. Development trend of industrial robot integration

Based on the diversification of its domestic robotics industry, China has entered a stage of high-quality development of robotics industry. In various sectors such as industry, healthcare, and services, system integrators have sprung up and are constantly developing new advanced integrated systems for their respective fields of expertise.

(1) The scale and development of the system integration industry

The automotive industry is the largest application market for industrial robots in China nowadays. With its increasing acceptance by the market, the robot products are applied not only in the automotive industry but the general industry. The research and development of integration for general industries will become more and easier, and robots for different industries will gradually be developed and applied to the corresponding industries, with system integration technology continue to be improved.

(2) The standardization level of projects

The standardization level of projects will continue to be improved. Therefore, the robot integration industry is facing a new round of adjustment under the allocation of market economy; for the system integration industry, the dominance of the automotive industry will be broken and other industries will have more applications of the system integration with more specific requirements, and system integrators need to master more industry processes to meet the needs of development of the industry. When only the robot bodies are standardized, the standardization level of the entire project will only be 30%–50%. Nowadays, many robot integrators are promoting the standardization of robot body processing technology, and the standardization level of robot integration projects in the future is expected to reach 75%.

(3) Digital intelligent factory

Digital intelligent factories appear when the information based development of modern factories enters a new stage, with digitalization as its core. Intelligence and digitalization will run through all aspects of production, reducing uncertainty from design to manufacturing, which shortens the conversion time from design to production of products, and improves their reliability and success rate. Robot integrators are developing to-

件方面的集成。

wards digital factories, and in the future, they will not only be able to have integration of hardware equipment, but more integration of top-level architecture design and software.

【课后巩固】

1. 简述工业机器人工作站的定义。
2. 简述工业机器人生产线的组成。工业机器人与工作站的区别在哪里？
3. 简述工业机器人工作站的分类。
4. 机器人输送线物流自动化系统主要由哪几部分组成？

[Consolidation after Class]

1. Briefly describe the definition of industrial robot workstations.
2. Briefly describe the composition of industrial robot production lines. What are the differences between industrial robots and workstations?
3. Briefly describe the classification of industrial robot workstations.
4. What are the main components of the automation system of the logistics of a robot conveying line?

任务 1.2　典型的工业机器人自动化生产线集成工作站
Task 1.2　Typical Integrated Workstations of Industrial Robot Automated Production Lines

【知识目标】

1. 熟悉典型工业机器人工作站的分类。
2. 掌握机器人工作站的应用。

【技能目标】

1. 能够识别不同种类的工业机器人工作站。
2. 能够说明各种工业机器人工作站的应用。

【素质目标】

1. 具有分析与决策能力。
2. 具有发现问题、解决问题的能力。
3. 具有团体协作能力。
4. 具有组织管理能力。

[Knowledge Objectives]

1. To be familiar with the classification of typical industrial robot workstations.
2. To master the applications of robot workstations.

[Skill Objectives]

1. To be able to identify different types of industrial robot workstations.
2. To be able to explain the applications of various industrial robot workstations.

[Competence Objectives]

1. To have the ability to conduct analysis and decision-making.
2. To have the ability to discover and solve problems.
3. To have the ability for teamwork.
4. To have organizational and management abilities.

【任务情景】

智能制造生产线系统以智能制造应用为核心，实现零部件加工、打磨、检测识别、分拣入库等生产工艺环节。认识机器人自动化生产线各工作站的组成和功能。

【任务分析】

智能制造生产线系统集智能仓储物流、工业机器人、数控加工、智能检查等模块为一体，利用物联网、工业以太网实现信息互联，依托 MES 实现数据采集和联控，满足轮毂的定制生产。智能制造生产线系统工作站整体布置如图 1-7 所示。

[Task Scenario]

The intelligent manufacturing production line system is centered around intelligent manufacturing applications to complete production processes such as component processing, polishing, detection and identification, sorting and storing. Understanding of the composition and functions of each workstation in the robot automated production line is required.

[Task Analysis]

The intelligent manufacturing production line system integrates modules such as intelligent warehousing and logistics, industrial robots, CNC machining, intelligent inspection to achieve information interconnection relying on the Internet of Things and industrial Ethernet, and achieve data collection and joint control relying on the MES, meeting the customized production of wheel hubs. The overall layout of the intelligent manufacturing production line system is shown in Figure 1-7.

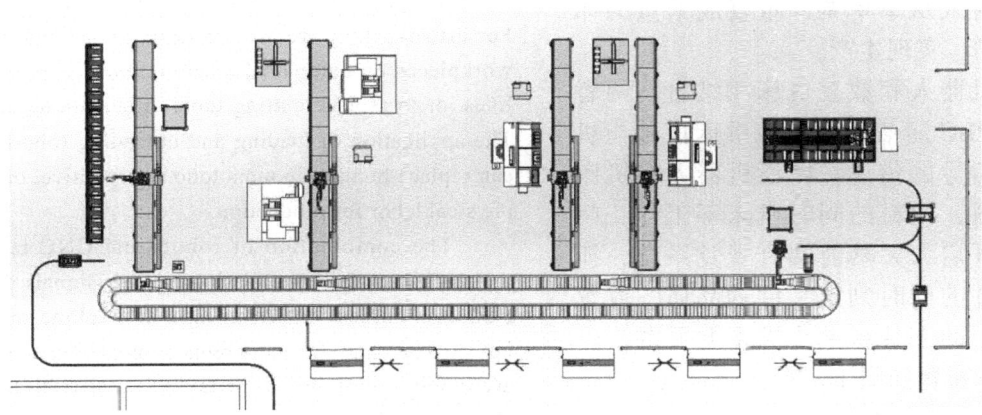

图 1-7 智能制造生产线系统工作站整体布置图

Figure 1-7 Overall Layout of Intelligent Manufacturing Production Line System

【知识准备】

1.2.1 机床上下料工作站

近年来，随着工业自动化的不断发展和用人成本的逐渐增加，许多工厂、数控加工中心对车床上下料机械

[Assumed Knowledge]

1.2.1 Machine Tool Loading and Unloading Workstation

With the continuous development of industrial automation and the gradual increase in labor costs in recent years, many factories and CNC machining

手的需求也越来越大。机械手具有动作快速、灵活、能适应危险和恶劣的环境、重复定位精度高以及可以长时间连续作业等优点,并且采用上下料机械手可减轻工人的劳动量,降低生产成本,提高生产效率。

工业机器人上下料工作站由上料机器人、数控机床、PLC控制柜、输送线等组成。数控机床上下料机械手是一种模拟人手操作的一种自动控制、可重复编程、多功能、多自由度的操作机(固定式的或是移动式的),用于搬运材料、工件、操持工具或检测装置,完成各种作业的自动化设备。

上下料机械手通常用作机床或其他机器的附加装置,如在自动机床或自动化生产线上装卸和传递工件,在加工中心中更换刀具等,一般没有单独的控制装置。应用上下料机械手可以代替人从事单调、重复或繁重的体力劳动,实现生产。

机器人和数控机床相结合,两者通过PLC通信实现数控机床信号与机器人信号的传递,保证机器人在机床内放料、取料的同时机床不工作,门保持开启。夹具前端有视觉系统,抓取工件的同时判定工件的品质,放置在不同的工件台上。机床上下料工作站整体布置如图1-8所示。

(1) 数控(CNC)机床

数控机床的任务是对工件进行加工,而工件的上下料则由工业机器人完成。

centers have an increasing demand for robotic arms for lathe loading and unloading. The robotic arms have the advantages of fast movement, flexibility, adaptability to dangerous and harsh environments, high precision in repeated positioning, and the ability to operate continuously for long periods of time. Moreover, the use of loading and unloading robotic arms can reduce the workload of workers and the production costs when improving production efficiency.

The industrial robot loading and unloading workstations are composed of loading and unloading robots, CNC machine tools, PLC control cabinets, conveying lines, etc. The loading and unloading robotic arms of CNC machine tools are operating machines (fixed or mobile) which simulate human hand operations, with automated control, repeatable programming, multi functions, and multi degree of freedom. They are automated equipment used to transport materials, workpieces, operate tools or detection devices to complete various tasks.

Loading and unloading robotic arms are commonly used as additional devices for machine tools or other machines, and have no separate control device generally. For instance, they are used to load, unload and transfer workpieces on automated machine tools or production lines, or to replace cutting tools in machining centers. The application of loading and unloading robotic arms can replace humans in monotonous, repetitive, or heavy physical labor for production.

The combination of robots and CNC machine tools achieves the transmission of their signals through PLC communication, ensuring that machine tools are not working, and the doors remain open when the robots are placing materials in or taking them from the machine tools. The front end of the fixture has a visual system which enables the determination of the quality of the workpieces when the fixture grasps them, and drops them on different workpiece tables. The overall layout of a machine tool loading and unloading workstation is shown in Figure 1-8.

(1) Computer Numerical Control (CNC) machine tools

The task of CNC machine tools is to process workpieces, while the loading and unloading of workpieces are completed by industrial robots.

图 1-8 机床上下料工作站整体布置图
Figure 1-8 Overall Layout of a Machine Tool Loading and Unloading Workstation

(2) 工业机器人及控制柜

数控机床加工的工件为圆柱体，重量≤1kg，机器人动作范围≤130°，故机床上下料机器人选用的是 FANUC M-20iA 机器人。末端执行器采用气动机械式二指单关节手爪夹持工件，控制手爪动作的电磁阀安装在机器人本体上。

(3) PLC 控制系统

PLC 控制柜用来安装断路器、PLC、开关电源、中间继电器、变压器等设备和组件。

(4) 上下料输送线

上下料输送线的功能是将载有待加工工件的托盘输送到上料工位，机器人将工件搬运至 CNC 机床进行加工，再将加工完成的工件搬运到托盘上，由输送线将加工完成的工件输送到装配工作站进行装配。

1.2.2 搬运工作站

搬运作业是指用一种设备握持工件，从一个加工位置移到另一个加工位置的过程。如果采用工业机器人来完成这个任务，整个搬运系统则构成

(2) Industrial robots and control cabinets

The workpieces processed by the CNC machine tools are cylindrical, with the weights of ≤ 1kg, and the robots are operating with a range of ≤ 130 degree. Therefore, the FANUC M-20iA robots are selected for the loading and unloading of the machine tools. The end effector adopts a pneumatic mechanical two-finger single-joint gripper to grip the workpieces, and the solenoid valve controlling the movement of gripper is installed on the robot body.

(3) PLC control system

The PLC control cabinet is used to install devices and components such as circuit breakers, PLCs, switching power supplies, intermediate relays, transformers.

(4) Loading and unloading conveying line

The function of the loading and unloading conveying line is to transport the pallet carrying the workpieces to be machined to the loading position. The robot transports the workpieces to the CNC machine tool for machining, and then transports the machined workpieces to the pallet. The conveying line transports the machined workpieces to the assembly workstation for assembly.

1.2.2 Handling Workstation

Handling operation refers to the process of using a device to hold a workpiece and move it from one machining position to another. If industrial robots are used to complete this task, the entire han-

了工业机器人搬运工作站。给搬运机器人安装不同类型的末端执行器，可以完成不同形态和状态的工件搬运工作。搬运机器人是可以进行自动化搬运作业的工业机器人。目前世界上使用的搬运机器人逾10万台，被广泛应用于机床上下料、冲压机自动化生产线、自动装配流水线、码垛搬运集装箱等的自动搬运。

搬运机器人工作站是一种集成化的系统，它包括工业机器人、控制器、PLC、机器人手爪、托盘等，并与生产控制系统相连接，以形成一个完整的集成化的搬运系统。搬运工作站整体布置如图1-9所示。

dling system constitutes an industrial robot handling workstation. Different types of end effectors can be installed on the handling robot to carry workpieces in different shapes and states. Handling robots are industrial robots that can perform automated handling operations. There are over 100,000 handling robots in use worldwide at present, and they are widely used for automated handling in loading and unloading of machine tools, automated production lines of stamping machines, automated assembly lines, and the palletizing and handling of containers, etc.

The handling robot workstation is an integrated system, which consists of industrial robot, controller, PLC, robot gripper, and tray, etc., and is connected with the production control system to form a complete integrated handling system. The overall layout of a handling workstation is shown in Figure 1-9.

图1-9　搬运工作站整体布置图
Figure 1-9　Overall Layout of a Handling Workstation

搬运机器人工作站一般具有以下特点：

1）应有物品的传送装置，其形式要根据物品的特点选用或设计。

The handling robot workstation generally has the following characteristics:

1) There shall be conveying devices for the items, and their forms shall be selected or designed according to the characteristics of the items.

2) 可使物品准确地定位，以便于机器人抓取。

3) 多数情况下设有物品托板，或机动或自动地交换托板。

4) 有些物品在传送过程中还要经过整形，以保证码垛质量。

5) 要根据被搬运物品设计专用末端执行器。

6) 应选用适合于搬运作业的机器人。

1.2.3 焊接工作站

焊接机器人是一种高度自动化的焊接设备，采用机器人代替手工焊接作业是焊接制造业的发展趋势，是提高焊接质量、降低成本、改善工作环境的重要手段。采用机器人进行焊接，光有一台机器人是不够的，还必须配备外围设备，即组成工作站系统。焊接机器人工作站系统广泛用于汽车及其零部件制造、摩托车、五金交电、工程机械、航空航天、化工等行业的焊接工程。

焊接机器人工作站系统以焊接机器人为系统核心，控制器、安全防护系统、操作台、回转工作台、变位机、焊接夹具、焊接系统(焊接电源、焊枪、自动送丝机构、水箱)等设备相结合。焊接系统结构合理，操作方便，适合大批量、高效率、高质量、柔性化生产。

焊接机器人是从事焊接的工业机器人。根据国际标准化组织(ISO)，工业机器人属于标准焊接机器人；工业机器人是一种多用途的、可重复编程的自动控制操作机，具有三个或更多可编程的轴，用于工业自动化领域。为了适应不同的用途，机器人最后一个轴的机械接口，通常是一个连接法

2) Items can be accurately positioned for robot grasping.

3) In most cases, there are item pallets, which can be exchanged mechanically or automatically.

4) Some items need to be reshaped during conveying to ensure stacking quality.

5) Specialized end effectors shall be designed based on the items being handled.

6) Robots suitable for handling operations shall be selected.

1.2.3 Welding Workstation

Welding robots are highly automated welding equipment, and it is the development trend in the welding manufacturing to adopt robot welding that replaces manual welding operations and serves as an important means to improve welding quality, reduce costs, and improve the working environment. Only a robot is not enough for robot welding, which requires peripheral equipment to form a workstation system. The welding robot workstation system is widely used for welding in such industries as automobile and its component manufacturing, motorcycles, hardware and electrical engineering, engineering machinery, aerospace, and chemical engineering.

The welding robot is the core of the welding robot workstation system, which combines equipment such as controllers, safety protection systems, operating consoles, rotary workbenches, positioners, welding fixtures, and welding systems (welding power supply, welding guns, automatic wire feeding mechanisms, water tanks). With reasonable structures and convenient operations, the welding system is suitable for large-scale, high-efficiency and quality, and flexible productions.

Welding robots are industrial robots engaged in welding. According to the International Organization for Standardization (ISO), industrial robots are defined as one kind of standard welding robots, and are versatile, reprogrammable automatically controlled operators with three or more programmable axes used in the field of industrial automation. In order to suit different purposes, the mechanical interface of the last axis of the robot is usually a

兰，可接装不同工具或末端执行器。焊接机器人就是在工业机器人末轴法兰处装接焊钳或焊（割）枪，使之能进行焊接、切割或热喷涂。其外形如图1-10所示。

connecting flange, which can be used to connect different tools or end effectors. Welding robots are equipped with welding clamps or welding (cutting) guns at the flange of the last axis of industrial robots, enabling them to perform welding, cutting, or thermal spraying. The appearance is shown in Figure 1-10.

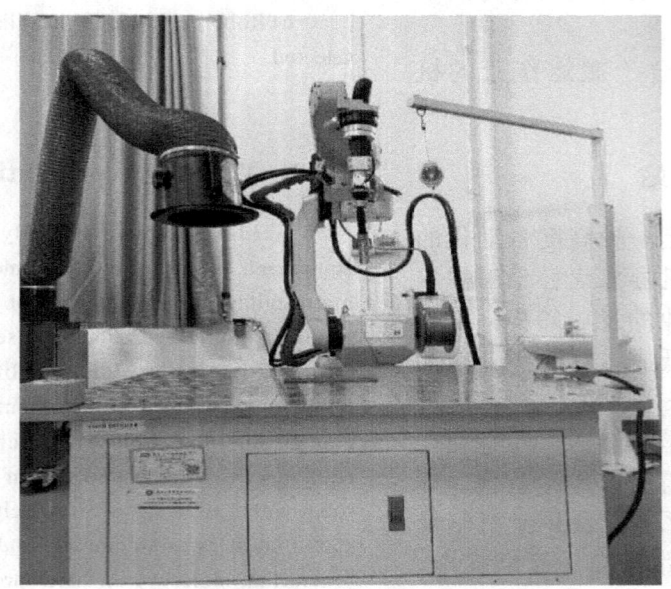

图1-10　焊接工作站整体布置图
Figure 1-10　Overall Layout of a Welding Workstation

焊接机器人可按用途、结构、受控运动方式、驱动方式等来进行分类。

1. 按用途分类

(1) 弧焊机器人

弧焊机器人是包括各种弧焊附属装置在内的柔性焊接系统，而不只是一台以规划的速度和姿态携带焊枪移动的单机，因而对其性能有着特殊的要求。

(2) 点焊机器人

汽车工业是点焊机器人系统一个典型的应用领域，在装配每台汽车车体时，大约60%的焊点是由机器人完成。最初，点焊机器人只用于增强焊作业（向已拼接好的工件上增加焊

Welding robots can be classified according to their purposes, structures, controlled motion modes, and driving modes, etc.

1. Classified by purpase

(1) Arc welding robot

An Arc welding robot is a flexible welding system including various arc welding accessories, rather than a single machine carrying a welding gun at a planned speed and posture. Therefore, special requirements are set for its performance.

(2) Spot welding robot

The automotive industry is a typical application field of spot welding robot systems, where approximately 60% of the welding points are completed by robots when assembling the body of a car. At first, spot welding robots were only used to enhance welding operations (adding welding points to spliced

点），后来为了保证拼接精度，又让机器人完成定位焊接作业性能。

2. 按结构坐标特点分类
1) 直角坐标机器人。
2) 圆柱坐标机器人。
3) 极坐标型机器人。
4) 多关节机器人。

3. 根据受控运动方式分类
(1) 点位 (PTP) 控制型

点位 (PTP) 控制型机器人受控运动方式为自一个点位目标移向另一个点位目标，只在目标点上完成操作。要求机器人在目标点上有足够的定位精度。相邻目标点间的运动方式之一是各关节驱动机以最快的速度趋近终点，各关节视其转角大小不同而到达终点有先有后；另一种运动方式是各关节同时趋近终点，由于各关节运动时间相同，所以角位移大的运动速度较高。点位控制型机器人主要用于点焊作业。

(2) 连续轨迹 (CP) 控制型

连续轨迹 (CP) 控制型机器人各关节同时作受控运动，使机器人终端按预期的轨迹和速度运动，为此各关节控制系统需要实时获取驱动机的角位移和角速度信号。连续控制主要用于弧焊机器人。

【课后巩固】

1. 简述搬运工作站的组成。
2. 简述焊接工作站的组成及分类。
3. 简述机床上下料的组成。

workpieces), but later, robots were used to perform positioning welding operations in order to ensure splicing precision.

2. Classified by structural coordinate characteristics
1) Cartesian coordinate robot.
2) Cylindrical coordinate robot.
3) Polar coordinate robot.
4) Multi-joint robot.

3. Classified by controlled movement mode
(1) Point to Point (PTP) control type

The controlled motion mode of robots of Point to Point (PTP) control type is to move from one point target to another, and only complete operations on the target points. The robots are required to have sufficient positioning precision at the target points. One of the motion modes between adjacent target points is that the joint driving machine approaches the endpoint at the fastest speed, and the joints reach the endpoint one after another depending on their angle sizes. The other motion mode is that all joints approach the endpoint at the same time and, due to the same movement time of all joints, the motion speed of the joint with large angular displacement is higher. Robots of Point to Point control type are mainly used for spot welding operations.

(2) Continuous Path (CP) control type

All joints of a robot of continuous path (CP) control type perform controlled movements simultaneously, making the robot terminal to move according to the expected path and speed. Therefore, the control system of the joints needs to obtain real-time angular displacement and speed signals of the driving machine. Continuous control is mainly used for arc welding robots.

[Consolidation after Class]

1. Briefly describe the composition of the handling workstations.
2. Briefly describe the composition and classification of welding workstations.
3. Briefly describe the composition of machine tool loading and unloading

任务 1.3　本书的学习方法
Task 1.3　Learning Methods for This Course

【知识目标】

掌握课程学习方法，包括硬件的选型、方案的编写以及软件的仿真等。

[Knowledge Objectives]

To master the learning methods of the course, including hardware selection, scheme writing, and software simulation.

【技能目标】

能够说出课程的结构，了解学习任务内容，掌握学习课程方法。

[Skill Objectives]

To be able to describe the structure of the course, understand the content of learning tasks, and master the methods of learning the course.

【素质目标】

1. 具有分析与决策能力。
2. 具有发现问题、解决问题的能力。
3. 具有团体协作能力。
4. 具有组织管理能力。

[Competence Objectives]

1. To have the ability to conduct analysis and decision-making.
2. To have the ability to discover and solve problems.
3. To have the ability for teamwork.
4. To have organizational and management abilities.

【任务情景】

通过课程学习方法的学习以及课程的结构思维导图，掌握课程的结构，更好地理解课程内容，完成学习任务。

[Task Scenario]

Through the study of the learning methods and the structural mind map of the course, one can master the structure of the course, and have better understanding of its course content to complete the learning tasks.

【任务分析】

本书主要介绍智能制造生产线系统的各部分组成以及实现的功能，包括工业机器人机床上下料工作站、搬运工作站、焊接工作站的系统设计与应用。最后进行数字化生产线整体运行与仿真指导，以便由点到面掌握智能制造生产线系统的性能。

[Task Analysis]

This course mainly introduces the components and functions of the intelligent manufacturing production line system, including the system design and application of the industrial robot machine tool loading and unloading workstations, handling workstations, and welding workstations. Finally, the overall operation and simulation of the digital production lines are conducted so that on can master the performance of the intelligent manufacturing production line system gradually and comprehensively.

【知识准备】　　　　　　　　[Assumed Knowledge]

1. 项目学习内容

项目1：认知工业机器人自动化生产线集成工作站

从本书的整体视角出发，将目前国内机器人的发展情况做初步分析，确定了机器人的发展前景。然后讲述了典型工业机器人工作站的组成与应用，概括了智能制造业的现状和未来的发展方向。最后，结合思维导图总结出教学方法和设计方案。

1. Learning contents of the programs

Program 1: Understanding of the Integrated Workstations of Industrial Robot Automated Production Lines

Starting from the overall perspective of this course, a preliminary analysis of the current development of robots in China has been conducted in the course to determine the development prospects of robots. Then, for the composition and application of typical industrial robot workstations, summarize the status quo and direction of future development of the intelligent manufacturing industry. Finally, teaching methods and design schemes are summarized based on the mind map of this course.

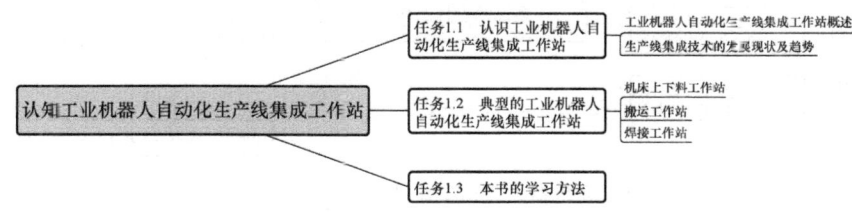

图1-11　认知工业机器人自动化生产线集成工作站知识导图

Figure 1-11　Knowledge Map of Understanding of the Integration Workstations of Industrial Robot Automated Production Line

项目2：典型工业机器人机床上下料工作站系统的设计及应用

该项目包括工艺要求分析及硬件选型、设计方案的编写、施工图的设计及绘制、机器人机床上下料工作站的仿真、工作站安装与调试及程序的编写及技术交底材料的整理和编写。

Program 2: Design and Application of a Typical Machine Tool Loading and Unloading Workstation System of the Industrial Robot

This Program includes process requirements analysis and hardware selection, compilation of design schemes, design and modeling of detailed drawings, simulation of machine tool loading and unloading workstation of robot, workstation, installation, commissioning and their programming, and organization and compilation of technical disclosure materials.

图1-12　典型工业机器人机床上下料工作站系统的设计及应用知识导图

Figure 1-12　Design and Application Knowledge Map of a Typical Machine Tool Loading and Unloading Workstation System of the Industrial Robot

项目3：典型工业机器人搬运工作站系统的设计及应用

该项目包括工艺要求分析及硬件选型、设计方案的编写、施工图的设计及建模、机器人搬运工作站的仿真、视觉系统的调试、工作站安装与调试及程序的编写、技术交底材料的整理和编写等。

Program 3: Design and Application of a Typical Industrial Robot Handling Workstation System

The Program includes process requirement analysis and hardware selection, compilation of design schemes, design and modeling of the detailed drawings, simulation of robot handling workstation, commissioning of the visual system, the writing, installation, commissioning and their programming, and organization and compilation of the technical disclosure materials.

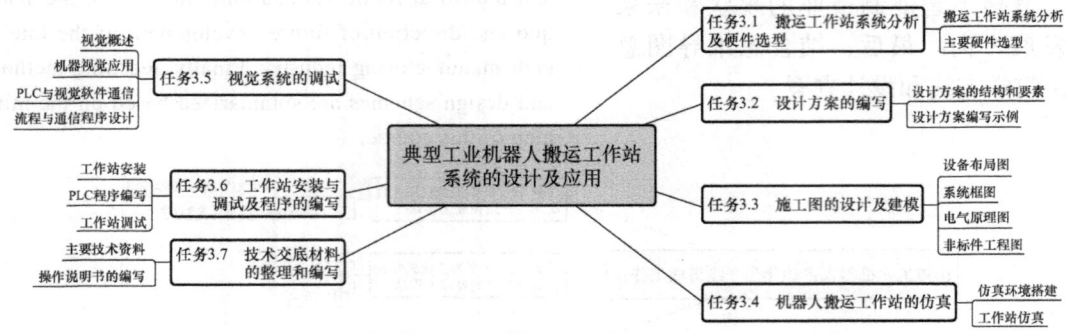

图 1-13　典型工业机器人搬运工作站系统的设计及应用知识导图

Figure 1-13　Design Knowledge Map of a Typical Industrial Robot Handling Workstation System

项目4：典型工业机器人弧焊工作站系统的设计及应用

该项目包括工艺要求分析及硬件选型，设计方案的编写，施工图的设计及建模，机器人弧焊工作站的仿真，程序编写、安装与调试，技术交底材料的整理和编写等。

Program 4: Design and Application of a Typical Industrial Robot Arc Welding Workstation System

This Program includes process requirements analysis and hardware selection, compilation of design schemes, design and modeling of detailed drawings, simulation of robot arc welding workstation, programming, installation and commissioning, and organization and compilation of technical disclosure materials.

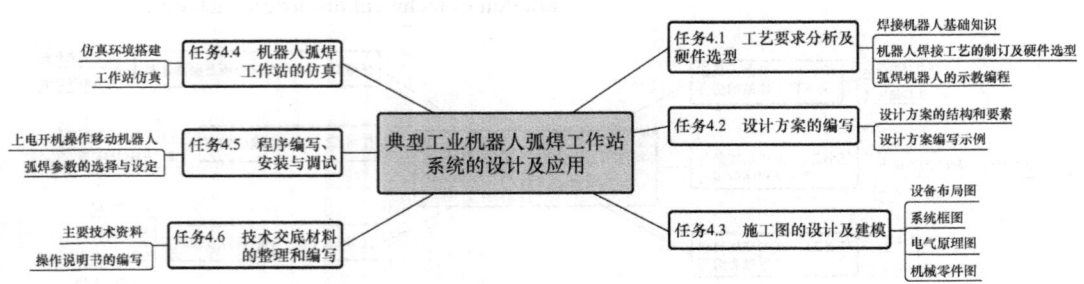

图 1-14　典型工业机器人弧焊工作站系统的设计及应用知识导图

Figure 1-14　Design and Application Knowledge Map of a Typical Industrial Robot Arc Welding Workstation System

项目 1　认知工业机器人自动化生产线集成工作站

项目 5：数字化生产线的构架及技术特点

Program 5: Architecture and Technical Characteristics of Digital Production Lines

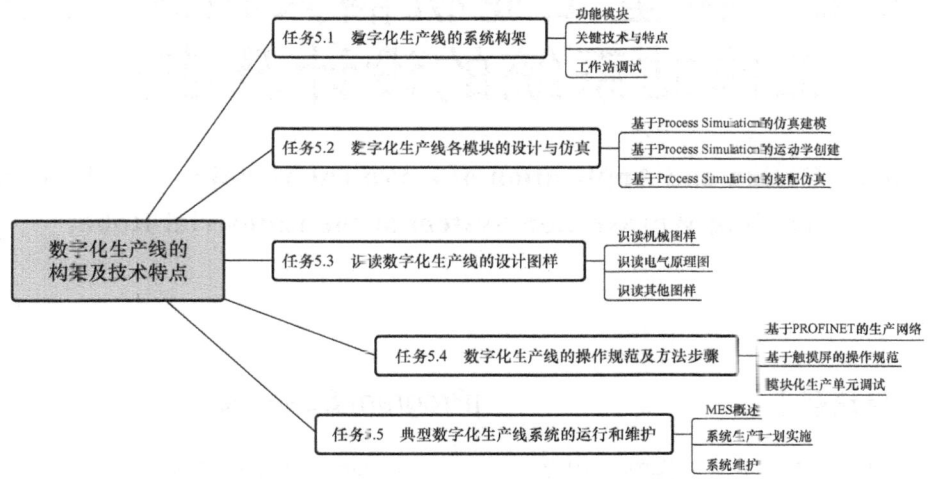

图 1-15　数字化生产线的构架及技术特点知识导图

Figure 1-15　Knowledge Map of the Architecture and Technical Characteristics of Digital Production Lines

2. 课程学习思路

2. Ideas of course learning

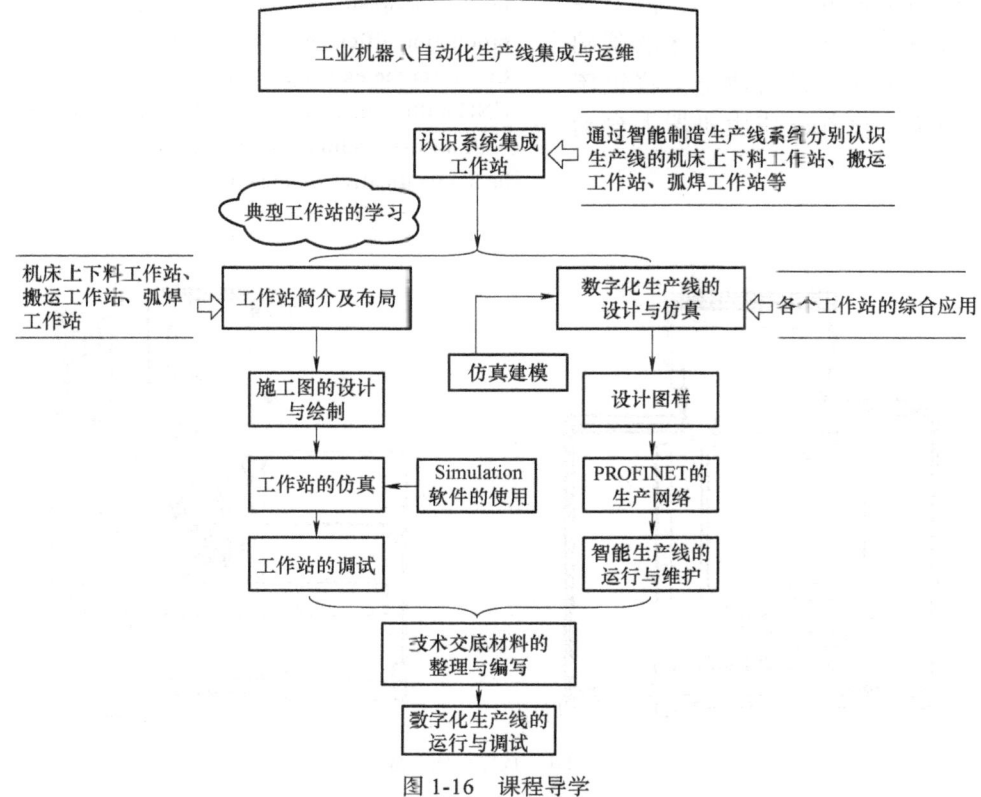

图 1-16　课程导学

Figure 1-16　Guidance of Course Learning

项目 2　典型工业机器人机床上下料工作站系统的设计及应用

Program 2　Design and Application of a Typical Machine Tool Loading and Unloading Workstation System of the Industrial Robot

【项目场景】

某企业有一数控加工工作站，由 2 台数控机床组成，如图 2-1 所示。该工作站需加工 3 种零件，如图 2-2 所示。当前该工作站机床的上下料工作全部由人工完成。为适应当今社会的发展需求，提高企业的生产效率，需要在车间现有数控铣床和数控车床的基础上，将现有工作站进行改造，将机床上下料的工作全部改成由机器人自动完成。

[Program Scenario]

An enterprise has a CNC machining workstation composed of 2 CNC machine tools, as shown in Figure 2-1. This workstation is required to machine three types of parts, as shown in Figure 2-2. At present, all the loading and unloading work of the machine tools at the workstation are done manually. In order to adapt to the development of society nowadays and improve the production efficiency of the enterprise, it is necessary to transform the existing workstation based on the existing CNC milling machines and CNC lathes in the workshop, with all the loading and unloading of the machine tools be automatically completed by robots.

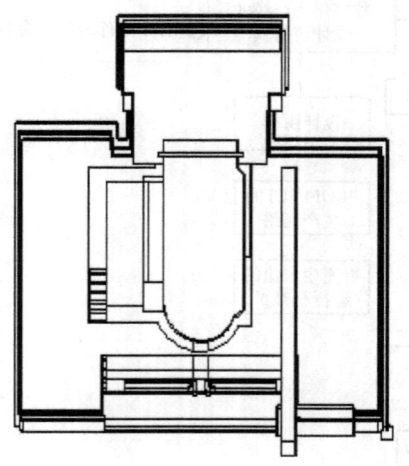 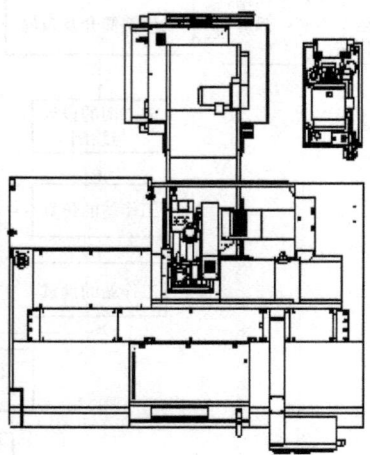

图 2-1　数控机床组成

Figure 2-1　Composition of a CNC Machine Tool

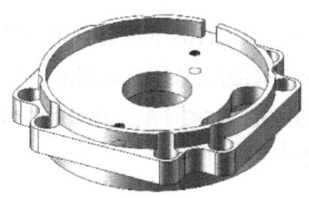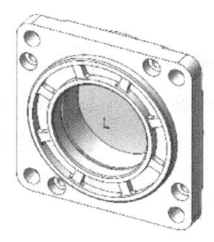

图 2-2　工作站加工零件

Figure 2-2　Parts Machined by the Workstation

【项目描述】

项目主要设备包含 2 台数控加工机床。现需要完成机器人机床上下料系统工作站的设计。本次项目设计的机床上下料工作站的主要功能是实现从触摸屏完成 3 个不同工件加工订单的下单。机床上下料工作站应该能够完成从滚筒线工位取料放入数控机床中进行加工，并将成品从数控机床下料后，放入滚筒线工位的这一自动化流程。

[Program Description]

Two CNC machine tools are included in the main equipment of the program. Now, it is needed to complete the design of the robot machine tool loading and unloading system workstation. The main function of the machine tool loading and unloading workstation designed in this program is to complete the order placement of the machining of 3 different workpieces from the touch screen. It shall be designed to complete automatically the flow of taking materials from the stations of the drum line and putting them into the CNC machine tool for machining, discharging the finished products from the CNC machine tool and then placing them into the stations of the drum line.

【知识目标】

1. 了解设计方案的结构和要素，学习设计方案的基本体例编制。

2. 掌握编写技术交底材料的方法和步骤。

[Knowledge Objectives]

1. To understand the structures and elements of design schemes, and learn the preparation of their basic styles.

2. To master the methods and steps for the compilation of technical disclosure materials.

【技能目标】

1. 学会根据任务要求分析硬件需求，并根据硬件需求进行硬件选型。

2. 依据硬件选型结果和设计方案，进行建模和仿真验证。

3. 能够运用所学知识，综合完成工业机器人机床上下料工作站的集成与调试。

[Skill Objectives]

1. To be able to analyze hardware demands according to task requirements, and conduct hardware selection based on the hardware demands.

2. To conduct modeling and simulation verification based on hardware selection results and design schemes.

3. To be able to complete comprehensively the integration and commissioning of the industrial robot machine tool loading and unloading workstation based on the knowledge learned.

【"工业机器人操作与运维职业技能等级标准"对中级的相关要求】

1.3.1 能对工业机器人的各轴进行归零调试和试运行功能调试。

1.3.2 能对工业机器人进行信号处理调试。

1.3.3 能对工业机器人及周边辅助设备(液压、气动、电气、夹具等)进行联调。

2.1.1 能使用工业机器人运动指令进行基础编程。

2.1.2 能完成工业机器人运动指令参数的设置。

2.1.3 能完成工业机器人手动程序调试。

2.1.4 能熟练应用中断程序,正确触发动作指令。

2.1.5 能通过编程完成对装配物品的定位、夹紧和固定。

2.1.6 能完成工业机器人的典型手动示教操作(矩形轨迹、三角形轨迹、曲线轨迹和圆弧轨迹等)。

2.1.7 能根据工业机器人典型应用(搬运码垛、装配)的任务要求,编写工业机器人程序。

[Relevant Requirements for Intermediate Level in the "Standard of Vocational Skill Level for Operation and Maintenance of Industrial Robot"]

1.3.1 Be able to perform the commissioning of return-to-zero and trial operation on all axis of industrial robots.

1.3.2 Be able to perform the commissioning of signal processing for industrial robots.

1.3.3 Be able to perform joint commissioning of industrial robots and peripheral auxiliary equipment (hydraulic, pneumatic, electrical, fixtures, etc.).

2.1.1 Be able to use motion commands of industrial robots for basic programming.

2.1.2 Be able to complete the setting of motion command parameters of industrial robots.

2.1.3 Be able to complete manual program commissioning of industrial robots.

2.1.4 Be proficient in applying interrupt programs and correctly triggering action instructions.

2.1.5 Be able to complete the positioning, clamping, and fixing of assembled items through programming.

2.1.6 Be able to complete typical manual teaching operations for industrial robots (rectangular, triangular, curve, and circular arc path, etc.).

2.1.7 Be able to write industrial robot programs based on the task requirements of typical applications (handling, stacking, and assembly) of industrial robots.

任务 2.1　工艺要求分析及硬件选型
Task 2.1　Process Requirement Analysis and Hardware Selection

【知识目标】

1. 掌握典型工业机器人机床上下料工作站硬件选型方法和步骤。

2. 掌握机床上下料工作站工艺要求分析方法。

[Knowledge Objectives]

1. To master the methods and steps of hardware selection for typical industrial robot machine tool loading and unloading workstations.

2. To master the analysis method of process requirements for machine tool loading and unloading workstations.

项目2 典型工业机器人机床上下料工作站系统的设计及应用

【技能目标】

1. 根据工艺要求完成标准部件和非标准部件的选型。
2. 能编写机床上下料工作站硬件选型方案。

[Skill Objectives]

1. To complete the selection of standard and non-standard components according to process requirements.
2. To be able to write hardware selection schemes for machine tool loading and unloading workstations.

【素质目标】

1. 树立乐观积极、务实进取的人生态度。
2. 加强专业技术应用能力、沟通协调能力和再学习能力。

[Competence Objectives]

1. To establish an optimistic, positive, pragmatic, and enterprising attitude towards life.
2. To strengthen the ability of professional technical application, communication and coordination, as well as relearning.

【任务情景】

某企业有一数控加工工作站，主要设备包含2台数控机床。该工作站需加工3种零件，当前该工作站机床的上下料工作全部由人工完成。为适应当今社会的发展需求，提高企业的生产效率，需要在车间现有数控铣床和数控车床的基础上，将现有工作站进行改造，将机床上下料的工作全部由机器人自动完成。为实现这一功能，现请你完成机床上下料工作站的工艺要求分析，并完成机床上下料工作站的硬件选型，编写选型报告。

[Task Scenario]

An enterprise has a CNC machining workstation composed of 2 CNC machine tools. This workstation is required to machine three types of parts, and all the loading and unloading work of the machine tools at the workstation are done manually. In order to adapt to the development of society nowadays and improve the production efficiency of the enterprise, it is necessary to transform the existing workstation based on the existing CNC milling machines and CNC lathes in the workshop, with all the loading and unloading of the machine tools be automatically completed by robots. To achieve this function, please complete the analysis of process requirement and the hardware selection for the machine tool loading and unloading workstation, and prepare a report of the hardware selection.

【任务分析】

目标：用机器人给2台数控机床自动上下料，全程无人参与加工。

加工工件：电动机转子、电动机前端盖、电动机后端盖。尺寸如图2-3～图2-5所示。

机床数量：2台数控机床。

工件的加工节拍：180s/件。

完成任务需要根据加工工件的加

[Task Analysis]

Objective: to load and unload automatically 2 CNC machine tools with robots, without humans participating in the machining during the entire process.

Workpieces to be machined: motor rotor, front end cover of motor, and rear end cover of motor, with dimensions shown in Figure 2-3 to 2-5.

Number of machine tools: 2 CNC machine tools.

The machining speed of the workpieces: 180s/piece.

To complete the task, it is necessary to design a

工要求设计工艺流程，根据设计出的工艺流程对整个工作站进行合理的布局和规划，最后对机器人、工控设备、行走轴等硬件设备进行硬件选型，撰写选型报告。

process flow based on the machining requirements of the workpieces, and design a reasonable layout and plan of the entire workstation according to the process flow designed, and finally complete the hardware selection of equipment such as robots, industrial control equipment, and walking axis, and prepare a report of the hardware selection.

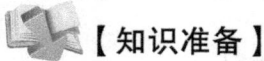
【知识准备】

[Assumed Knowledge]

2.1.1 工艺要求分析

2.1.1 Process Requirement Analysis

本次项目设计的机床上下料工作站的主要功能是实现从触摸屏完成3个不同工件加工订单的下单。机床上下料工作站应该能够完成从传送带取料放入数控机床进行加工，并将成品从数控机床中下料后，放入传送带的这一自动化流程。为这一流程的实现形成了一整套的工序，工序需要多种元件配合实现，因此可以根据要完成的工序选用相应的硬件，并进行电气电路连接以及编程调试。为避免危险的工作环境，可采用仿真软件设计机床上下料工作站，构造虚拟工作环境，进行验证工作。电动机转子（轴）工艺过程卡片见表2-1，工程图如图2-3所示；电动机前端盖工艺过程卡片见表2-2，工程图如图2-4所示；电动机后端盖工艺过程卡片见表2-3，工程图如图2-5所示。工位布局如图2-6所示，具体工艺流程如图2-7所示。

The main function of the machine tool loading and unloading workstation designed in this program is to complete the order placement of the machining of 3 different workpieces from the touch screen. It shall be designed to complete automatically the flow of taking materials from the conveyor belt and putting them into the CNC machine tool for machining, discharging the finished products from the CNC machine tool and then placing them into conveyor belt. To achieve this flow, a complete set of work procedures has been formed, which requires multiple components to cooperate. Therefore, corresponding hardware can be selected based on the work procedures to be completed, and electrical circuit connections and programming commissioning shall be carried out. To avoid dangerous working environments, simulation software can be used to design the machine tool loading and unloading workstation, construct a virtual working environment to perform verification. The technological process card of the motor rotor (shaft) is shown in Table 2-1, and the engineering is shown in Figure 2-3; the technological process card of the front end cover of the motor is shown in Table 2-2, and the engineering is shown in Figure 2-4; the technological process card for the rear end cover of the motor is shown in Table 2-3, and the engineering is shown in Figure 2-5. The design of station layout is shown in Figure 2-6, with the specific process flow shown in Figure 2-7.

项目2　典型工业机器人机床上下料工作站系统的设计及应用

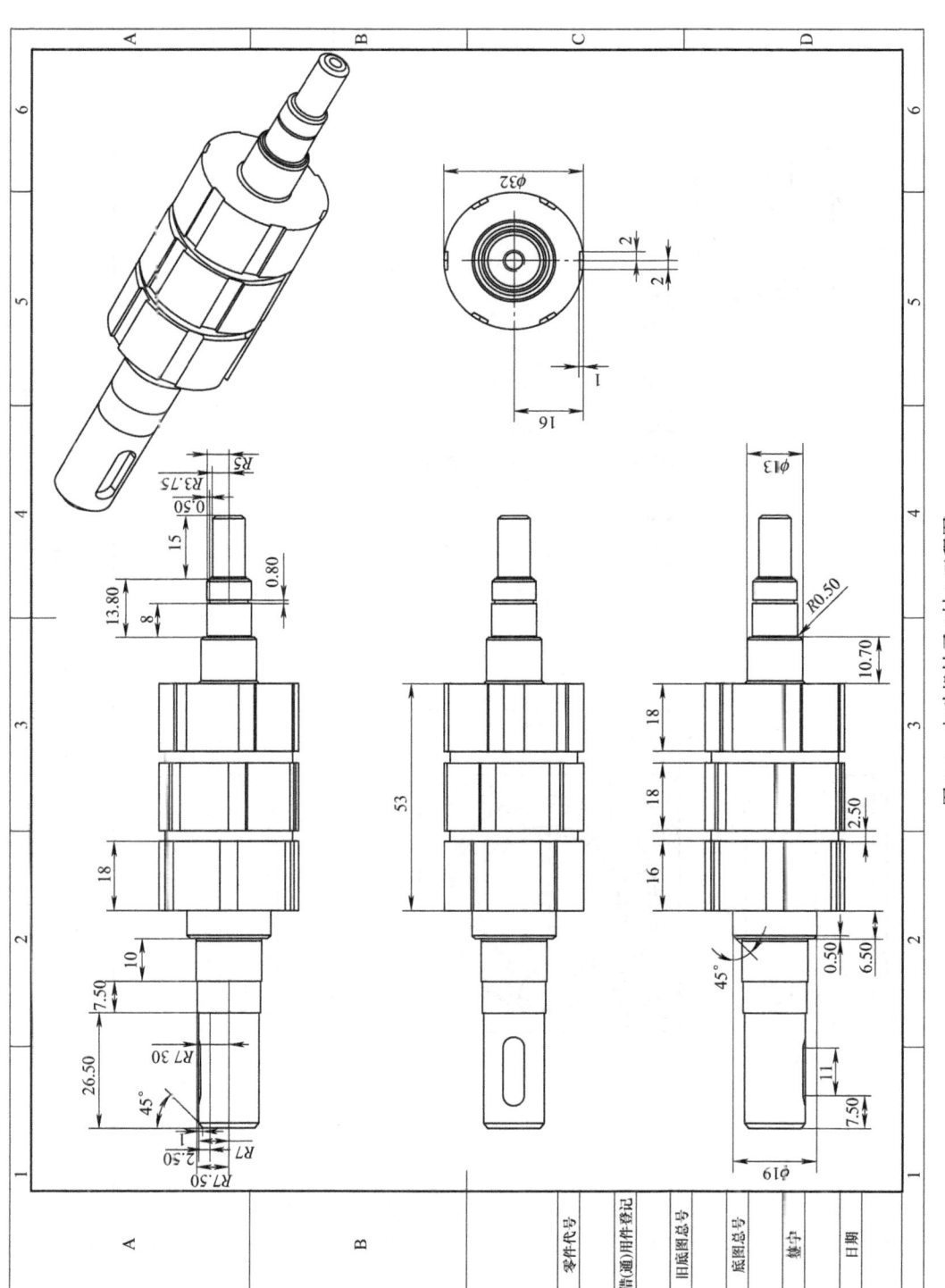

图 2-3　电动机转子（轴）工程图
Figure 2-3　Engineering Drawing of a Motor Rotor (Shaft)

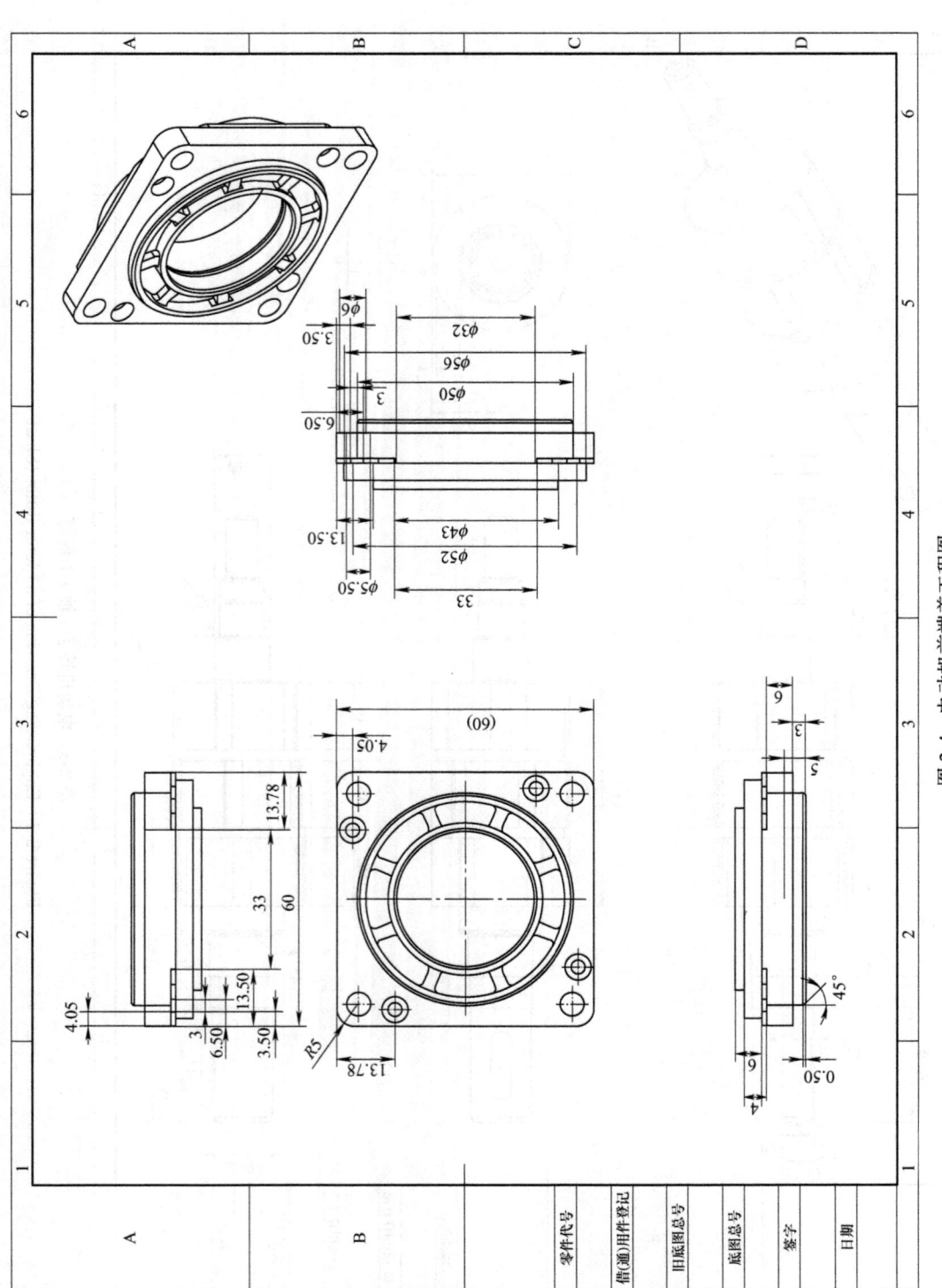

图 2-4　电动机前端盖工程图

Figure 2-4　Engineering Drawing of a Front End Cover of Motor

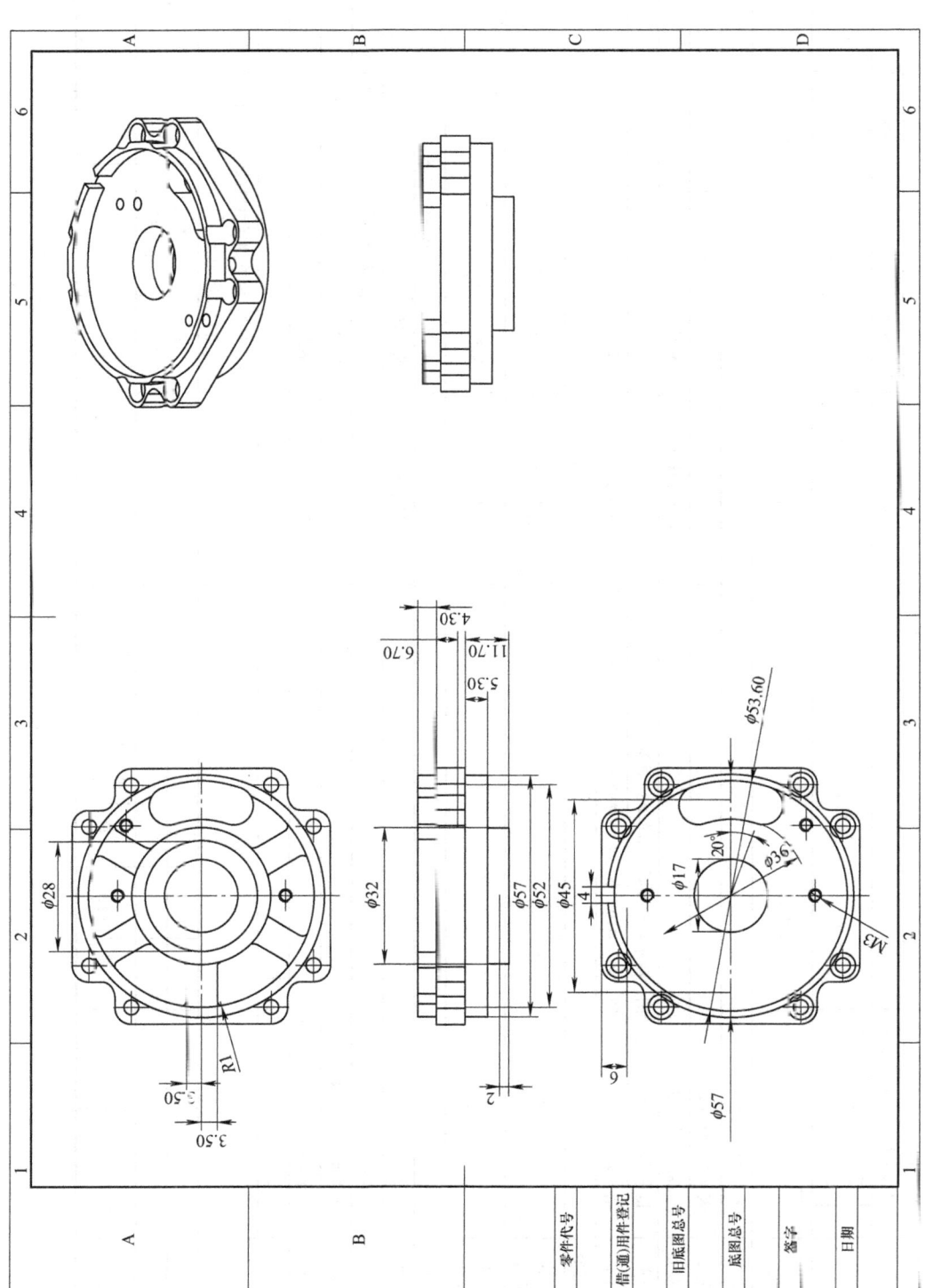

图 2-5 电动机后端盖工程图

Figure 2-5 Engineering Drawing of a Rear End Cover of Motor

表 2-1 电动机转子（轴）工艺过程卡片

Table 2-1 Technological Process Card of a Motor Rotor (Shaft)

材料牌号	机械加工工艺过程卡片		产品型号		零（部）件图号			
			产品名称		零（部）件名称		共（ ）页 第（ ）页	
	毛坯种类	毛坯外形尺寸		$\phi 35 \times 180$	每个毛坯可制件数	电动机轴 1	每台件数 1	备注
序号	工序名称	工序内容			设备	工艺装备		
1	上料	毛坯 $\phi 35 \times 180$			机器人 1	夹 $\phi 35$		
2	装夹	夹 $\phi 35$，伸出 120			斜轨车床 1	自动卡盘（夹持长度 60）		
3	车端面	车端面，深度 3			斜轨车床 1	偏刀		
4	粗车	粗车 $\phi 14$、$\phi 14.6$、$\phi 15$ 外圆			斜轨车床 1	偏刀		
5	精车	精车 $\phi 14$、$\phi 14.6$、$\phi 15$ 外圆，车转子槽			斜轨车床 1	偏刀		
6	铣键槽				斜轨车床 1	铣刀		
7	铣转子槽				斜轨车床 1	铣刀		
8	钻孔攻牙	$M5 \times 25$			斜轨车床 1	钻头、丝锥		
9	下料				机器人 1	夹 $\phi 32$，放置到托盘中转架		
10	上料				机器人 2	夹 $\phi 35$		
11	装夹	夹 $\phi 32$，伸出 70			斜轨车床 2	自动卡盘，需改造定位（夹转子部位）		
12	车端面	车端面，深度 34			斜轨车床 2	偏刀		
13	粗车	粗车 $\phi 7.5$、$\phi 10$、$\phi 13$ 外圆			斜轨车床 2	偏刀		
14	精车	精车 $\phi 7.5$、$\phi 10$、$\phi 13$ 外圆，车槽			斜轨车床 2	偏刀		
15	下料				机器人 2	夹 $\phi 10$，放置到托盘成品架，如果成品架有干涉，则需要进行二次中转		
					设计（日期）	审核（日期）	标准化（日期）	会签（日期）
标记	处数	更改文件号	签字	日期	标记	处数	更改文件号	签字 日期

表 2-2　电动机前端盖工艺过程卡片

Table 2-2　Technological Process Card of a Front End Cover of Motor

机械加工工艺过程卡片		产品型号		零(部)件图号		共( )页	第( )页
		产品名称		零(部)件名称	前端盖		
材料牌号	毛坯种类	毛坯外形尺寸		每个毛坯可制件数	每台件数		备注
铝合金	型材	68×68×25		1	1		
序号	工序名称	工序内容		设备	工艺装备		
1	上料	68×68×25		机器人3	机器人对心抓手V形夹对角(68×68)		
2	装夹	夹两侧底部高度8，上面剩余17		CNC1或CNC2	机床自动对心夹具V形夹对角(68×68)		
3	钻孔	钻内孔φ32～φ20		CNC1或CNC2	φ20钻头		
4	铣顶端面	铣顶端面		CNC1或CNC2	φ16立铣刀		
5	粗铣	粗铣四周至61×61，粗铣φ56外圆，粗铣φ43外圆，粗铣内孔φ32		CNC1或CNC2	φ16立铣刀		
6	铣槽	键铣刀铣槽到零件图尺寸		CNC1或CNC2	φ4键铣刀		
7	精铣	精铣四周至60×60，精铣φ36外圆，精铣φ43外圆，精铣内孔φ32		CNC1或CNC2	φ16立铣刀		
8	铣台阶	换刀，铣台阶到零件图尺寸		CNC1或CNC2	φ5立铣刀，专用夹具		
9	倒角	全部孔和外圆都倒角		CNC1或CNC2	0.5倒角刀		
10	下料			机器人3	机器人对心抓手撑内孔(φ32)		
11	中转			中转架	把半成品翻转180°		
12	上料			机器人3	机器人对心抓手V形夹对角(60×60)		
13	夹紧	夹方轮廓高度6.5，上面剩余9		CNC1或CNC2	机床对心夹具V形夹对角(60×60)		
14	找基准	自动找头找圆心		汉默欧			
15	粗铣端面	粗铣端面		CNC1或CNC2	φ16立铣刀，专用夹具		
16	粗铣	粗铣φ50外圆		CNC1或CNC2	φ16立铣刀，专用夹具		
17	铣槽	键铣刀铣槽到零件图尺寸		CNC1或CNC2	φ5键铣刀，专用夹具		
18	精铣	精铣φ50外圆		UNC1或CNC2	φ16立铣刀，专用夹具		
19	钩槽	换刀，钩内孔槽钩到尺寸		CNC1或CNC2	1.2钩槽刀，专用夹具		
20	钻孔	钻孔φ5.5(四处)，钻孔φ3.4(四处)，沉孔φ6(四处)，深度3.4		CNC1或CNC2	φ5.5钻头、φ3.4钻头、φ6立铣刀		
21	倒角	全部孔和外圆都倒角		CNC1或CNC2	0.5倒角刀，专用夹具		
22	下料			机器人3	机器人对心抓手撑内孔(φ32)		
		设计(日期)	审核(日期)		标准化(日期)		会签(日期)

表 2-3　电动机后端盖工艺过程卡片

Table 2-3　Technological Process Card of a Rear End Cover of Motor

机械加工工艺过程卡片				产品型号		零(部)件图号		共( )页	
材料牌号 铝合金	毛坯种类 型材	毛坯外形尺寸 68×68×25		产品名称		零(部)件名称 后端盖	每个毛坯可制件数 1	第( )页	
序号	工序名称	工序内容			设备		工艺装备	每台件数 1	备注
1	上料	68×68×25			机器人3		机床自动对心夹具V形夹对角(68×68)		
2	装夹	夹两侧底部高度7,上面剩余18			CNC1或CNC2		机床自动对心夹具V形夹对角(68×68)		
3	钻孔	钻内孔 φ17～φ16			CNC1或CNC2		φ16钻头		
4	铣顶端面				CNC1或CNC2		φ16立铣刀		
5	粗铣	粗铣四周至61×61,粗铣φ57外圆,粗铣φ32外圆,粗铣内孔 φ26,粗铣腰孔			CNC1或CNC2		φ16立铣刀		
6	铣槽	键槽铣刀铣槽到零件图尺寸			CNC1或CNC2		φ4键铣刀		
7	精铣	精铣四周至60×60(按图轮廓),精铣腰孔,外圆精铣内孔 φ26,精铣φ56外圆,精铣φ43			CNC1或CNC2		φ16立铣刀		
8	钩槽	钩内孔槽钩到尺寸			CNC1或CNC2		1.2钩槽刀		
9	倒角	全部孔和外圆都倒角			CNC1或CNC2		0.5倒角刀		
10	下料				机器人3		机器人对心抓手撑内孔(φ26)		
11	中转				中转架		把半成品翻转180°		
12	上料				机器人3		机器人对心抓手撑内孔(φ17)		
13	夹紧	夹方轮廓高度6.2,上面剩余8.5			CNC1或CNC2		机床对心夹具V形夹对角(60×60)		
14	找基准	自动测头找圆心			汉默欧				
15	粗铣端面	粗铣端面			CNC1或CNC2		φ16立铣刀		
16	粗铣	粗铣φ57外圆,粗铣φ54内圆			CNC1或CNC2		φ16立铣刀		
17	铣槽	键槽铣刀铣槽到零件图尺寸			CNC1或CNC2		φ4键铣刀		
18	精铣	精铣φ50外圆			CNC1或CNC2		φ16立铣刀		
19	钻孔	钻孔沉头			CNC1或CNC2				
20	倒角	全部孔和外圆都倒角			CNC1或CNC2		0.5倒角刀,专用夹具		
21	下料				机器人3		机器人对心抓手撑内孔(φ17)		
		设计(日期)		审核(日期)		标准化(日期)		会签(日期)	

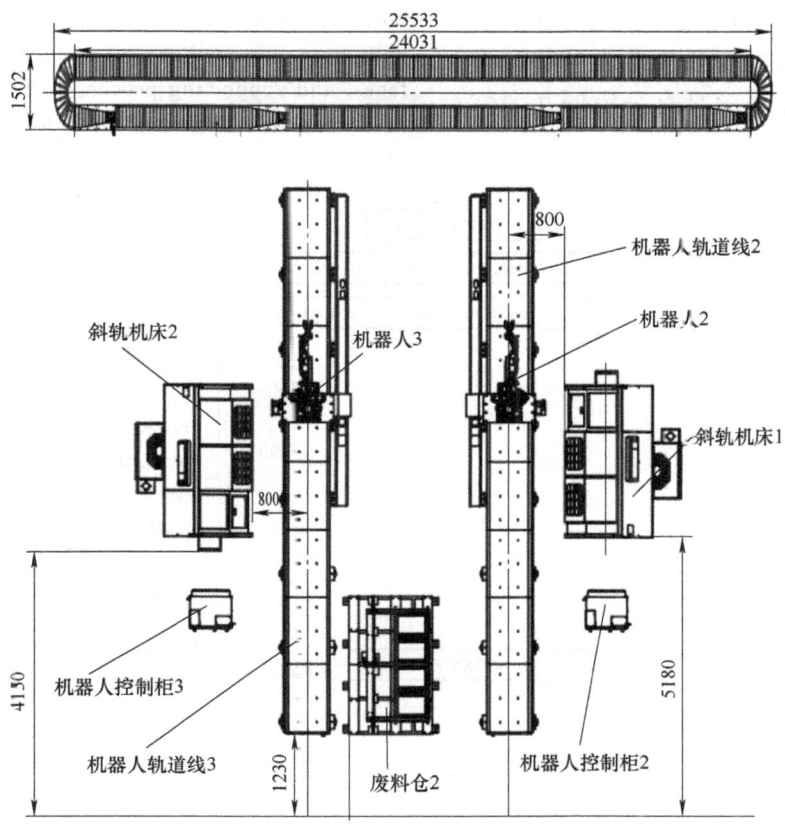

图 2-6 机器人机床上下料工作站工位布局

Figure 2-6　Equipment Layout Diagram of a Machine Tool Loading and Unloading Workstation of the Robot

2.1.2 主要硬件选型

1. 工业机器人选型

工业机器人是本系统的核心设备，其选型尤为重要。首先，由于不同品牌工业机器人的技术特点、擅长领域各不相同，所以首先根据工作任务的工艺要求、项目的预算来确定工业机器人的品牌；其次，根据工作任务、操作对象以及工作环境等因素决定所需工业机器人的负载、最大运动范围、防护等级等性能指标，确定工业机器人的型号；再次，详细考虑如系统先进性、配套工艺软件、I/O 接口、总线通信方式、外部设备配合等问题。在满足工作任务要求的前提下，尽量选

2.1.2　Main Hardware Selection

1. Selection of industrial robots

Industrial robots are the core equipment of this system, and their selection is particularly important. Firstly, due to their differences in the technical characteristics and areas of expertise, industrial robots of different brands shall be determined based first on the process requirements of the work task and the budget of project; secondly, determine the models of the industrial robot based on the performance indicators such as load, maximum range of motion, and protection level of the industrial robot which are determined based on factors such as work tasks, operating objects, and work environment; then, consider such issues in details as system advancement, supporting process software, I/O interface, bus communication mode, and external equipment cooperation. When the requirements of work tasks are met, choose as for as

用控制系统更先进、I/O 接口更多、有配套工艺软件的工业机器人品牌和型号，以利于使系统具有一定的冗余性和扩充性。

possible the brands and models of industrial robot with more advanced control systems, more I/O interfaces, and supporting process software, in order to ensure a certain degree of redundancy and scalability of the system.

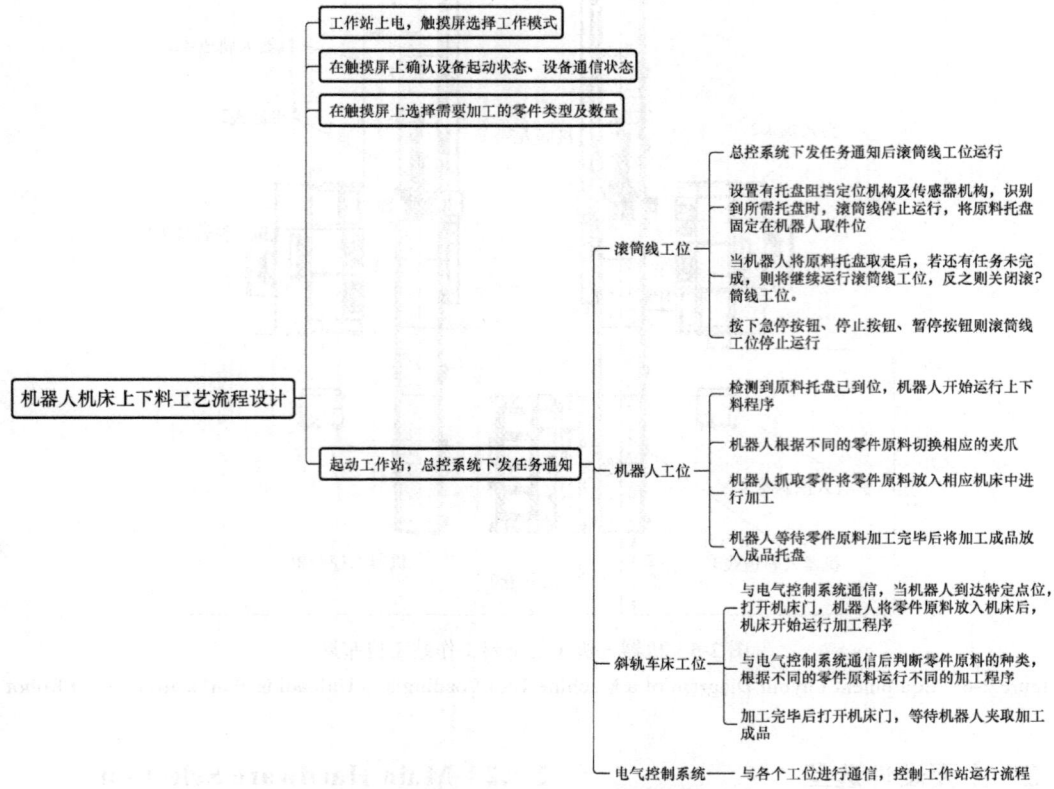

图 2-7　机器人机床上下料工作站工艺流程

Figure 2-7　Process Flow Diagram of a Machine Tool Loading and Unloading Workstation of the Robot

本项目工件毛坯最大质量不超过 5kg，夹具质量不超过 5kg，机器人工作半径要求不小于 1500mm。考虑到本项目中根据工件加工流程、加工机床及设备功能和布局选择提供 2 个工业机器人选择方案，分别为：FANUC 机器人，型号为 M-20iA/12L，负载 12kg，可达半径为 2009mm，如图 2-8 和图 2-9 所示；华数机器人，型号为 HSR-JR620L，负载 20kg，工作范围 1848mm，如图 2-10 和图 2-11 所示。所选的两款机器人都可满足本工作站的工艺要求。

In this program, the maximum weight of the workpiece blank does not exceed 5kg, that of the fixture does not exceed 5kg, and the working radius of the robot shall not be less than 1500mm. Two options of industrial robot are provided based on the workpiece machining process, machine tools, and equipment functions and layout in the program, namely: FANUC Robot, with the model of M-20iA/12L, load of 12kg, and achievable radius of 2009mm, as shown in Figure 2-8 and 2-9; Huashu Robot, with the model of HSR-JR620L, load of 20kg, working range of 1848mm, as shown in Figure 2-10 and 2-11. Robots of both options can meet the process requirements of this workstation.

项目2 典型工业机器人机床上下料工作站系统的设计及应用

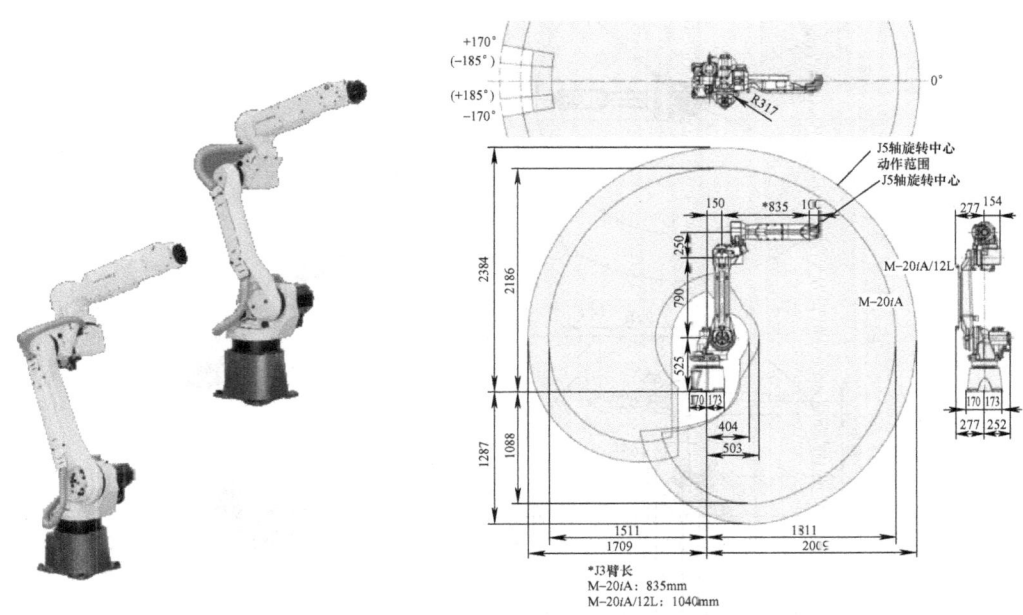

图 2-8　FANUC M-20iA/12L 机器人工作范围
Figure 2-8　Working Range of FANUC Robot M-20iA/12L

项目		规格			
		M-20iA	M-20iA/12L	M-10iA/10M	M-10iA/10MS
控制轴数		6轴（J1、J2、J3、J4、J5、J6）			
可达半径		1811mm	2009mm	1422mm	1101mm
安装方式		地面安装、顶吊安装、倾斜角安装			
动作范围（最高速度）	J1轴旋转	340°/370°（选项）（195°/s）5.93 rad/6.45 rad(选项)(3.40 rad/s)	340°/370°（选项）(225°/s) 5.93 rad/6.45 rad(选项)(3.49 rad/s)	340°/360°（选项）(200°/s) 5.93 rad/6.28 rad(选项)(3.93 rad/s)	340°/360°（选项）(290°/s) 5.93 rad/6.28 rad(选项)(5.06 rad/s)
	J2轴旋转	260°（175°/s）4.54 rad (3.05 rad/s)		250°（205°/s）4.36 rad (3.58 rad/s)	250°（280°/s）4.36 rad (4.89 rad/s)
	J3轴旋转	458°（180°/s）8.00 rad (3.14 rad/s)	460°（190°/s）8.04 rad (3.32 rad/s)	445°（225°/s）7.76 rad (3.93 rad/s)	341°（315°/s）5.95 rad (5.50 rad/s)
	J4轴手腕旋转	400°（360°/s）6.98 rad (6.28 rad/s)	400°（430°/s）6.98 rad (7.50 rad/s)	380°（420°/s）6.63 rad (7.33 rad/s)	
	J5轴手腕摆动	360°（360°/s）6.28 rad (6.28 rad/s)	360°（430°/s）6.28 rad (7.50 rad/s)	280°（420°/s）6.63 rad (7.50 rad/s)	
	J6轴手腕旋转	900°（550°/s）15.71 rad (9.60 rad/s)	900°（630°/s）15.71 rad (11.0 rad/s)	720°（700°/s）12.57 rad (12.22 rad/s)	720°（720°/s）12.57 rad (12.57 rad/s)
手腕部可搬运质量		20 kg	12 kg	10 kg	
手腕允许负载转矩	J4轴	44.0 N·m	22.0 N·m	22 N·m	
	J5轴	44.0 N·m	22.0 N·m	22 N·m	
	J6轴	22.0 N·m	9.8 N·m	10 N·m	
手腕允许负载转动惯量	J4轴	1.04 kg·m²	0.65 kg·m²	0.50 kg·m²	
	J5轴	1.04 kg·m²	0.65 kg·m²	0.50 kg·m²	
	J6轴	0.28 kg·m²	0.17 kg·m²	0.20 kg·m²	
重复定位精度		±0.03 mm			
机器质量		250 kg		130 kg	
安装条件		环境温度：0～45℃ 环境湿度：通常在75%RH以下（无结露现象） 短期95%RH以下（1个月内） 振动加速度：4.9m/s²(0.5g)以下			

图 2-9　FANUC M-20iA/12L 机器人性能参数
Figure 2-9　Performance Parameter of FANUC Robot M-20iA/12L

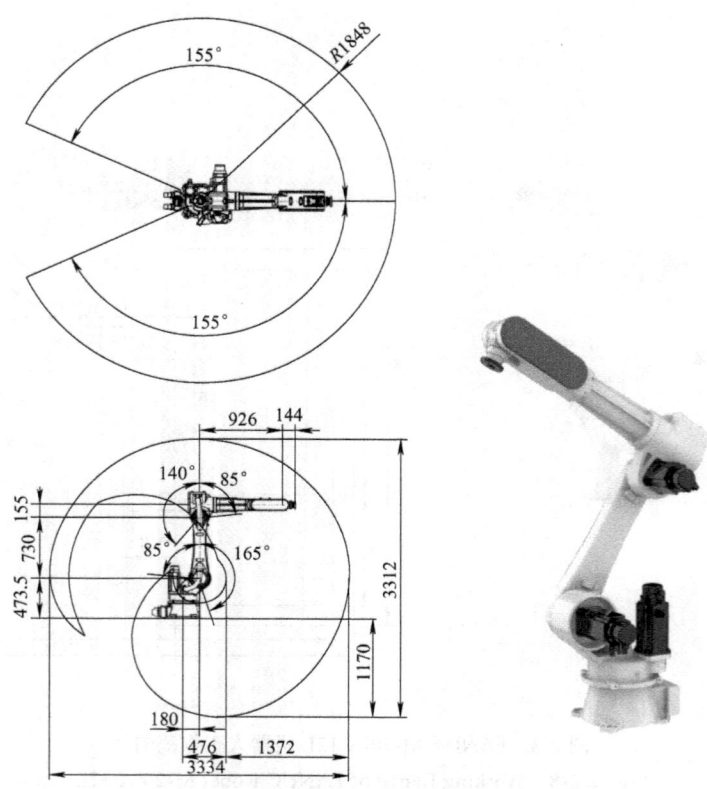

图 2-10　华数 HSR-JR620L 机器人工作范围
Figure 2-10　Working Range of Huashu Robot HSR-JR620L

产品型号		HSR-JR620L
自由度		6
最大负载		20kg
最大工作半径		1848mm
重复定位精度		±0.06mm
运动范围	J1轴	±160°
	J2轴	-175°/+75°
	J3轴	+40°/+265°
	J4轴	±180°
	J5轴	±125°
	J6轴	±360°
额定速度	J1轴	1.73rad/s, 99°/s
	J2轴	1.52rad/s, 87°/s
	J3轴	2.51rad/s, 144°/s
	J4轴	3.14rad/s, 180°/s
	J5轴	3.14rad/s, 180°/s
	J6轴	3.92rad/s, 225°/s
容许惯性矩	J6轴	0.8kg·m²
	J5轴	3.3kg·m²
	J4轴	10.9kg·m²
容许扭矩	J6轴	30.7N·M
	J5轴	73.4N·M
	J4轴	140.4N·M
	温度	0~45℃
	湿度	20%~80%
	其他	避免与易燃易爆或腐蚀性气体、液体接触,远离高电子噪声源(等离子)
防护等级		IP54
安装方式		地面安装
本体质量		305kg

图 2-11　华数 HSR-JR620L 机器人性能参数
Figure 2-11　Performance Parameters of Huashu Robot HSR-JR620L

2. PLC 选型

本项目需要与触摸屏、数控机床进行网络通信。通过分析整个控制流程以及具体的硬件连接，实现整个动作流程需要输入点数为50，输出点数为40。整个工作台的控制都采用数字量控制，没有模拟量的计算，整个项目控制并不复杂。因此，对于控制器的选择只需要选用小型的 PLC，并且通信网络与数控机床和触摸屏需要兼容。通过对机床的分析得知，机床支持 PROFINET 通信协议。考虑到整个系统后续的可扩展性和网络接口备用性，因此选择了本体自带有2个 PROFINET 通信接口的西门子 S7-1215C DC/DC 型 PLC，其性能参数如图 2-12 所示。但是该类型 PLC 只有14个输入和10个输出，因此需要加入3个含有16个输入和16个输出的扩展模块 SM 1223 DC/RLY。

3. 机器人行走轴选型

根据测算，两个数控机床之间的工件抓取工作点布局间隔为3.2m，机器人的质量为305kg，因此依据此数据作为行走轴选择的主要依据，具体参数见表2-4。行走轴安装后的具体效果如图2-13所示。

4. 数控机床选型

(1) 确定典型加工工件
- 箱体类零件应选择卧式加工中心。
- 板类零件应选择立式加工中心。
- 轴类零件应选择车削加工中心。

(2) 选择数控机床规格
- 机床工作台面积应大于典型零件尺寸以便于安装夹具。
- 机床行程应大于典型零件加工范围以便于出刀。

2. Selection of PLCs

It requires network communication with touch screens and CNC machine tools in this program. Based on the analysis of the entire control process and specific hardware connections, the entire action process requires 50 input points and 40 output points. The control of the entire program is not complex, as the control of the entire workbench adopts digital control without analog calculation. Therefore, PLC of small capacities can be selected as the controller, and the communication network needs to be compatible with CNC machine tools and touch screens. Based on the analysis of the machine tools, it is found that the machine tools support the PROFINET communication protocol. When the scalability of the entire system and network interface backup are considered, the Siemens S7-1215C DC/DC type PLC with two PROFINET communication interfaces is selected, and its performance parameters are shown in Figure 2-12. However, this type of PLC only has 14 inputs and 10 outputs. Therefore, it is necessary to add 3 expansion modules of SM 1223 DC/RLY with 16 inputs and 16 outputs.

3. Selection of robot walking axis

Based on calculations, the interval of work points for workpiece gripping between 2 CNC machine tools is 3.2m, and the weight of the robot is 305kg. The figures above are used as the main basis for selecting the walking axis, as shown in Table 2-4. The installed walking axis is rendered as Figure 2-13.

4. Selection of CNC machine tools

(1) Determine typical workpiece for machining
- Horizontal machining centers shall be selected for box parts.
- Vertical machining centers shall be selected for plate parts.
- Turning centers shall be selected for axis parts.

(2) Select the specifications of CNC machine tools
- The area of the machine tool workbench shall be larger than the size of typical parts for easier installation of fixtures.
- The machine tool stroke shall be greater than the machining range of typical parts to facilitate cutting.

型号		CPU 1211C	CPU 1212C	CPU 1212FC	CPU 1214C	CPU 1214FC	CPU 1215C	CPU 1215FC	CPU 1217C
外观		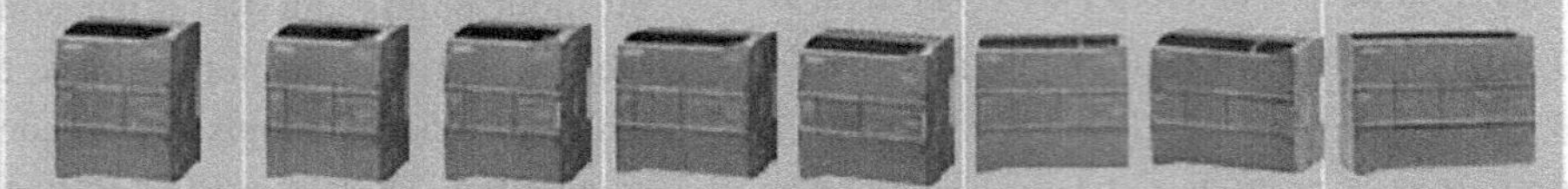							
标准CPU		DC/DC/DC, AC/DC/RLY, DC/DC/RLY							DC/DC/DC
故障安全CPU		—	DC/DC/DC, DC/DC/RLY						—
物理尺寸		90mm×100mm×75mm			110mm×100mm×75mm		130mm×100mm×75mm		150mm×100mm×75mm
用户存储器									
• 工作存储器		• 50KB	• 75KB	• 100KB	• 100KB	• 125KB	• 125KB	• 150KB	• 150KB
• 装载存储器		• 1MB	• 2MB	• 2MB	• 4MB	• 4MB	• 4MB	• 4MB	• 4MB
• 保持性存储器		• 10KB	• 10KB	• 10KB	• 10KB	• 10KB	• 10KB	• 10KB	• 10KB
本体集成I/O									
• 数字量		• 6点输入/4点输出	• 8点输入/6点输出		• 14点输入/10点输出		• 14点输入/10点输出		
• 模拟量		• 2路输入	• 2路输入		• 2路输入		• 2路输入/2路输出		
过程映像大小		1024字节输入(1)和1024字节输出(Q)							
位存储器(M)		4096个字节			8192个字节				
信号模块扩展		无	2		8				
信号板		1							
最大本地I/O-数字量		14	82		284				
最大本地I/O-模拟量		3	19		67		69		
通信模块		3(左侧扩展)							
高速计数器	总计	最多可组态6个使用任意内置输入或SB输入的高速计数器							
	差分1MHz	—							Ib.2～Ib.5
	100/80kHz	Ia.0～Ia.5							
	30/20kHz	—			Ia.6～Ia.7		Ia.6～Ib.5		Ia.6～Ib.1
		使用SB 1223 DI 2×24 V DC, DQ 2×24 V DC时可达30/20kHz							
	200/160kHz	使用SB 1221 DI 4×24 V DC, 200 kHz, SB 1221 DI 4×5 V DC, 200 kHz, SB 1223 DI 2×24 V DC/DQ 2×24 V DC, 200 kHz SB 1223 DI 2×5 V DC/DQ 2×5 V DC, 200 kHz时最高可达200/160kHz							
脉冲输出	总计	最多可组态4个使用DC/DC/DC CPU任意内置输出或SB输出的脉冲输出							
	差分1MHz	—							Qa.0～Qa.3
	100kHz	Qa.0～Qa.3							Qa.4～Qb.1
	20kHz	—			Qa.4～Qa.5		Qa.4～Qb.1		—
		使用SB 1223 DI 2×24 V DC, DQ 2×24 V DC时可达20kHz							
	200kHz	使用SB 1222 DQ 4×24 V DC, 200 kHz, SB 1222 DQ 4×5 V DC, 200 kHz, SB 1223 DI 2×24 V DC/DQ 2×24 V DC, 200 kHz, SB 1223 DI 2×5 V DC/DQ 2×5 V DC, 200 kHz时最高可达200kHz							
存储卡		SIMATIC存储卡(选件)							
实时时钟保持时间		通常为20天，40℃时最少12天							
PROFINET		1个以太网通信端口，支持PROFINFT通信					2个以太网端口，支持PROFINFT通信		
实数数学运算执行速度		2.3μs/指令							
布尔运算执行速度		0.08μs/指令							

图 2-12 S7-1200 PLC 性能参数

Figure 2-12 Performance Parameter of S7-1200 PLC

表 2-4 行走轴选型具体参数
Table 2-4 Specific Parameters for Selection of Walking Axis

序号 S/N	项目 Item	参数 Parameter	备注 Remarks
1	宽度 width	920mm	
2	工作面高度 Working face height	350mm	
3	有效长度 Effective length	3.8m	整个长度不超过5m The entire length does not exceed 5m
4	驱动方式 Driving mode	伺服电动机+减速机 Servo motor+reducer	
5	传动方式 Transmission mode	齿轮齿条 Gear and rack	
6	控制方式 Control mode	机器人示教器控制 Controlled by robot teach pendant	
7	线速度 Linear velocity	最大速度大于1.5m/s Maximum velocity greater than 1.5m/s	
8	润滑方式 Lubrication method	手动润滑泵 Manual lubrication pump	
9	负载 Load	大于500kg Greater than 500kg	
10	重复定位精度 Repetitive positioning precision	±0.2mm	
11	安装后导轨平面度 Flatness of the installed guide rail	±0.3mm	

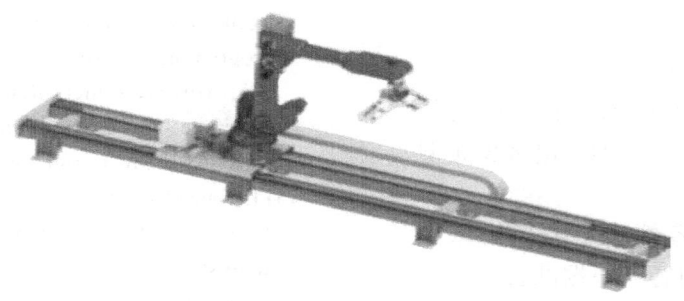

图 2-13 行走轴效果
Figure 2-13 Rendering of Walking Axis

- 机床工作台承重能力应大于零件和夹具的质量。
- 主电动机功率：主电动机功率越大，其每分钟可切除的金属余量就越多，表明机床切削能力越强，刚性也越高。

(3) 选择机床的精度
- 应统一使用 ISO 标准衡量机床的定位和重复定位精度值。
- 机床的重复定位精度反映了该控制轴在行程内任意定位点的定位稳定性，这是衡量该控制轴能否稳定可靠的基本指标。
- 零件在单轴上移动加工两孔的孔距精度约为机床单轴定位精度的 2 倍左右，双轴移动则为机床单轴定位精度的 3 倍左右。

(4) 数控系统的选择
- 铣削应选择铣床系统，车削应选择车床系统，钻削应选择钻床系统。
- 进口系统性能稳定，价格高；国产系统可靠性差，价格低。
- 系统基本功能都已固化，都必须选择，特殊选项价格特别贵，可根据实际需要适当选择，如 FANUC-0I-MC 铣床系统中图形显示功能、自动编程软件、刀具测量系统、工件测量系统、以太网接口及通信功能可增加成本 20 多万元。

(5) 估算工时和节拍
- 选择机床时必须做可行性分析：一年内该机床能加工出多少典型零件。
- 根据典型零件确定数控机床加工工序的内容，根据准备给机床配置的刀具种类和数量来确定切削用量，并计算每道工序的切削

- The load-bearing capacity of the machine tool workbench shall be greater than the weight of the parts and fixtures.
- The power of main motor: the higher power of the main motor, the more metal allowance it can cut off per minute and, as it indicates, the better cutting ability and rigidity of the machine tool.

(3) Select the precision of the machine tools
- ISO standards shall be uniformly used to measure the positioning and the precision value of repeated positioning of machine tools.
- The repeated positioning precision of the machine tool reflects the positioning stability of the control axis at any positioning point within its stroke, which is a basic indicator to determine whether the control axis is stable and reliable.
- The precision of the hole spacing between two holes processed by moving the parts on a single axis is about twice that of the single axis positioning precision of the machine tool, and about 3 times that of the double axis positioning precision of the machine tool.

(4) Selection of CNC systems
- The milling machine system shall be selected for milling, when the turning system for turning, and the drilling system for drilling.
- The imported systems have stable performance and cost more, while the domestic systems have poor reliability and cost less.
- The basic functions of the system have been solidified and must be selected. Special options are particularly expensive and can be selected according to actual needs. For example, more than 200,000 yuan will be increased when selecting the graphic display function, automated programming software, tool measurement system, workpiece measurement system, Ethernet interface, and communication function in the FANUC-0I-MC milling machine system.

(5) Estimation of working hours and speed
- When selecting a machine tool, a feasibility analysis has to be conducted to determine how many typical parts it can machine within a year.
- Determine the content of the machining procedures of CNC machine tools based on typical

时间及相应的辅助时间，一般换刀时间按 10s 计算。

(6) 刀库选择及刀柄配置
- 刀库容量越大，价格越贵，故障率也越高，加工中心 50% 以上的故障都与刀库有关。
- 立式车床选用 20 把左右刀具容量的刀库、卧式车床选用 40 把左右刀具容量的刀库基本能满足要求。
- 根据典型零件要加工的工序内容确定刀柄的种类和数量，用户不太熟悉时可由机床厂家或刀具供应商协助分析加工工艺，并制定刀具和刀柄的选配方案。
- 选用复合式刀具预调仪：为提高数控机床的开动率，加工前刀具的准备工作尽量不要占用机床工时。为提高预调仪的利用率，最好是一台预调仪为多台机床服务，把其纳入数控机床的技术准备工作中，作为一个重要环节。

(7) 数控机床驱动及电动机的选择
- 数控机床的控制系统、驱动放大器、驱动电动机最好选择同一家配套系统，其性能才能得到最佳匹配和发挥。如 FANUC 控制系统，则应选择 FANUC 推荐的 FANUC 驱动放大器和 FANUC 驱动电动机。同一控制系统，驱动不同、电动机不同则性能相差悬殊，价格也相差悬殊。
- 进给电动机功率和转矩越大，其响应能力（即起动、制动、加减速、准确定位、换向）越强，定位能力越高，价格也越高。
- 电动机功率越大，其每分钟可切除的金属余量就越多，表明机床承受切削的能力越强，刚性也

parts, the cutting amount based on the type and quantity of tools to be configured for the machine tools, and calculate the cutting time and corresponding auxiliary time for each procedure. Generally, the tool changing time is set as 10 seconds.

(6) Selection of tool magazine and configuration of tool handle
- The larger the capacity of the tool magazine, the more expensive the price, and the higher the failure rate. More than 50% of the faults in the machining center are related to the tool magazine.
- Requirements are basically met when a tool magazine with a capacity of about 20 tools is selected for the vertical lathe and a tool magazine with a capacity of about 40 tools for the horizontal one.
- Determine the type and quantity of tool holders based on the content of machining procedures of typical parts. If the user is not familiar on this regard, the machine tool manufacturer or tool supplier can assist in analyzing the machining process and developing a selection plan for tools and their holders.
- Select composite tool presetters: in order to improve the efficiency of CNC machine tools, try to best to make the preparation of tools done before machining to avoid occupying machine hours. In order to improve the utilization rate of the presetter, it would be better to have one presetter serve multiple machine tools and include it in the technical preparation of CNC machine tools and treat it as an important procedure.

(7) Selection of drives and motors for CNC machine tools
- It would be better to choose the supporting systems of the same brand of the control system, drive amplifier, and drive motor for CNC machine tools in order to achieve the best matching and performance. For instance, if the FANUC control system is used, the FANUC drive amplifier and FANUC drive motor recommended by FANUC should be selected. For the same control system, different drives or motors will result in significant differences in performance

越高。

综上所述，再根据零件加工工艺要求进行分析，得出以下机床选型：

1) 型号：T2C-500。生产公司：沈阳一机。机床具体参数见表2-5和表2-6。

2) 型号为Viva T2C/500或型号为Viva T2Cm/500的斜轨车床。两种型号的车床参数基本一致，Viva T2Cm/500比Viva T2C/500增加了2个动力头，是铣削复合式车床。具体参数见表2-7。

➤ The higher the power and torque of the feed motor, the stronger its response ability (i.e. starting, braking, acceleration and deceleration, accurate positioning, and commutation), the higher its positioning ability and price.

➤ The power of motor: the higher power of the motor, the more metal allowance it can cut off per minute and, as it indicates, the better cutting ability and rigidity of the machine tool.

In summary, based on the analysis of the machining process requirements of the parts, selection of machine tools is as follows:

1) Model: T2C-500; manufacturer: Shenyang Yiji CNC Machine Tool Equipment Co., Ltd., specific parameters of the machine tool are shown in Table 2-5 and 2-6.

2) Inclined rail lathe with model Viva T2C/500 or Viva T2Cm/500. The parameters of the lathes of the two models are basically the same, with the Viva T2Cm/500 one having two additional power heads compared to the Viva T2C/500 one, making the former a compound milling lathe, and its specific parameters are shown in Table 2-7.

表2-5 T2C-500型号机床运行参数

Table 2-5 Operating Parameters of T2C-500 Machine Tool

项目 Item	单位 Unit	规格 Specification	备注 Remarks
床身上最大回转直径 Maximum swing diameter on the lathe	mm	560	
最大切削长度 Maximum cutting length	mm	500	
最大切削直径 Maximum cutting diameter	mm	280	
滑板上最大回转直径 Maximum turning diameter on the sliding plate	mm	350	
主轴端部型式及代号 Type and code of the main axle end		A2-6	
主轴孔直径 Hole diameter of the main axle	mm	65	
最大通过棒料直径 Maximum diameter of bar passing through	mm	50	

（续）

项目 Item		单位 Unit	规格 Specification	备注 Remarks
单主轴主轴箱 Main axle box for single main axle	主轴转速范围 Speed range of the main axle/ 主轴最大输出扭矩 Maximum output torque of the main axle	r/min, N·m	50～4500, 235	FANUC 0i mate-TD
主电动机输出功率 Main motor output power	30min/连续 (continuous)	kW	15/11	βiIP22/6000
标准卡盘 Standard chuck	卡盘直径 Chuck diameter	in	8	
X轴快移速度 X-axis quick movement speed		m/min	30	滚动导轨 Rolling guide rail
Z轴快移速度 Z-axis quick movement speed		m/min	30	滚动导轨 Rolling guide rail
X轴行程 X-axis stroke		mm	200	
Z轴行程 Z-axis stroke		mm	560	
尾座行程 Tailstock stroke		mm	450	
尾座主轴锥孔锥度 Tailstock main axle taper		莫氏 (Morse)	5号	
标准刀架形式 Standard tool holder form			卧式8工位	
刀具尺寸 Tool size	外圆刀 Outer cylindrical knife	mm	25×25	
	镗刀杆直径 Boring bar diameter	mm	φ40, φ32, φ25, φ20	
刀盘可否就近选刀 Can the cutter head be selected nearby			可 (can)	
机床质量 Machine tool weight	总质量 Total weight	kg	4000	
最大承重 Max bearing capacity	盘类件 Disk parts	kg	200	
	轴类件 Shaft parts	kg	500	

表 2-6 T2C-500 型号机床加工精度
Table 2-6 Machining Precision of T2C-500 Machine Tool

检验项目 Inspection items		工厂标准 Factory standards
加工精度 Machining precision		IT6
加工工件圆度 Processing workpiece roundness		0.0025mm/ϕ75
加工工件圆柱度 Processing workpiece cylindricity		0.010mm/150mm
加工工件平面度 Processing workpiece flatness		0.010mm/ϕ200mm
加工工件表面粗糙度 Processing workpiece surface roughness		Ra1.25μm
定位精度 Positioning precision	X 轴 X-axis	0.010mm
	Z 轴 Z-axis	0.010mm
重复定位精度 Repetitive positioning precision	X 轴 X-axis	0.004mm
	Z 轴 Z-axis	0.005mm

表 2-7 Viva T2C/500 斜轨车床参数
Table 2-7 Parameters of Viva T2C/500 Inclined Rail Lathe

名称 Name		规格 Specifications	单位 Unit
工作台 Workbench	工作台尺寸 Workbench size	1000×500	mm
	允许最大荷重 Maximum allowable load	600	kg
	T 形槽尺寸 T-groove size	18×5	mm×个
加工范围 Machining range	工作台最大行程 –X 轴 Maximum stroke of workbench –X-axis	850	mm
	滑座最大行程 –Y 轴 Maximum stroke of sliding seat –Y-axis	560	mm

（续）

名称 Name			规格 Specifications	单位 Unit
加工范围 Machining range	主轴最大行程 –Z 轴 Maximum stroke of main axle –Z-axis		650	mm
	主轴端面至工作台面距离 Distance from the main axle end face to the workbench face	最大 maximum	800	mm
		最小 minimum	150	mm
	主轴中心到导轨基面距离 Distance from the main axle ceter to the guide rail base face		665	mm
主轴 Main axle	锥孔 Conical hole		HSK–A63	
	转数范围 Revolution range		50～18000	r/min
	最大输出转矩 Maximum output torque		29.4/43.6	N·m
	主轴电动机功率 Main axle motor power		18.5/26	kW
	主轴传动方式 Main axle transmission mode		电主轴形式 Mortorized main axle	
刀具 Tools	刀柄型号 Knife handle model		HSK63A	
进给 Feed	快速移动 Quick movement	X 轴 X-axis	32	m/min
		Y 轴 Y-axis	32	m/min
		Z 轴 Z-axis	30	m/min
	三轴拖动电动机功率 (X/Y/Z)		2.5/2.5/3	kW
	三轴拖动电动机转矩 (X/Y/Z)		20/20/27	N·m
	进给速度 Feed rate		1～20000	mm/min

（续）

名称 Name		规格 Specifications		单位 Unit
刀库 Tool magazine	刀库形式 Tool magazine form	机械手 manipulator		
	选刀方式 Tool selection method	双向就近选刀 Bidirectional nearest knife selection		
	刀库容量 Magazine capacity	24		把
	最大刀具长度 Maximum tool length	300		mm
	最大刀具质量 Maximum tool weight	7		kg
	最大刀盘直径 Maximum cutter diameter	满刀 Full tool	$\phi 80$	mm
		相邻空刀 Empty of tool adjacent	$\phi 150$	mm
	换刀时间 Tool changing time	2.5		s
		JISB6336-4:2000	GB/T 18400.4—2010	
定位精度 Positioning precision	X 轴 X-axis	0.016	0.016	mm
	Y 轴 Y-axis	0.012	0.012	mm
	Z 轴 Z-axis	0.012	0.012	mm
重复定位精度 Repetitive positioning precision	X 轴 X-axis	0.010	0.010	mm
	Y 轴 Y-axis	0.008	0.008	mm
	Z 轴 Z-axis	0.008	0.008	mm
	机床质量 Machine tool weight	6800		kg
	电气总容量 Total electrical capacity	25		kV·A
	机床轮廓尺寸（长×宽×高） Outline dimension (length × width × height) of machine tool	4355 × 2275 × 2920		mm

5. 其余零部件选型

除主要设备外，根据工艺需要，其余设备选型清单见表 2-8，软件配置清单见表 2-9。

5. Selection of other parts and components

Except for the main equipment and based on the process requirements, the list of other equipment is shown in Table 2-8, and that of software configuration in Table 2-9.

表 2-8 项目其余设备清单

Table 2-8 List of Other Equipment for the Program

序号 S/N	名称 Name	数量 Qty	单位 Unit	备注 Remarks
1	机器人夹具 Robot fixture	1	套 set	3 种夹具，配快换系统 Three types of fixtures, with quick change system
2	立体仓库 Warehouse	1	套 set	设 5 层 6 列共 30 个仓位，含安全门、按钮、RFID 检测设备、光电开关及指示灯 Set up a total of 30 bins on 5 layers and 6 columns, including safety doors, switch buttons, RFID detection equipment, photoelectric switches, and indicator lights
3	在线测量装置 Online measurement device	1	套 set	用于加工中心 For machining centers
4	雄克零点定位系统 SCHUNK zero-point clamping system	1	套 set	用于加工中心 For machining centers
5	触摸屏 Touch screen	1	台 piece	TP900 精致面板，TFT 显示屏，PROFINET/工业以太网接口（2 个端口） TP900 exquisite panel, TFT display screen, PROFINET/industrial Ethernet interface (2 ports)
6	安全防护系统 Safety protection system	1	套 set	防止意外闯入、保护人员安全 Prevent accidental entry and protect personnel safety
7	电气控制柜 Electrical control cabinet	1	套 set	用于放置电气元件和电气设备 For placement of electrical components and equipment

表 2-9 软件配置清单

Table 2-9 List of Software Configuration

序号 S/N	软件名称 Software name	数量 Qty	单位 Unit	基本功能 Basic function
1	TIA 软件 TIA software	1	套 set	负责周边设备及机器人控制，实现智能制造单元的流程和逻辑总控 Being responsible for controlling peripheral equipment and robots to achieve the flow and overall logic control of intelligent manufacturing units
2	机器人仿真软件 Robot simulation software	1	套 set	单元设备模拟，虚拟安装调试，布局优化 Unit device simulation, virtual installation and commissioning, layout optimization
3	SOLIDWORKS 或 NX 软件 SOLIDWORKS or NX software	1	套 set	三维模型设计和编程，编制零件加工工艺 3D model design and programming, and parts machining process compiling

【课后巩固】

1. 要满足项目的生产需求，除了 FANUC M-20iA/12L 机器人与华数 HSR-JR620L 机器人之外还可以选用什么型号的机器人呢？

2. 工业机器人有几种切换手爪的方式呢？

[Consolidation after Class]

1. What other models of robots can be selected besides FANUC M-20iA/12L and Huashu HSR-JR620L to meet the production needs of the program?

2. How many ways are there for industrial robots to switch their grippers?

任务 2.2　设计方案的编写
Task 2.2　Compilation of Design Schemes

【知识目标】

1. 了解工业机器人机床上下料工作站的工作流程及控制要求。

2. 了解工业机器人机床上下料工作站的设备清单。

3. 了解工业机器人机床上下料工作站设计方案的结构和要素。

[Knowledge Objectives]

1. To understand the workflow and control requirements of the industrial robot machine tool loading and unloading workstation.

2. To understand the equipment list of the industrial robot machine tool loading and unloading workstation.

3. To understand the structure and elements of the design scheme of the industrial robot machine tool loading and unloading workstation.

【技能目标】

1. 能根据工艺要求说明工作流程及控制要求。

2. 能编写工业机器人机床上下料工作站的设计方案。

[Skill Objectives]

1. To be able to explain workflow and control requirements according to process requirements.

2. To be able to prepare design schemes for the industrial robot machine tool loading and unloading workstation.

【素质目标】

1. 培养良好的自主学习习惯。
2. 培养团队协作精神。

[Competence Objectives]

1. To cultivate good independent study habits.
2. To cultivate a teamwork spirit.

【任务情景】

某企业需要将一数控加工工作站进行改造，将机床上下料的工作全部设计由机器人自动完成，目前已确定工业机器人机床上下料工作站的工作

[Task Scenario]

An enterprise needs to transform a CNC machining workstation, with all the loading and unloading work of the machine tool to be automatically completed by robots. Currently, the workflow and control requirements of the industrial robot machine tool loading and

流程及控制要求，根据给出的工作流程、工作布局图、硬件选型方案，请完成机器人机床上下料工作站设计方案的编写。

【任务分析】

1. 了解设计方案所需的结构和要素。

2. 将设计方案中所需的结构、要素用流畅的语言描述出来。

【知识准备】

2.2.1 设计方案的结构和要素

在项目实施的过程中，必须要编写出设计方案交给客户。设计方案必须详细地叙述出项目实施的优势，项目能够给企业带来的利益和生产效率的提升，项目实施过程中的设备选型和布局，项目的工程预算等。一个好的设计方案对于项目的推进和实施有着重要的意义。

常见的设计方案目录结构如图2-14所示，主要包括工作站简介及布局、加工工件说明、工艺动作流程、工业机器人机床上下料工作站主要设备清单。

2.2.2 设计方案编写示例

这里以"一、工作站简介及布局"的内容编写为例，为大家演示如何进行方案文档编写。方案的其他内容请同学们按要求自己编写，具体任务要求及文档编写规则请按照本教材配套的活页式学材的方法和步骤实施。

unloading workstation have been determined. Please prepare design scheme of the robot machine tool loading and unloading workstation based on the provided workflow, work layout, and the scheme of the hardware selection.

[Task Analysis]

1. Understand the structure and elements required for the design scheme.

2. Describe the structure and elements required in the design scheme in fluent language.

[Assumed Knowledge]

2.2.1 Structure and Elements of Design Schemes

During the program implementation process, it is necessary to prepare a design scheme and submit it to the client. The design scheme must provide a detailed description of the advantages of the program, the benefits that the program can bring to the enterprise, the improvement of production efficiency, the selection and layout of equipment during the program implementation process, the engineering budget of the program, etc. A good design scheme has important significance for the launch and implementation of the program.

The structure and elements of common design schemes are shown in Figure 2-14, mainly including workstation introduction and layout, workpiece machining instructions, process flow, and a list of main equipment for the industrial robot machine tool loading and unloading workstation.

2.2.2 An Example of Design Scheme Compilation

Here, the compilation of the content of " I. Workstation Introduction and Layout" is taken as an example to demonstrate how to compile a scheme. Other content of the scheme shall be compiled by the student according to the requirements. For specific task requirements and document preparation rules, please follow the methods and steps of the loose-leaf learning materials matched with this textbook.

图 2-14 设计方案目录结构
Figure 2-14 Structure of the Table of Contents of a Design Scheme

示例：

工作站简介及布局

目前在机床加工行业中，要求加工精度高、批量加工速度快使生产线自动化程度要有很大的提升，首先一点就是针对机床方面进行全方位自动化处理，使人力从中解放出来。

在机器人机床上下料工作站中，机床几乎要 24 小时运行。在一些欧美国家早已采用机器手来自动上料和下料。要根据加工零件的形状及加工工艺的不同，来采用不同的手爪抓取系统。而完成抓取、搬运和取走过程的运动机构就是大型直角坐标机器人，它们通常就是一个水平运动轴（X 轴）和上下运动轴（Z 轴）。本方案机器人采用 FANUC 品牌的 M-20iA/12L 型号的工业机器人来完成机床上下料的工作。根据被加工零件形状和重量不同，所采用的手爪形状及结构也不

Example:

Workstation Introduction and Layout

At present, in the machine tool machining industry, the requirements for high machining precision and fast mass machining have led to a significant improvement in the automation level of production lines. The first is to carry out comprehensive automation machining for machine tools, freeing manpower from it.

In the robot machine tool loading and unloading workstation, the machine tool runs almost continuously for 24 hours. Robotic arms have long been used to automatically load and unload materials in some European and American countries. Different gripper grasping systems shall be used based on the shape and machining technology of the parts to be machined. The motion mechanism that completes the process of grasping, handling and taking is a large Cartesian coordinate robot, which usually consists of a horizontal motion axis (X axis) and an up and down motion axis (Z axis). FANUC industrial robots of the model M-20iA/12L are adopted in the scheme to complete the loading and unloading work of the machine tool. The shape and structure of the grippers used vary depending on the shape and weight

同。手爪的类型及尺寸要根据具体的零件及加工工艺来定。

在上下料过程中,机器人要与机床工作台运动及卡盘张紧精确协调,严格按信号流顺序来控制上下料过程,在放下加工好的零件和取要加工的新零件时也必须与其配套的设施精确同步协调。在考虑工作站的工作流程、安全保障事项后,设计出的工业机器人机床上下料工作站布局如图2-15、图2-16所示,包含数控车床、在线检测单元、六轴多关节机器人、行走轴、物料传送系统和中央控制系统等。

of the parts to be machined The type and size of the grippers need to be determined according to the specific parts and machining technology.

During the loading and unloading process, the robot needs to coordinate accurately with the workbench movement and chuck tension of the machine tool, strictly controlling the loading and unloading process according to the signal flow sequence. When placing processed parts and taking new parts for machining, it has to coordinate accurately with its supporting facilities. Considering the workflow and safety assurance of the workstation, the layout of the industrial robot machine tool loading and unloading workstation is designed as shown in Figure 2-15 and 2-16, including CNC lathes, online testing units, six-axis multi-joint robots, walking axis, material conveying systems, and central control systems.

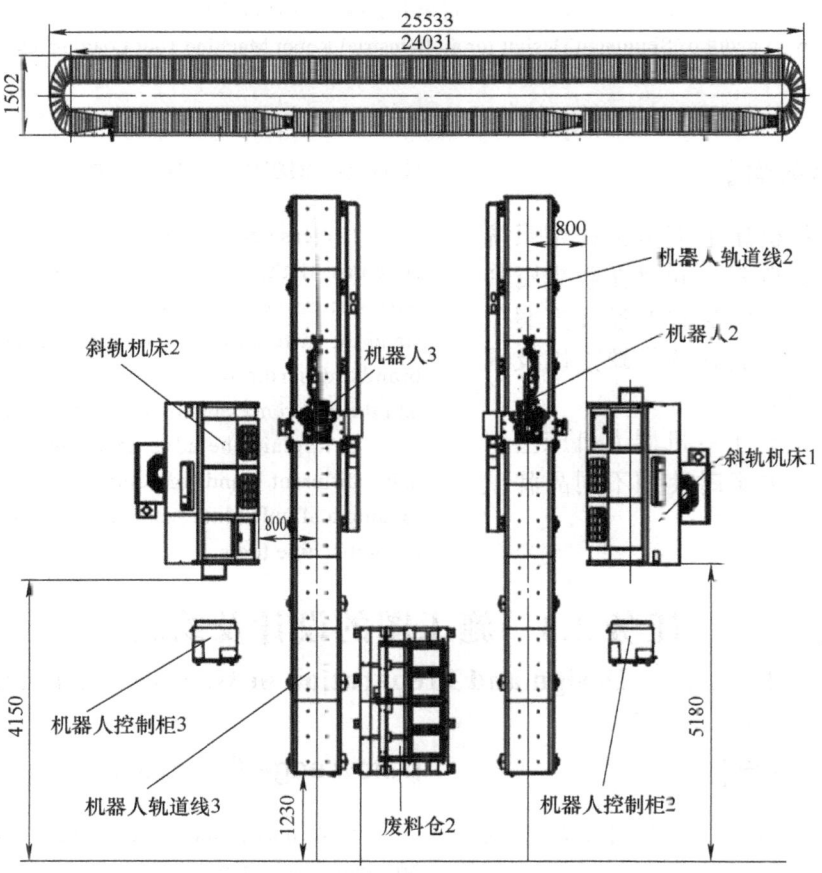

图 2-15 工业机器人机床上下料工作站布局

Figure 2-15 Layout of an Industrial Robot Machine Tool Loading and Unloading Workstation

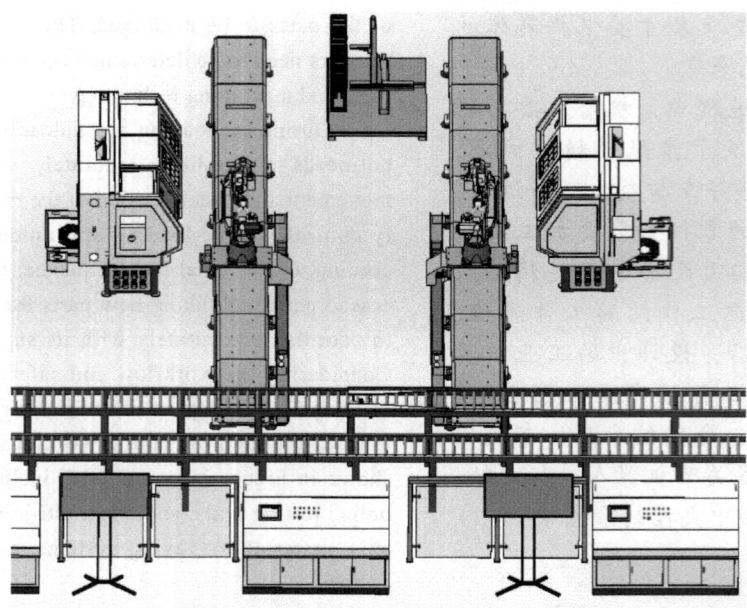

图 2-16 工业机器人机床上下料工作站仿真设计布局

Figure 2-16 Layout of Simulated Design for an Industrial Robot Machine Tool Loading and Unloading Workstation

【课后巩固】

1. 请收集机床上下料工作站系统设计方案相关资料，同时在收集资料过程中，了解不同品牌的设计方案的区别，说出不同品牌的优缺点以及国产和国外方案区别。

2. 说明设计工业机器人机床上下料工作站系统方案时使用不同品牌设备的优缺点。

[Consolidation after Class]

1. Please collect information related to the design schemes of the machine tool loading and unloading workstation system and, during this process, understand the differences in the design schemes of different brands, describe their advantages and disadvantages, and the differences in domestic and foreign schemes.

2. Explain the advantages and disadvantages of using different brands of equipment when designing the loading and unloading workstation system for industrial robot machine tools.

任务 2.3　施工图的设计及绘制
Task 2.3　Design and Preparation of Detailed Drawings

【知识目标】

1. 掌握设备布局图设计与绘制基本方法。

2. 掌握系统框图设计与绘制基本方法。

[Knowledge Objectives]

1. To master the basic methods of designing and preparing equipment layouts.

2. To master the basic methods of designing and preparing system charts.

3. 掌握电气原理图设计与绘制基本方法。

4. 掌握非标件工程图设计与绘制基本方法。

【技能目标】

1. 能够根据任务要求绘制设备布局图。

2. 能够根据任务要求绘制系统框图。

3. 能够根据任务要求绘制电气原理图。

4. 能够根据任务要求绘制非标件工程图。

【素质目标】

1. 培养学生善于观察、归纳总结的能力。

2. 培养学生严谨的工作态度、潜心研究的敬业精神。

【任务情景】

某企业需要将一数控加工工作站进行设计和改造，将机床上下料的工作全部设计与机器人自动完成，目前已确定工业机器人机床上下料工作站的工作流程及控制要求，根据给出的机器人机床上下料工作站设计方案，请你完成工作站施工图的设计及绘制。

【任务分析】

1. 掌握设备布局图、系统框图、电气原理图、非标件工程图设计与绘制基本方法。

3. To master the basic methods of designing and preparing electrical schematic diagrams.

4. To master the basic methods of designing and preparing non-standard parts engineering drawings.

[Skill Objectives]

1. To be able to prepare equipment layouts according to task requirements.

2. To be able to prepare system charts according to task requirements.

3. To be able to prepare electrical schematic diagrams according to task requirements.

4. To be able to prepare non-standard parts engineering drawings according to task requirements.

[Competence Objectives]

1. To cultivate students the ability to observe and summarize.

2. To cultivate students a rigorous attitude to work and a dedicated spirit of research.

[Task Scenario]

An enterprise needs to design and transform a CNC machining workstation, with all the loading and unloading work of the machine tool to be automatically completed by robots. Currently, the workflow and control requirements of the industrial robot machine tool loading and unloading workstation have been determined. Please complete the design and preparation of the detailed drawing of the workstation based on the provided design scheme for the robot machine tool loading and unloading workstation.

[Task Analysis]

1. To master the basic methods of designing and preparing equipment layouts, system charts, electrical schematic diagrams, and non-standard parts engineering drawings.

2. 根据任务要求绘制设备布局图、系统框图、电气原理图、非标件工程图。

2. To prepare equipment layouts, system charts, electrical schematic diagrams, and non-standard parts engineering drawings according to task requirements.

【知识准备】

[Assumed Knowledge]

2.3.1 设备布局图

工业机器人机床上下料工作站布局如图 2-17 所示，包含数控车床、加工中心、在线检测单元、六轴多关节机器人、工业机器人导轨和中央控制系统。

2.3.1 Equipment Layout Diagram

The layout of the industrial robot machine tool loading and unloading workstation is shown in Figure 2-17, including CNC lathes, machining centers, online testing units, six-axis multi-joint robots, industrial robot guide rails, and central control systems.

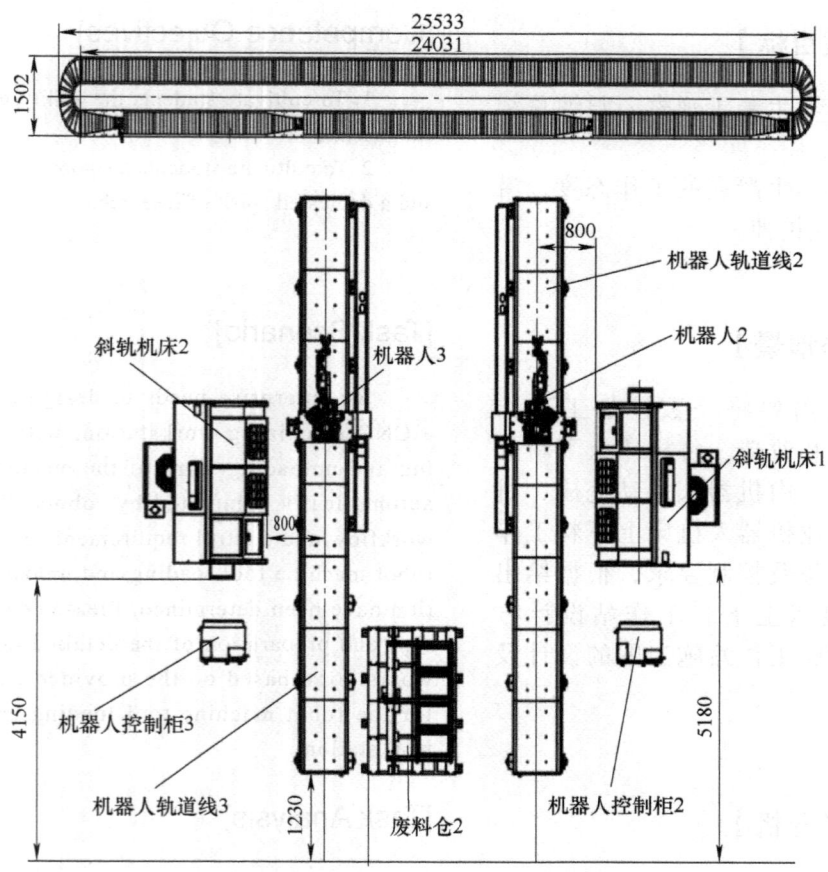

图 2-17 工业机器人机床上下料工作站布局
Figure 2-17　Layout of a Machine Tool Loading and Unloading Workstation of the Industrial Robot

2.3.2 系统框图

工作站各主要设备之间的连接关系及控制关系的系统框图如图 2-18 所示，为后续电气原理图的绘制提供基础。

2.3.2 System Chart

The system chart of the connection and control relationships among the main equipment of the workstation is shown in Figure 2-18, which provides a basis for the subsequent drawing of electrical schematic diagrams.

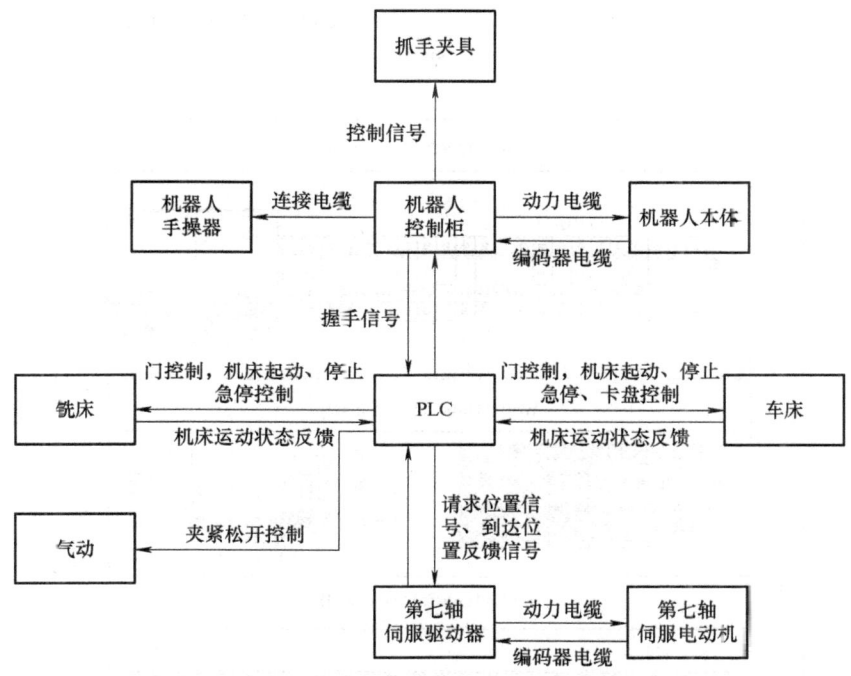

图 2-18 工业机器人机床上下料工作站系统框图

Figure 2-18 System Chart of a Machine Tool Loading and Unloading Workstation of the Industrial Robot

2.3.3 电气原理图

电气原理图是用来表明设备电气的工作原理及各电气元件的作用相互之间的关系的一种表示方式。运用电气原理图的方法和技巧，对于分析电气线路、排除电路故障、程序编写是十分有益的。电气原理图一般由电气元件分布图、主回路电气原理图、开关电源电路原理图、信号分配电气原理图、接线端子与现场信号接线原理图等组成。

这里将部分电气原理图展示给大家，为大家演示如何绘制电气原理图，

2.3.3 Electrical Schematic Diagram

The electrical schematic diagram is a representation used to indicate the working principle of equipment electrification and the functions of various electrical components, and their relationships. The use of electrical schematic methods and techniques is very beneficial for analyzing electrical circuits, circuit troubleshooting, and programming. The electrical schematic diagram generally consists of such parts as electrical component distribution diagram, electrical schematic diagram of main circuits, circuit schematic diagram of switch power supply, signal distribution electrical schematic diagram, wiring terminal and on-site signal wiring schematic diagram.

Here, some electrical schematic diagrams are

未绘制的图纸请同学们按要求自行绘制，具体任务要求及图纸绘制规则请按照与教材配套的活页式学材的方法和步骤实施。

1. 电气元件分布图

图 2-19 所示为电气元件分布图。

shown to demonstrate how to prepare electrical schematic diagrams. For drawings that have not been prepared, please prepare them by yourself according to the requirements. For specific task requirements and rules for preparation, please follow the methods and steps of the loose-leaf learning materials matched with this textbook.

1. Electrical component distribution diagram

The electrical component distribution diagram is shown in Figure 2-19.

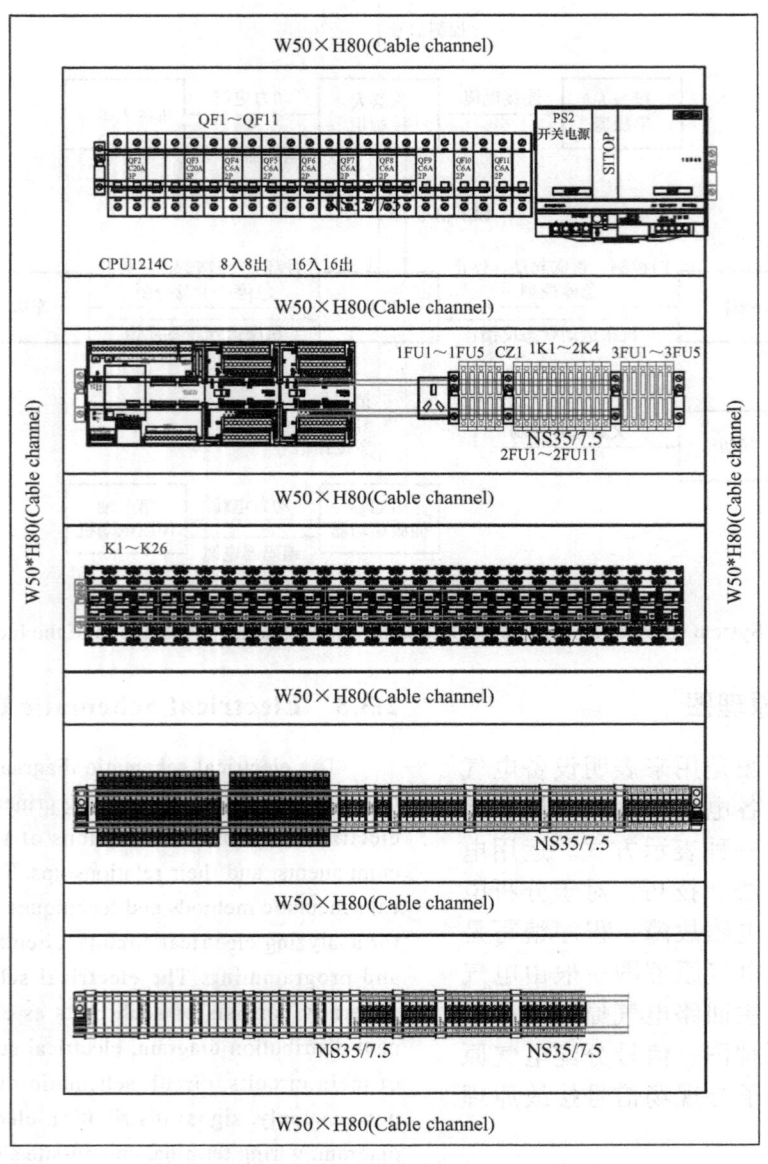

图 2-19　电气元件分布图

Figure 2-19　Electrical Component Distribution Diagram

第一层为断路器和开关电源，第二层为PLC模块和熔断器，第三层为继电器，第四、第五层为信号接线端子。

The first layer is for circuit breakers and switch power supplies, the second layer is for PLC modules and fuses, when the third layer is for relays, and the fourth and fifth layers are for signal wiring terminals.

2. 主回路电气原理图

1) 电柜进线经过柜门总断路器进入柜内第一层断路器，这是整个电柜的主电源进线，如图2-20所示。

2) 柜内断路器分别对变频器机器人、开关电源等分配控制电源，部分柜内断路器电源分配如图2-21所示。

2. Electrical schematic diagram of the main circuit

1) The incoming lines of the electrical cabinet enter the circuit breaker on the first layer inside the cabinet through the main circuit breaker of the cabinet door, which are the incoming lines of the main power supply of the entire electrical cabinet, as shown in Figure 2-20.

2) The circuit breakers inside the cabinet distribute control power to the frequency converter robots, switch power supplies, etc. Some of the power distribution of the circuit breakers inside the cabinet is shown in Figure 2-21.

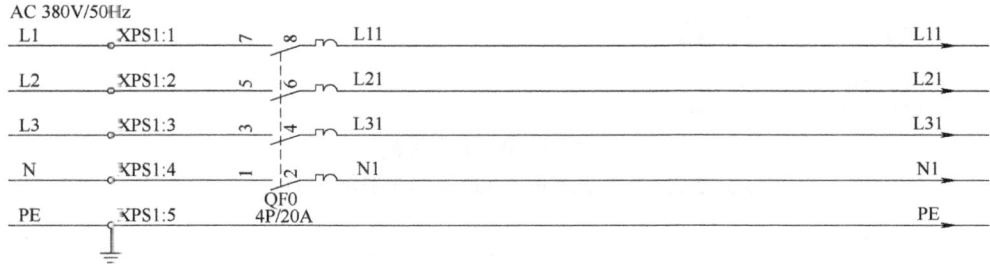

图 2-20　电柜进线

Figure 2-20　Incoming Line Diagram of the Electric Cabinet

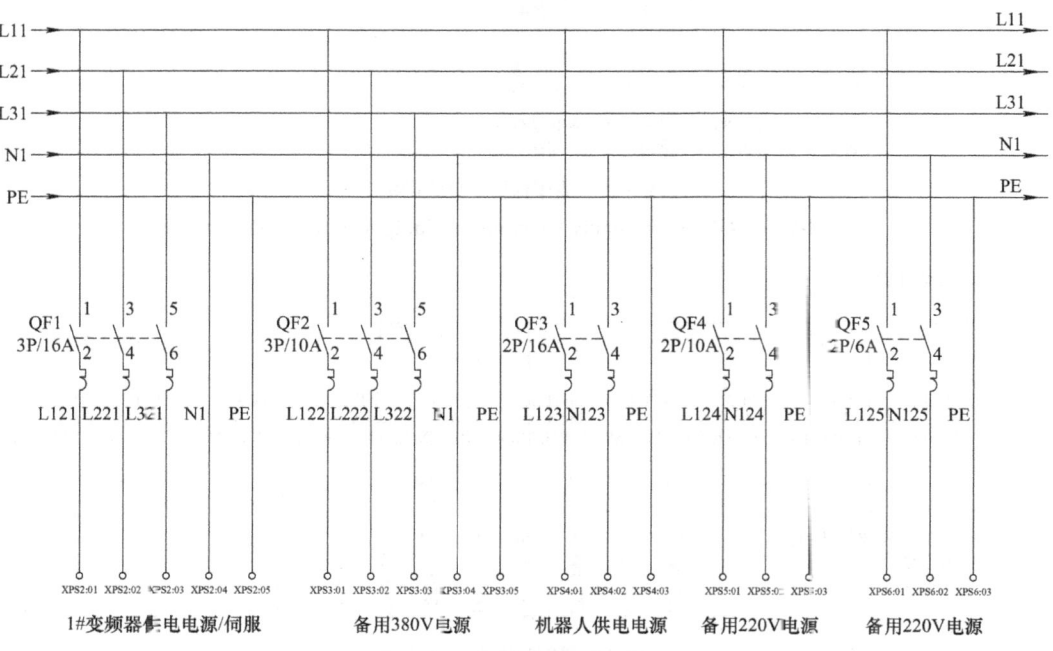

图 2-21　柜内断路器电源分配

Figure 2-21　Power Distribution Diagram of circuit breakers inside the Cabinet

3. 开关电源电路原理图

1) 开关电源分配如图 2-22 所示。

2) 柜内设备 24V 供电如图 2-23 所示。

3) 柜外设备 24V 供电如图 2-24 所示。

3. Schematic diagram of switch power supply circuit

1) The switch power supply distribution is shown in Figure 2-22.

2) The 24V power supply for the equipment inside the cabinet is shown in Figure 2-23.

3) The 24V power supply for the equipment outside the cabinet is shown in Figure 2-24.

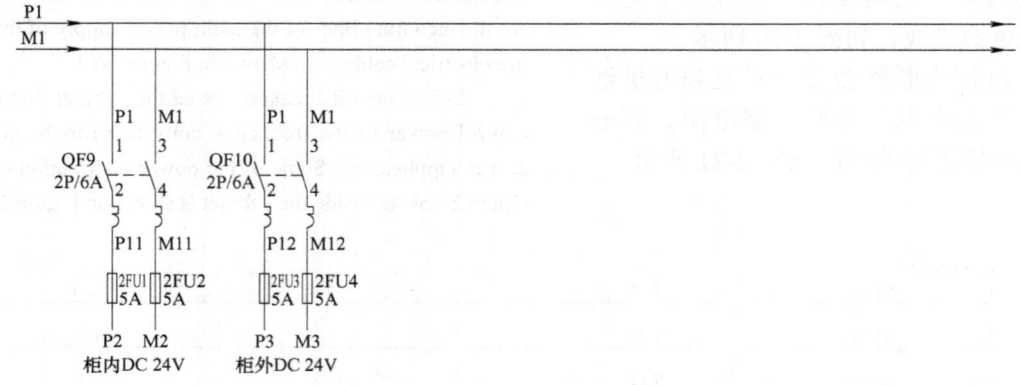

图 2-22 开关电源分配

Figure 2-22 Switch Power Supply Distribution Diagram

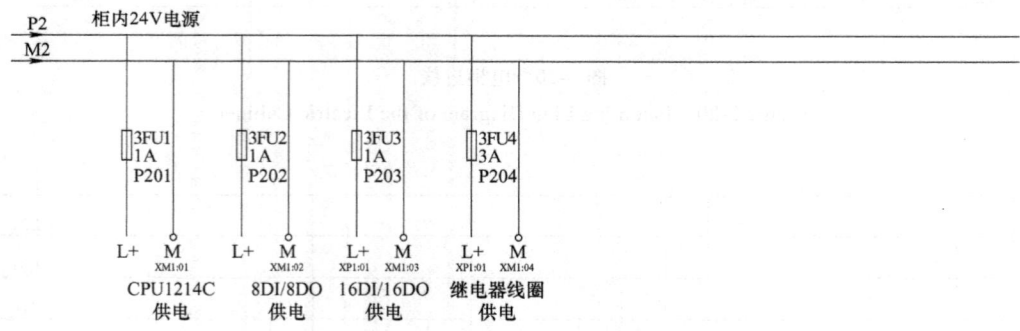

图 2-23 柜内设备 24V 供电

Figure 2-23 24V Power Supply Diagram for Equipment inside the Cabinet

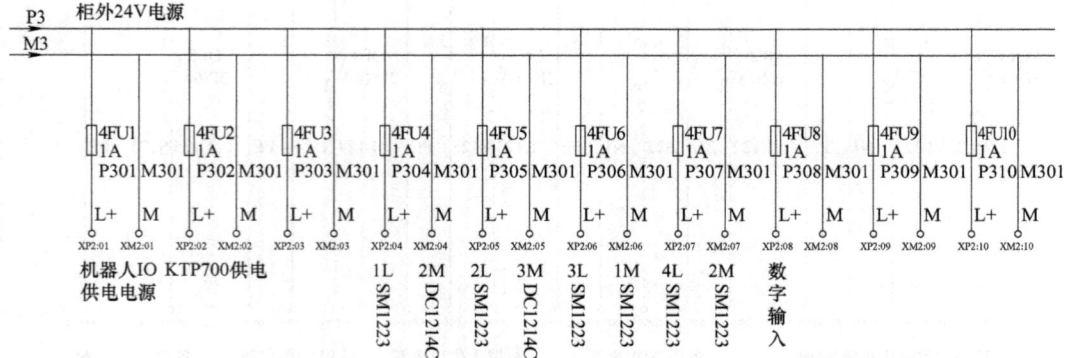

图 2-24 柜外设备 24V 供电

Figure 2-24 24V Power Supply Diagram for Equipment outside the Cabinet

4. 信号分配电气原理图

1) PLC 的信号输入全部由 PLC 模块端子接到电柜第四层对应的接线端子，外围信号接入到接线端子然后完成信号采集。数字量信号输入接线如图 2-25 所示。

2) PLC 输出信号通过控制继电器控制现场设备，继电器的一路常开触点接到电柜第五层相应的接线端子。数字量信号输出接线如图 2-26 所示。

4. Electrical schematic diagram of signal distribution

1) All the signal inputs of the PLC are connected from the PLC module terminals to the corresponding wiring terminals on the fourth layer of the electrical cabinet, and the peripheral signals are connected to the wiring terminals to complete signal acquisition. The digital signal input wiring is shown in Figure 2-25.

2) The PLC output signal controls the on-site equipment through a control relay, the normally open points of which is connected to the corresponding wiring terminal on the fifth layer of the electrical cabinet. The digital signal output wiring is shown in Figure 2-26.

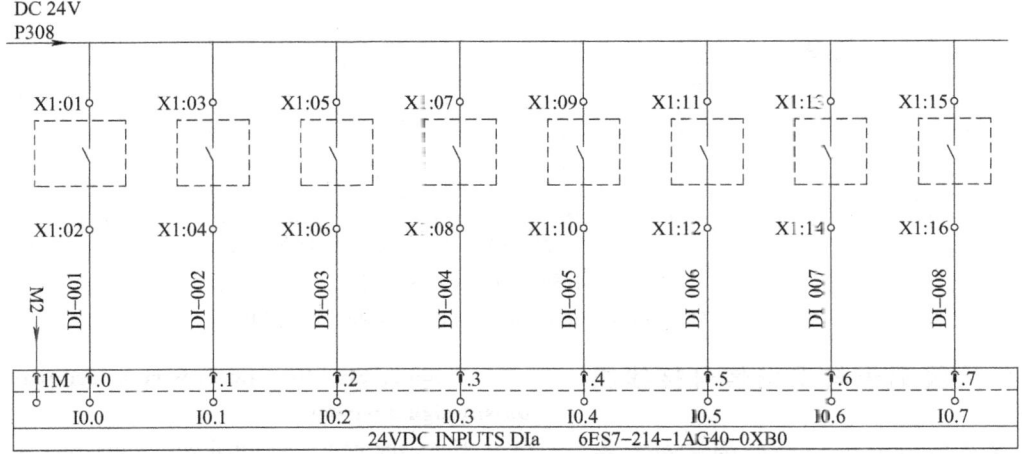

图 2-25 数字量信号输入接线

Figure 2-25 Digital Signal Input Wiring Diagram

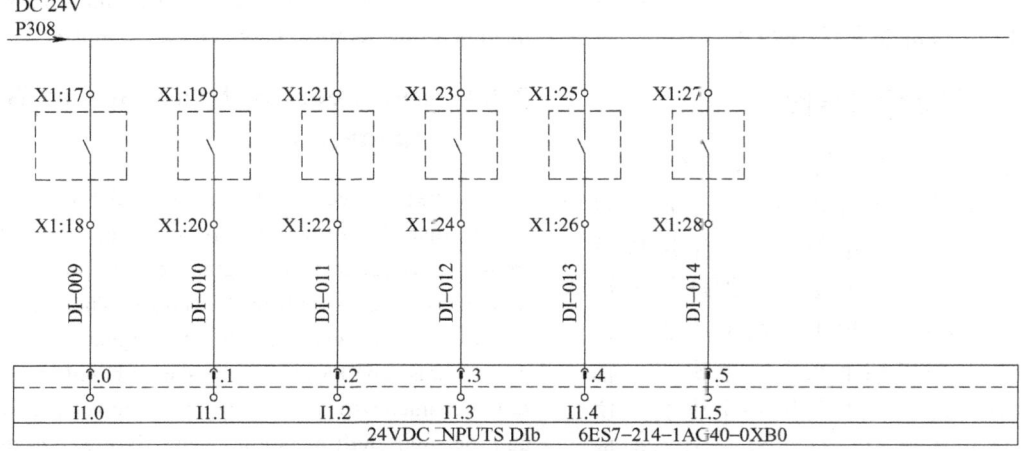

图 2-26 数字量信号输出接线

Figure 2-26 Digital Signal Output Wiring Diagram

3) PLC 的 16 入 16 出扩展模块为继电器输出模块,模块的输出直接与机器人输入相连接。部分机器人与 PLC 信号接线如图 2-27 所示。

3) The 16 inputs and 16 outputs expansion module of PLC is a relay output module. The output of the module is directly connected with the robot input. The connection of signals between some robots and PLC is shown in Figure 2-27.

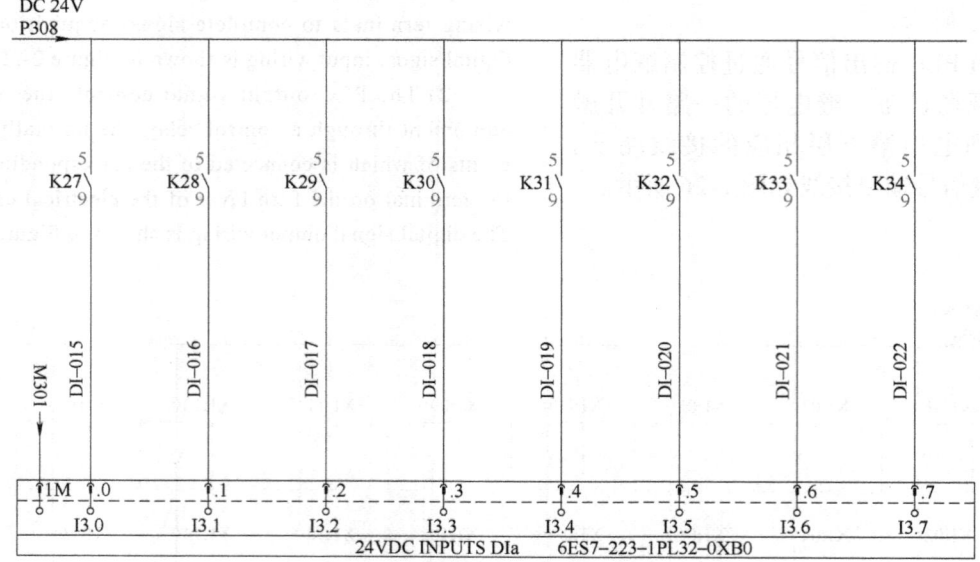

图 2-27　机器人与 PLC 信号接线
Figure 2-27　Connection of Signals between Robots and PLC

5. 接线端子与现场信号接线原理图

这里以电源端子接线图为例,展示接线端子接线原理图的绘制方法,如图 2-28 所示。请按示例绘制接线端子的现场信号输入图、PLC 信号输入图、PLC 与机器人信号图。

5. Schematic diagram of wiring terminals and on-site signal wiring

Here, the power terminal wiring diagram is taken as an example to demonstrate the drawing method for the schematic diagram of the terminal wiring, as shown in Figure 2-28. Please draw the field signal input diagram, PLC signal input diagram, PLC and robot signal diagram of the terminal based on the examples.

2.3.4　非标件工程图

在本项目中,需要针对 3 个工件,设计出安装在工业机器人第 6 轴法兰盘上的夹具。在本项目中,采用快换的方式进行夹具的更换,因此需要针对 3 个物料设计出 1 个快换夹具,其中根据上端盖和下端盖、电机转子轴的外形特征,可以共用一个夹具。这些夹具和快换工具的设计,因为根据工件的不同,需要特别的设计,并没

2.3.4　Engineering Drawings of Non-standard Parts

A fixture needs to be designed for 3 workpieces and installed on the flange at the 6th axis of the industrial robot in this program, and a quick-change method is used to replace the fixture. Therefore, one quick-change fixture needs to be designed for 3 kinds of materials. The fixture can be shared based on the external characteristics of the upper and lower end covers and the rotor shaft of motor. These fixtures and quick-change tool requires special design according to the different workpieces, as there are no standard parts.

项目2　典型工业机器人机床上下料工作站系统的设计及应用

有标准件，因此需要用到三维软件根据工件特性进行设计，大家可采用之前所学过的SOLIDWORKS软件或者NX软件进行非标产品的设计。

Therefore, three-dimensional software is needed to design according to the characteristics of the workpieces. The SOLIDWORKS software or NX software which you have learned before can be used for the design of non-standard products.

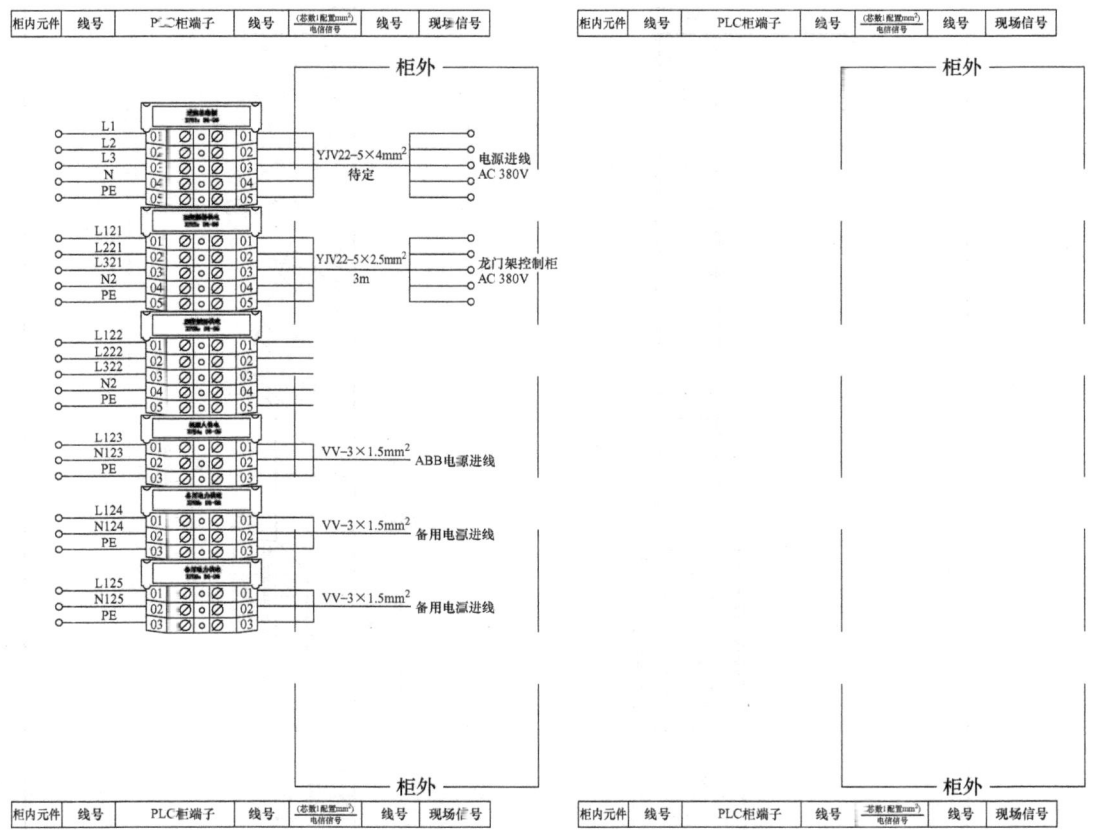

图 2-28　电源端子接线图

Figure 2-28　Power Terminal Wiring Diagram

本项目除了需要设计夹具外，还需要设计固定在机器人法兰盘上的快换固定装置。

这里以小圆形棒料夹具设计为例讲述夹具设计的具体步骤及方法，其他两个夹具的设计需要大家按要求自行绘制。

1) 首先设计出圆棒料夹具夹头，设计这个夹具夹头要考虑工件毛坯的尺寸、抓取位置、放置位置、气缸安装尺寸等因素，根据以上因素设计出圆棒料夹具夹头工程图如图2-29所示。

In addition to the design of fixtures, it also requires the design of a quick change-fixing device fixed to the flange of the robot in this program.

Here, the design of a small round bar fixture chuck is taken as an example to explain the specific steps and methods of fixture design. Please design the other two fixtures according to the requirements.

1) Firstly, design the round bar fixture chuck. For the design of this fixture chuck, one should consider factors such as the size of the workpiece blank, picking position, dropping position, and cylinder installation size, based on which a round bar fixture chuck is designed as shown in Figure 2-29.

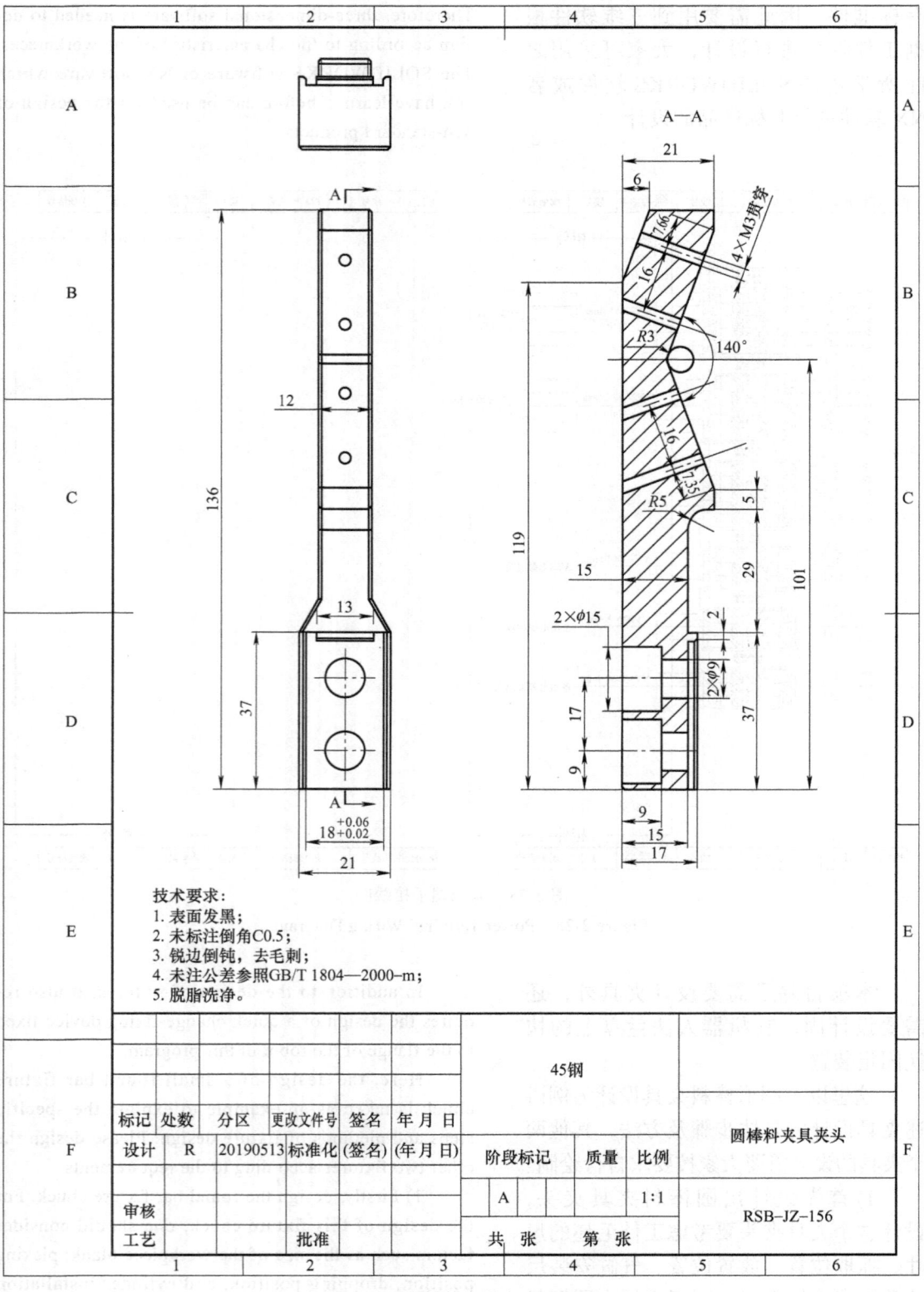

图 2-29　圆棒料夹具夹头工程图

Figure 2-29　Engineering Drawing of a Round Bar Fixture Chuck

2) 接着设计机器人夹具法兰，设计夹具法兰要考虑快换头连接尺寸、气缸连接尺寸以及夹具总体长度等，具体工程图如图2-30所示。

2) Next, design the robot fixture flange, for which one should take into account the connection size of the quick-change chuck, connection size of the cylinder, and overall length of the fixture. The specific engineering is shown in Figure 2-30.

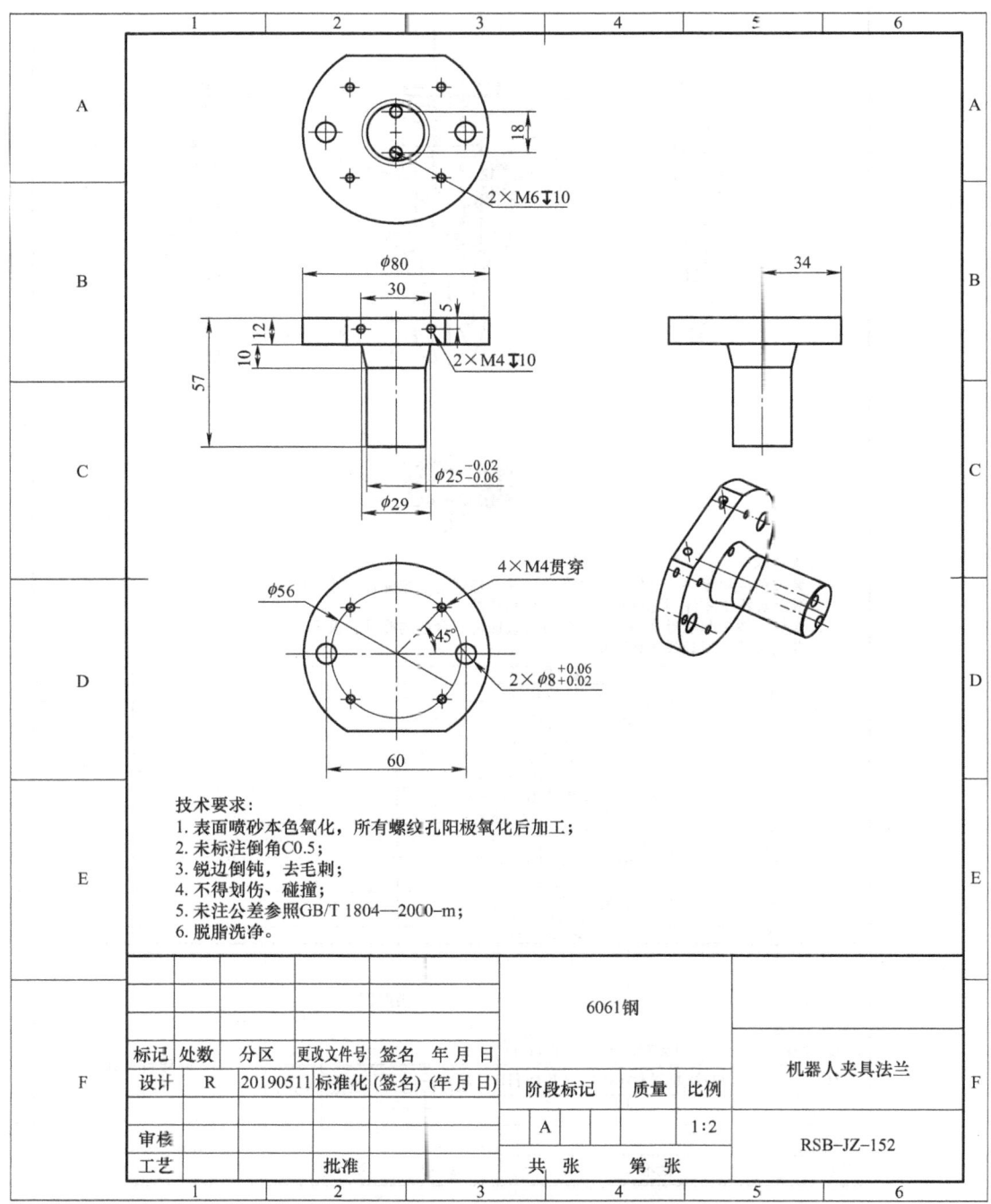

图 2-30 机器人夹具法兰工程图

Figure 2-30　Engineering Drawing of Robot Fixture Flange

3) 最后完成装配图，如图 2-31 所示。

3) Finally, complete the assembly diagram, as shown in Figure 2-31.

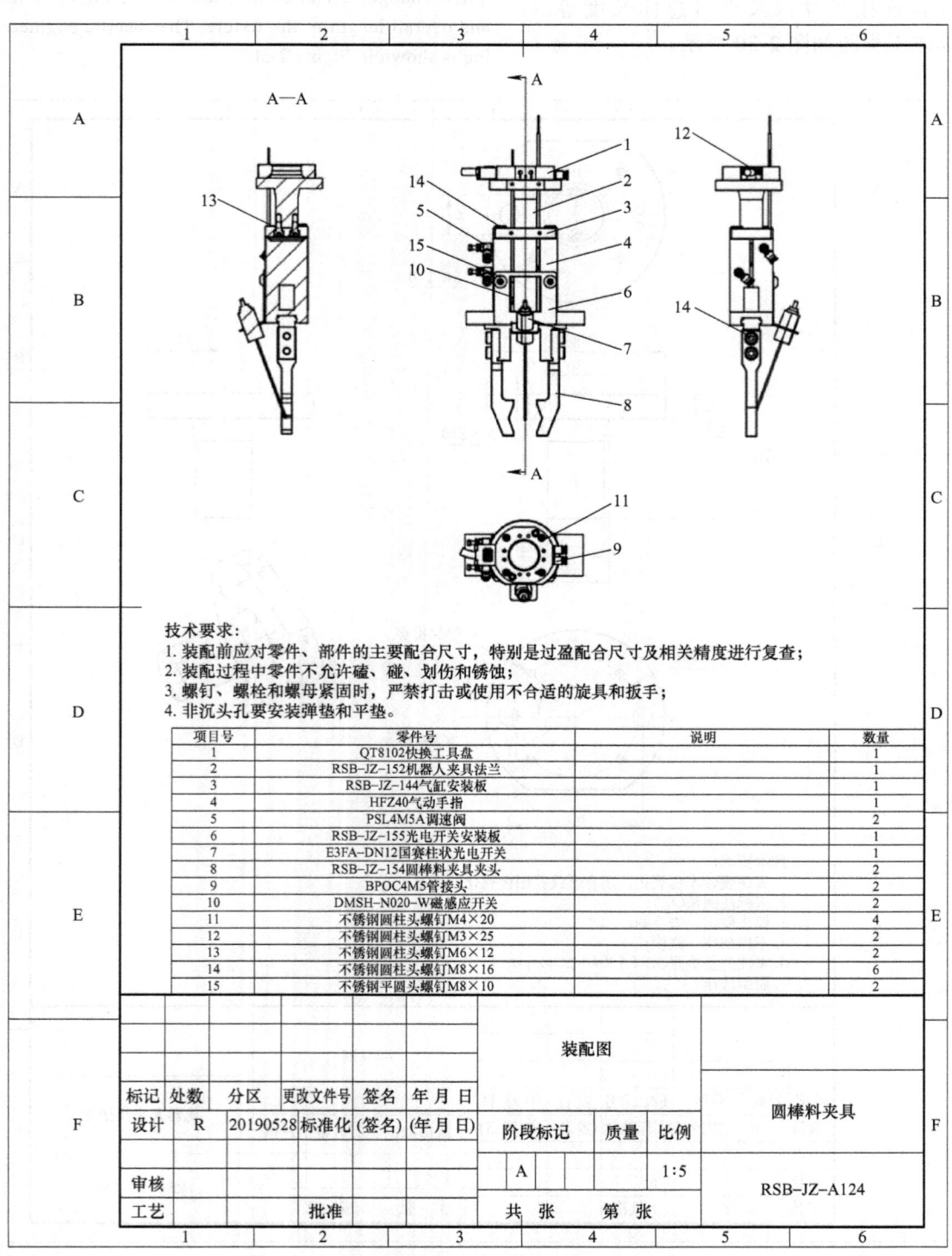

图 2-31 小圆棒料夹具装配图

Figure 2-31 Assembly Diagram of Small Round Bar Fixture

【课后巩固】

1. 依据以上部件的设计方法自行绘制工业机器人夹具支架的设计。

2. 针对特定的加工工件（上下端盖、电动机转子轴）是否可以采用吸盘式夹具呢？请说明你的理由。

[Consolidation after Class]

1. Prepare the design of the industrial robot fixture holder based on the design method of the above components.

2. Can suction cup fixtures be used for the machining of specific workpieces (upper or lower end covers, rotor shaft of motor)? Please explain your reasons.

任务 2.4　机器人机床上下料工作站的仿真
Task 2.4　Simulation of Machine Tool Loading and Unloading Workstation of Robot

【知识目标】

1. 掌握在 ROBOGUID 软件中导入三维模型的基本方法。

2. 掌握工业机器人仿真程序编写的基本方法。

3. 掌握机器人机床上下料系统仿真的基本方法。

[Knowledge Objectives]

1. To master the basic method of importing 3D models to the ROBOGuid software.

2. To master the basic methods for programming of industrial robot simulation.

3. To master the basic methods for the simulation of robot machine tool loading and unloading system.

【技能目标】

1. 能够根据任务要求、工艺要求进行机器人程序编写。

2. 能够正确导入/导出机器人仿真程序。

3. 能够准确区分工业机器人机床上下料工作站各部件在仿真软件中的类型及仿真布局。

[Skill Objectives]

1. To be able to program robots according to task and process requirements.

2. To be able to correctly import/export robot simulation programs.

3. To be able to accurately distinguish the types and simulation layouts of various components of the industrial robot machine tool loading and unloading workstation in the simulation software.

【素质目标】

1. 培养学生自主学习、自主探索的能力。

2. 培养学生善于观察、总结归纳的能力。

[Competence Objectives]

1. To cultivate students' ability to learn and explore independently.

2. To cultivate students' ability to observe and summarize.

【任务情景】

某企业需要将一数控加工工作

[Task Scenario]

An enterprise needs to design and transform a

站进行设计和改造，将机床上下料的工作全部由机器人自动完成，目前已确定工业机器人机床上下料工作站的工作流程及控制要求，根据给出的机器人机床上下料工作站设计方案，完成机器人机床上下料工作站的仿真项目。

CNC machining workstation, with all the loading and unloading work of the machine tool to be automatically completed by robots. Currently, the workflow and control requirements of the industrial robot machine tool loading and unloading workstation have been determined. Please complete the simulation of the robot machine tool loading and unloading workstation based on the provided design scheme of the robot machine tool loading and unloading workstation.

【任务分析】

1. 根据工作站布局图及现场环境情况搭建相应的仿真环境。

2. 根据任务要求完成机器人机床上下料仿真程序的编写。

3. 根据任务要求完成机器人机床上下料系统的仿真。

[Task Analysis]

1. Build a simulation environment based on the corresponding workstation layout and on-site environmental conditions.

2. Complete the programming of the robot machine tool loading and unloading simulation according to the task requirements.

3. Complete the simulation of the robot machine tool loading and unloading system according to the task requirements.

【知识准备】

在完成系统硬件的建模和工程图纸的设计后，正式设备投产和开展项目安装和调试前，还需要对设计出来的模型以及整个系统的功能进行仿真，通过仿真检验前期设计方案是否可行，因此仿真是项目正式开展现场施工前一个非常重要的环节。在本项目中，我们可以用FANUC机器人配套的软件ROBOGUIDE进行仿真，也可以用VISIUALONE和PROCESS软件进行仿真，具体选用哪款软件，可根据自身条件进行选用，这里以ROBOGUIDE为例进行讲解。

[Assumed Knowledge]

When completing the hardware modeling and the engineering design of the system, and before the formal operation of equipment, installation and commissioning of the project, it is necessary to simulate the designed model and the functions of the entire system. It is possible to verify the feasibility of the designed scheme through simulation, which is a very important step before the start of on-site construction of the project. In this program, we can use either the ROBOGUIDE software that matches up with the FANUC robots, or the VISIUALONE or PROCESS software for simulation, based on the actual conditions. Here, the ROBOGUIDE is taken as a case for explanation.

2.4.1 仿真环境搭建

根据工作站布局图2-17及现场环境情况搭建出相应的机器人机床上下料工作站仿真环境，如图2-32所示。

2.4.1 Establishment of Simulation Environment

Establish a simulation environment for the robot machine tool loading and unloading workstation based on the corresponding workstation layout of diagram 2-17 and the on-site environment, as shown in Figure 2-32.

项目2 典型工业机器人机床上下料工作站系统的设计及应用

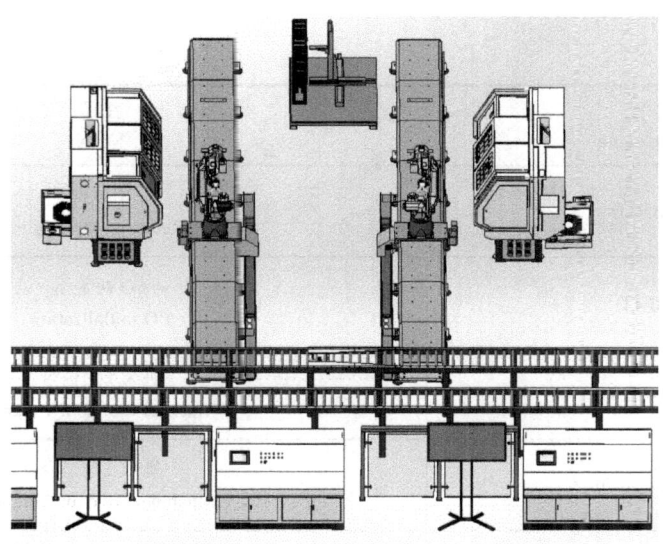

图 2-32 机器人机床上下料工作站仿真环境
Figure 2-32 Simulation Environment of the Robot Machine Tool Loading and Unloading Workstation

2.4.2 工作站仿真

1. 仿真工作站机器人程序编写

仿真环境建立后，必须在仿真软件中按照工作站工艺流程编写相应的机器人程序才能够进行运动仿真。这里将参考程序展示给大家，其中表 2-10 为程序介绍。

2.4.2 Workstation Simulation

1. Programming of the simulation of workstation robot

When the simulation environment is built, a robot program must be written in the simulation software according to the process flow of the workstation in order to perform motion simulation. Here, a reference program is shown, with the program introduction in the Table 2-10.

表 2-10 机器人程序介绍
Table 2-10 Introduction to Robot Programs

程序名 Program name	注释 Notes
CNC1-FEED	工件放置机床 1 Workpiece feeding CNC machine 1
CNC1-TAKE	工件抓取机床 1 Workpiece taking CNC machine 1
CNC2-FEED	工件放置机床 2 Workpiece feeding CNC machine 2
CNC2-TAKE	工件抓取机床 2 Workpiece taking CNC machine 2

（续）

程序名 Program name	注释 Notes
GO HOME	回原点 Return to the origin
I/O_INIT	I/O 初始化 I/O initialization
DOOR1-DETE	机床 1 门信号 Machine tool door 1 signal
DOOR2-DETE	机床 2 门信号 Machine tool door 2 signal
FEED-001	工件放置工位 1 Workpiece feeding station 1
FEED-002	工件放置工位 2 Workpiece feeding station 2
FEED-003	工件放置工位 3 Workpiece feeding station 3
GET-DATA	接收数据 Receive data
READ-ID-01	读取 RFID 信号 1 Read RFID signal 1
TAKE-001	抓取工位 1 Taking station 1
TAKE-002	抓取工位 2 Taking station 2
TAKE-003	抓取工位 3 Taking station 3
VISI/ON	视觉程序 Visual program
RSR001	主程序 Main program
SENDDATA	发送数据 Send data
SENDEVNT	发送结果 Send results

项目2　典型工业机器人机床上下料工作站系统的设计及应用

具体程序及程序注解如下：　　　　　　The specific program and its annotations are as follows:

```
 1:  UTOOL_NUM=9 ;                                                      坐标系9
 2:  UFRAME_NUM=1 ;                                                     坐标系1
 3:  WAIT RI[2:OFF:Jig1 jiawei]=ON AND DI[16:OFF:lingjia OK]=ON   ;     等待放松到位和料架停止到位
 4:  WAIT R[3]>=121 AND R[3]<=150   ;                                   等待料架层数
 5:  WAIT DI[16:OFF:lingjia OK]=ON   ;                                  料架停止到位
 6:  DO[1:ON :Rack1 Action LN]=OFF  ;                                   禁止料架旋转打开
 7:L PR[11:lian jia WEI] 200mm/sec CNT30   ;                            移动
 8:J PR[31:lianjia FANG] 50% CNT30   ;                                  移动
 9:  DO[1:ON :Rack1 Action LN]=OFF  ;                                   禁止料架旋转打开
10:L P[1] 400mm/sec FINE   ;                                            料架入口
11:L P[2] 50mm/sec FINE   ;                                             上升位置
12:  RO[1:OFF:jiazhuao1 ]=OFF  ;                                        夹爪放松
13:  WAIT RI[1:ON :Jig1 songwei]=ON   ;                                 夹爪放松等待时间
14:  WAIT    .20(sec)   ;                                               等待时间
15:L P[3] 30mm/sec FINE   ;                                             移动
16:L P[4] 50mm/sec FINE   ;                                             移动
17:  DO[1:ON :Rack1 Action LN]=ON  ;                                    料架可旋转
18:L PR[31:lianjia FANG] 400mm/sec CNT30   ;                            移动
19:J PR[11:lian jia WEI] 50% CNT30   ;                                  移动
20:L P[5] 300mm/sec CNT30   ;                                           移动
21:    ;                                                                结束
22:    ;                                                                结束
23:  END   ;                                                            结束
```

```
 1:  UTOOL_NUM=9 ;                                                      坐标系9
 2:  UFRAME_NUM=7 ;                                                     坐标系7
 3:  WAIT R[4]=1   ;                                                    机床号
 4:  IF R[3]<>0,JMP LBL[1]   ;                                          判断是否直接放料
 5:L P[2] 300mm/sec CNT30   ;                                           移动
 6:J P[3] 50% FINE   ;                                                  移动
 7:  WAIT   .50(sec)  ;                                                 等待时间
 8:  WAIT RI[3:ON :jia1 guan wei]=ON AND DI[14:OFF:CNC1 kai men OK]=ON  ;  等待和门打开到位
 9:L P[4] 400mm/sec FINE   ;                                            移动
10:L P[5] 400mm/sec FINE   ;                                            移动
11:  LBL[1] ;                                                           标签
12:L P[8] 300mm/sec FINE   ;                                            移动
13:L P[9] 100mm/sec FINE   ;                                            移动
14:  DO[3:OFF:CNC1 kapan kai]=ON   ;                                    卡盘打开
15:  RO[2:OFF:jiazhuang2]=OFF   ;                                       夹爪放到位
16:  WAIT RI[4:ON :Jig2 songwei]=ON   ;                                 等待夹爪放到位
17:  WAIT   .50(sec)   ;                                                等待时间
18:  DO[10:OFF:CNC1 chuiqi]=ON   ;                                      吹气
19:L P[1] 50mm/sec FINE   ;                                             移动
20:L P[10] 50mm/sec FINE   ;                                            移动
21:  DO[3:OFF:CNC1 kapan kai]=OFF   ;                                   卡盘关闭
22:L P[7] 300mm/sec FINE   ;                                            移动
23:L P[11] 350mm/sec FINE   ;                                           移动
24:  DO[10:OFF:CNC1 chuiqi]=ON   ;                                      吹气打开
25:L P[12] 300mm/sec FINE   ;                                           移动
26:    ;
27:  WAIT DI[1:OFF:CNC1-kapan-ok]=ON   ;                                气密检测OK
28:  DO[10:OFF:CNC1 chuiqi]=OFF   ;                                     吹气关闭
```

```
 1:  UTOOL_NUM=9 ;                                                      坐标系9
 2:  UFRAME_NUM=1 ;                                                     坐标系1
 3:  WAIT R[4]=2   ;                                                    机床号
 4:  IF R[3]<>0,JMP LBL[2]   ;                                          判断是否直接放料
 5:L PR[22:CNC2 ruo kou] 200mm/sec FINE   ;                             移动
 6:J P[2] 50% FINE   ;                                                  移动
 7:  WAIT RI[3:ON :jia1 guan wei]=ON AND DI[9:OFF:CNC2 kai men OK]=ON  ;  等待开门到位
 8:L P[3] 400mm/sec FINE   ;                                            移动
 9:  WAIT   .80(sec)   ;                                                等待时间
10:  WAIT RI[3:ON :jia1 guan wei]=ON   ;                                等待开门到位
11:L P[1] 300mm/sec FINE   ;                                            移动
12:    ;
13:  LBL[2] ;                                                           标签
14:L P[4:fang lian kou] 250mm/sec FINE   ;                              移动
15:L P[11] 150mm/sec FINE   ;                                           移动
16:  DO[4:OFF:CNC2 kapan kai]=ON   ;                                    卡盘打开
17:  RO[2:OFF:jiazhuang2]=OFF   ;                                       夹爪放到位
18:  WAIT RI[4:ON :Jig2 songwei]=ON   ;                                 等待夹爪放到位
19:  WAIT   .20(sec)   ;                                                等待时间
20:  DO[11:OFF:CNC2 chuiqi]=ON   ;                                      吹气
21:L P[5:xia jiang wei] 50mm/sec FINE   ;                               移动
22:L P[6:hou tui wei] 70mm/sec FINE   ;                                 移动
23:  DO[4:OFF:CNC2 kapan kai]=OFF   ;                                   卡盘关闭
24:L P[12] 350mm/sec FINE   ;                                           移动
25:L P[7] 350mm/sec FINE   ;                                            移动
26:L P[8] 300mm/sec FINE   ;                                            移动
27:  DO[11:OFF:CNC2 chuiqi]=ON   ;                                      吹气打开
28:  WAIT DI[2:OFF:CNC2 kapan OK]=ON   ;                                气密检测OK
```

2. 仿真验证

在创建的工作站仿真环境中，依据工作站工艺流程进行结构仿真和运动仿真，仿真流程如图 2-33 ～图 2-37 所示，用以完成以下工作：

（1）机器人抓取毛坯及放到工作台卡盘上的过程

毛坯料通常由链条式传送带运输到指定的位置，由气动或电动定位机构进行初步定位，保证每次机器人从同一位置抓取零件。当 X 轴向右运动到毛坯料前方时停止运动，Z 轴向下运动使张开的手爪刚好能抓住毛坯件。这时闭合手爪抓住毛坯。然后 Z 轴向上运动到指定高度后（不会发生碰撞），X 轴向左运动到工作台卡盘正上方，然后 Z 轴向下运动把毛坯装入卡盘或工装内。然后卡盘夹紧，Z 轴上升到超出机床防护罩上方，X 轴再运动到毛坯上方或等待卡盘上方。

（2）从工作台卡盘取下零件及放置到特定位置的过程

当 X 轴运动到卡盘的正上方后，Z 轴向下运动使手爪刚好能抓住工件，然后给气压使手爪合并抓住工件。这时机械手的控制系统控制液压卡盘松开，当控制系统得到卡盘松开信号后，Z 轴向上运动到出来机床防护板，然后 X 轴向左运动（取决于放下料的位置）把工件运动到放料位置正上方。这时 Z 轴下降到放料件上，再张开手爪及提升 Z 轴，从而完成取料及放料过程。

2. Simulation verification

In the workstation simulation environment established, structural simulation and motion simulation are carried out based on the workstation process flow. The simulation process is shown in Figure 2-33 to 2-37 to complete the following tasks:

(1) The process of the robot grabbing the workpiece blanks and placing them on the workbench chuck

Workpiece blanks are usually conveyed to designated positions by chain conveyors, and are initially positioned by pneumatic or electric positioning mechanisms to ensure that the robot grabs the parts from the same position every time. When the X-axis moves rightward to front of the blank and stops, the Z-axis moves downward until the open gripper can just grasp the blank and, at this moment, close the gripper to grab the blank. When the Z-axis moves up to the specified height (without collision), the X-axis moves leftward to right above the workbench chuck, and then the Z-axis moves downward to load the blank into the chuck or fixture. Then, the chuck clamps, the Z-axis rises above the protective cover of the machine tool, and the X-axis moves to the top of the blank or the waiting chuck.

(2) The process of removing parts from the workbench chuck and placing them in a specific position

When the X-axis moves right above the chuck, the Z-axis moves downward until the gripper is just able to grasp the workpiece, and then applies air pressure to make the gripper close and grasp the workpiece. At this time, the control system of the robotic arm controls the hydraulic chuck to release. When the control system receives the signal of chuck release, the Z-axis moves up out of the protective plate of the machine tool, and then the X-axis moves left (depending on the position where the workpiece is placed) to move the workpiece directly above the feeding position. At this time, the Z-axis descends to the feeding piece, opens the gripper and rises, which completes the process of workpiece taking and feeding.

图 2-33 仿真流程 1
Figure 2-33　Simulation Process 1

图 2-34 仿真流程 2
Figure 2-34　Simulation Process 2

图 2-35 仿真流程 3
Figure 2-35　Simulation Process 3

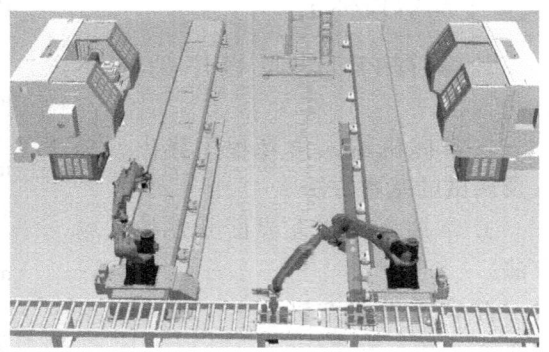

图 2-36 仿真流程 4
Figure 2-36　Simulation Process 4

图 2-37 仿真流程 5
Figure 2-37　Simulation Process 5

【课后巩固】

[Consolidation after Class]

1. 能导入 Process Simulate 仿真软件的模型一般都有哪些格式？

2. ROBOGUIDE 仿真软件是哪个公司开发的？

3. Process Simulate 仿真软件可以仿真哪些品牌的机器人？

1. What are the formats in general for models that can be imported into Process Simulate simulation software?

2. Which company developed the ROBOGUIDE simulation software?

3. What brands of robots can the Process Simulate software simulate?

任务 2.5　工作站安装与调试及程序的编写
Task 2.5　Workstation Installation, Commissioning and Their Programming

【知识目标】

[Knowledge Objectives]

1. 掌握 PLC 编写程序的基本方法。

2. 掌握机器人仿真程序导出、导入及调试的基本方法。

3. 掌握人机交互界面开发的基本方法。

4. 掌握机器人机床上下料系统电气接线的基本方法。

5. 掌握系统联调的基本方法。

1. To master the basic methods of PLC programming.

2. To master the basic methods of the exporting, importing and commissioning of robot simulation programs.

3. To master the basic methods of developing human-computer interaction interfaces.

4. To master the basic methods of electrical wiring for the robot machine tool loading and unloading system.

5. To master the basic methods of joint commissioning of the system.

【技能目标】

[Skill Objectives]

1. 能够根据任务要求、工艺要求编写工作站的 PLC 程序。

2. 能够将仿真项目中的机器人程序导出并重新示教点位。

3. 能够完成人机交互界面的开发。

4. 能够完成机器人机床上下料系统的电气接线。

5. 能够完成机器人机床上下料的系统联调。

1. To be able to write PLC programs for workstations according to task and process requirements.

2. To be able to export robot programs and reteach the points in simulation programs.

3. To be able to develop human-computer interaction interfaces.

4. To be able to complete the electrical wiring of the robot machine tool loading and unloading system.

5. To be able to complete joint commissioning of the robot machine tool loading and unloading system.

【素质目标】

1. 培养学生务实求真、自主探索的能力。

2. 培养学生大胆探索、敢于创新的能力。

[Competence Objectives]

1. To cultivate students' ability to be pragmatic, seek truth, and explore independently.

2. To cultivate students' ability to boldly explore and innovate.

【任务情景】

某企业需要将一数控加工工作站进行改造，将机床上下料的工作全部改成由机器人自动完成，目前已确定工业机器人机床上下料工作站的工作流程及控制要求，请你根据给出的机器人机床上下料工作站的施工图、设计方案、仿真项目，完成机器人机床上下料工作站的安装与调试。

[Task Scenario]

An enterprise needs to transform a CNC machining workstation, with all the loading and unloading work of the machine tool to be automatically completed by robots. The workflow and control requirements of the industrial robot machine tool loading and unloading workstation have been determined. Please complete the installation and commissioning of the robot machine tool loading and unloading workstation based on the provided detailed drawings, design schemes, and simulation programs.

 【任务分析】

1. 根据工作站的设计方案及仿真项目，编写工作站的 PLC 程序。

2. 将仿真项目中的机器人程序导出并重新示教点位。

3. 根据任务要求完成机器人机床上下料工作站的人机交互界面。

4. 根据任务要求完成机器人机床上下料系统的电气接线。

5. 根据任务要求完成机器人机床上下料的系统联调。

[Task Analysis]

1. Write the PLC program for the workstation based on the design scheme and simulation program of the workstation.

2. Export the robot program in the simulation program and re-teach the points.

3. Complete the human-machine interaction interface of the robot machine tool loading and unloading workstation according to the task requirements.

4. Complete the electrical wiring of the robot machine tool loading and unloading system according to the task requirements.

5. Complete the joint commissioning of the robot machine tool loading and unloading system according to the task requirements.

 【知识准备】

前面通过仿真已经对前期设计好的图纸和模型进行可行性论证，并对仿真出来的不合理的设计进行修改和调整。在完成了系统的仿真后下一步即可进入工作站的安装与调试环节，工作站的安装调试通过前期设计好的

[Assumed Knowledge]

The feasibility of the previously designed drawings and models has been verified through simulation, and unreasonable designs found by simulation have been modified and adjusted. The next step after the system simulation is the installation and commissioning of the workstation, which are required to be conducted according to the relevant detailed drawings previously

相关施工图样进行规范施工即可。在本项目中，在进行现场硬件的安装与调试的同时，即可进行 PLC 程序的编写和触摸屏工程的设计。在完成硬件安装和调试后，即可进行系统联调。

done. In this program, when the on-site hardware installation and commissioning are under progress, the PLC program can be written and the design of touch screen engineering conducted simultaneously. The joint commissioning of the system can be conducted when the hardware installation and commissioning are completed.

2.5.1 工作站安装

2.5.1 Workstation Installation

1. 设备及部件安装

根据施工图样完成各设备及部件的安装，工作站总体效果图如图 2-38 和图 2-39 所示。

2. 电气安装

按照电气原理图及硬件安装位置，完成电气设备安装及接线，总体效果图如图 2-40 所示。

1. Equipment and components installation

Complete the installation of various equipment and components according to the detailed drawings, and the rendering of the overall workstation is shown in Figure 2-38 and 2-39.

2. Electrical installation

Complete the installation and wiring of electrical equipment according to the electrical schematic diagrams and hardware installation locations, as shown in Figure 2-40.

图 2-38 机器人机床上下料工作站安装效果图 1

Figure 2-38 Rendering Drawing 1 of the Installation of Robot Machine Tool Loading and Unloading Workstation 1

图 2-39 机器人机床上下料工作站安装效果图 2

Figure 2-39 Rendering Drawing 2 of the Installation of Robot Machine Tool Loading and Unloading Workstation 2

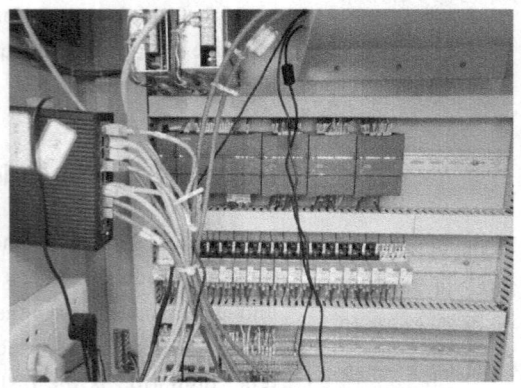

图 2-40 工作站电气安装总体效果图

Figure 2-40 Rendering Drawing of the Overall Electrical Installation of the Workstation

2.5.2 PLC 程序编写及交互界面设计

1. I/O 通信变量表

首先根据规划创建 PLC 的 I/O 变量表，见表 2-11，然后分别编写各部分程序。

2. PLC 程序编写

依据 I/O 通信变量表及工艺流程要求编写机器人程序。

1) 程序结构如图 2-41 所示。
2) 主程序如图 2-42 所示。

2.5.2 PLC Programming and Interactive Interface Design

1. Table of I/O communication variable

Firstly, create the I/O variable table for the PLC according to the plan, as shown in Table 2-11, and then write separate programs for each part.

2. PLC programming

Write robot programs based on the table of I/O communication variable and process requirements.

1) The program structure is shown in Figure 2-41.
2) The main program is shown in Figure 2-42.

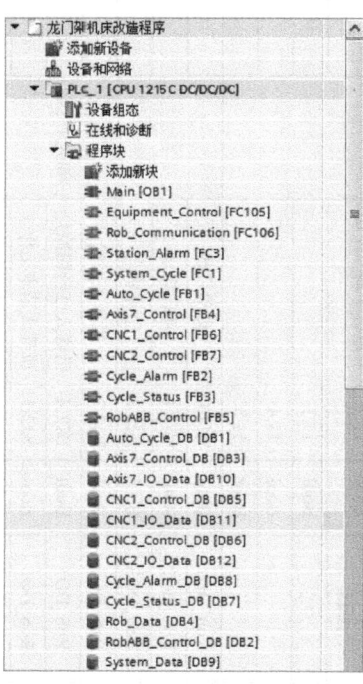

图 2-41　PLC 程序结构图

Figure 2-41　PLC Program Structure Diagram

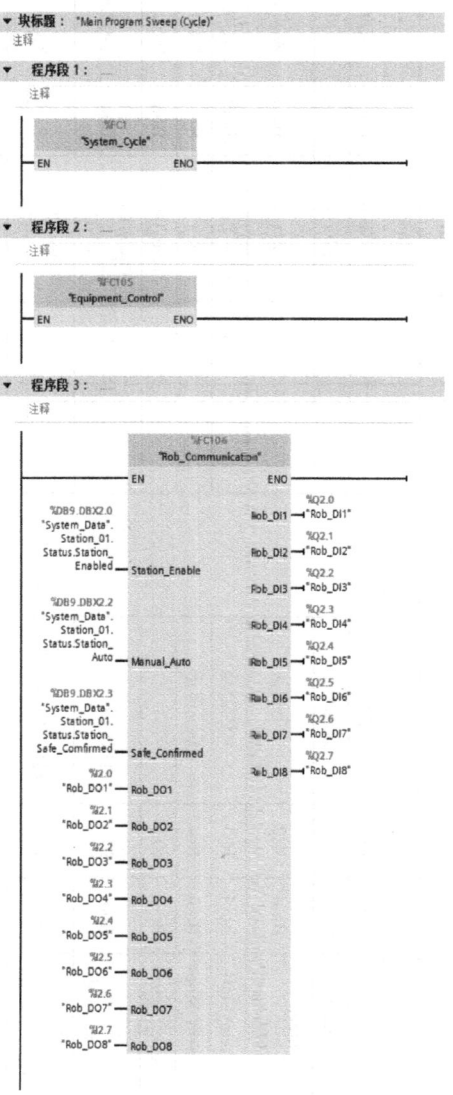

图 2-42　PLC 主程序

Figure 2-42　PLC Main Program

表 2-11　PLC 的 I/O 变量表
Table 2-11　I/O variable Table of PLC

序号	符号	格式	地址	注释	槽号	端子位号	序号	符号	格式	地址	注释	槽号	端子位号
1	ESTOP	Bool	I0.0	急停			1	LIG-R	Bool	%Q0.0	机床2卡盘松开		
2	Enable	Bool	I0.1	输出使能			2	LIG-G	Bool	%Q0.1	机床2卡盘夹紧		
3	Safety_Confirm	Bool	I0.2	安全确认			3	LIG-O	Bool	%Q0.2			
4	System_Start	Bool	I0.3	系统启动			4	LIG-Buzzer	Bool	%Q0.3	机床1液压松开		
5	System_Stop	Bool	I0.4	系统停止	1214C		5	Hyd_Close	Bool	%Q0.4	机床1液压夹紧	1214C	
6	Check_Signal	Bool	I0.5				6	Hyd_Open	Bool	%Q0.5			
7	CNC1_DoorOpend	Bool	I0.6	机床门1开到位			7	OpenDoor_CNC1	Bool	%Q0.6	机床门1开门		
8	CNC1_DoorClosed	Bool	I0.7	机床门1关到位			8	CloseDoor_CNC1	Bool	%Q0.7	机床门1关门		
9	CNC2_DoorOpend	Bool	I1.0	机床门2开到位			9	OpenDoor_CNC2	Bool	%Q1.0	机床门2开门		
10	CNC2_DoorClosed	Bool	I1.1	机床门2关到位			10	CloseDoor_CNC2	Bool	%Q1.1	机床门2关门		
11	KP_Open_Back	Bool	I1.2	卡盘打开动作反馈			11		Bool				
12		Bool	I1.3				12		Bool				
13		Bool	I1.4				13		Bool				
14		Bool	I1.5				14		Bool				
15	Rob_DO1	Bool	%I2.0	自动启动			15	Rob_DI1	Bool	%Q2.0	机器人上电		
16	Rob_DO2	Bool	%I2.1	循环启动			16	Rob_DI2	Bool	%Q2.1	机器人启动		
17	Rob_DO3	Bool	%I2.2	急停状态			17	Rob_DI3	Bool	%Q2.2	机器人停止		
18	Rob_DO4	Bool	%I2.3	机器人报错	16DI/16DO		18	Rob_DI4	Bool	%Q2.3	机器人急停	16DI/16DO	
19	Rob_DO5	Bool	%I2.4	机器人已经回原位			19	Rob_DI5	Bool	%Q2.4	机器人复位		
20	Rob_DO6	Bool	%I2.5	机器人请求工作1			20	Rob_DI6	Bool	%Q2.5	工作1完成		
21	Rob_DO7	Bool	%I2.6	机器人请求工作2			21	Rob_DI7	Bool	%Q2.6	工作2完成		
22	Rob_DO8	Bool	%I2.7				22	Rob_DI8	Bool	%Q2.7			
23	CNC1_Work	Bool	%I3.0	机床1作业中			23	CNC1_Start	Bool	%Q3.0	机床1启动		

项目2 典型工业机器人机床上下料工作站系统的设计及应用

(续)

序号	符号	格式	地址	注释	槽号	端子位号
24	CNC1_Stoped	Bool	%I3.1	机床1已停止	16DI/16DO	
25	CNC1_Fault	Bool	%I3.2	机床1故障		
26	CNC2_Work	Bool	%I3.3	机床2作业中		
27	CNC2_Stoped	Bool	%I3.4	机床2已停止		
28	CNC2_Fault	Bool	%I3.5	机床2故障		
29		Bool	%I3.6			
30		Bool	%I3.7			
15	Rob_DI1	Bool	DI0	机器人上电	DSQC651	
16	Rob_DI2	Bool	DI1	机器人启动		
17	Rob_DI3	Bool	DI2	机器人停止		
18	Rob_DI4	Bool	DI3	机器人急停		
19	Rob_DI5	Bool	DI4	机器人复位		
20	Rob_DI6	Bool	DI5	工作1完成		
21	Rob_DI7	Bool	DI6	工作2完成		
22	Rob_DI8	Bool	DI7			
23	Rob_ClampClosed	Bool	DI8	机器人夹具夹紧到位		
24	Rob_ClampOpend	Bool	DI9	机器人夹具打开到位		
25	Rob_7_Pos0	Bool	DI10	机器人在龙门架原点		
26	Rob_7_Pos1	Bool	DI11	机器人在龙门架位置1		
27	Rob_7_Pos2	Bool	DI12	机器人在龙门架位置2		
28		Bool	DI13			
29		Bool	DI14			
30		Bool	DI15			

序号	符号	格式	地址	注释	槽号	端子位号
24	CNC1_Stop	Bool	%Q3.1	机床1停止	16DI/16DO	
25	CNC1_Estop	Bool	%Q3.2	机床1急停		
26	CNC2_Start	Bool	%Q3.3	机床2启动		
27	CNC2_Stop	Bool	%Q3.4	机床2停止		
28	CNC2_Estop	Bool	%Q3.5	机床2急停		
29	Axis7_Estop	Bool	%Q3.6	天轨急停		
30		Bool	%Q3.7			
15	Rob_DO1	Bool	DO0	自动启动	16DI/16DO	
16	Rob_DO2	Bool	DO1	循环启动		
17	Rob_DO3	Bool	DO2	急停状态		
18	Rob_DO4	Bool	DO3	机器人报错		
19	Rob_DO5	Bool	DO4	机器人已经回原位		
20	Rob_DO6	Bool	DO5	机器人请求工作1		
21	Rob_DO7	Bool	DO6	机器人请求工作2		
22	Rob_DO8	Bool	DO7			
23	Rob_ClampClose	Bool	DO8	机器人夹具夹紧		
24	Rob_ClampOpen	Bool	DO9	机器人夹具打开		
25	Ask_Rob_7_Pos0	Bool	DO10	请求到龙门架原点		
26	Ask_Rob_7_Pos1	Bool	DO11	请求到龙门架位置1		
27	Ask_Rob_7_Pos2	Bool	DO12	请求到龙门架位置2		
28		Bool	DO13			
29		Bool	DO14			
30		Bool	DO15			

3) 功能程序仅展示工位状态处理程序，如图 2-43 所示。其他功能程序需要大家依据要求自行编写。

3) The functional program only displays the station status processing program, as shown in Figure 2-43. Other functional programs need to be written by yourself according to the requirements.

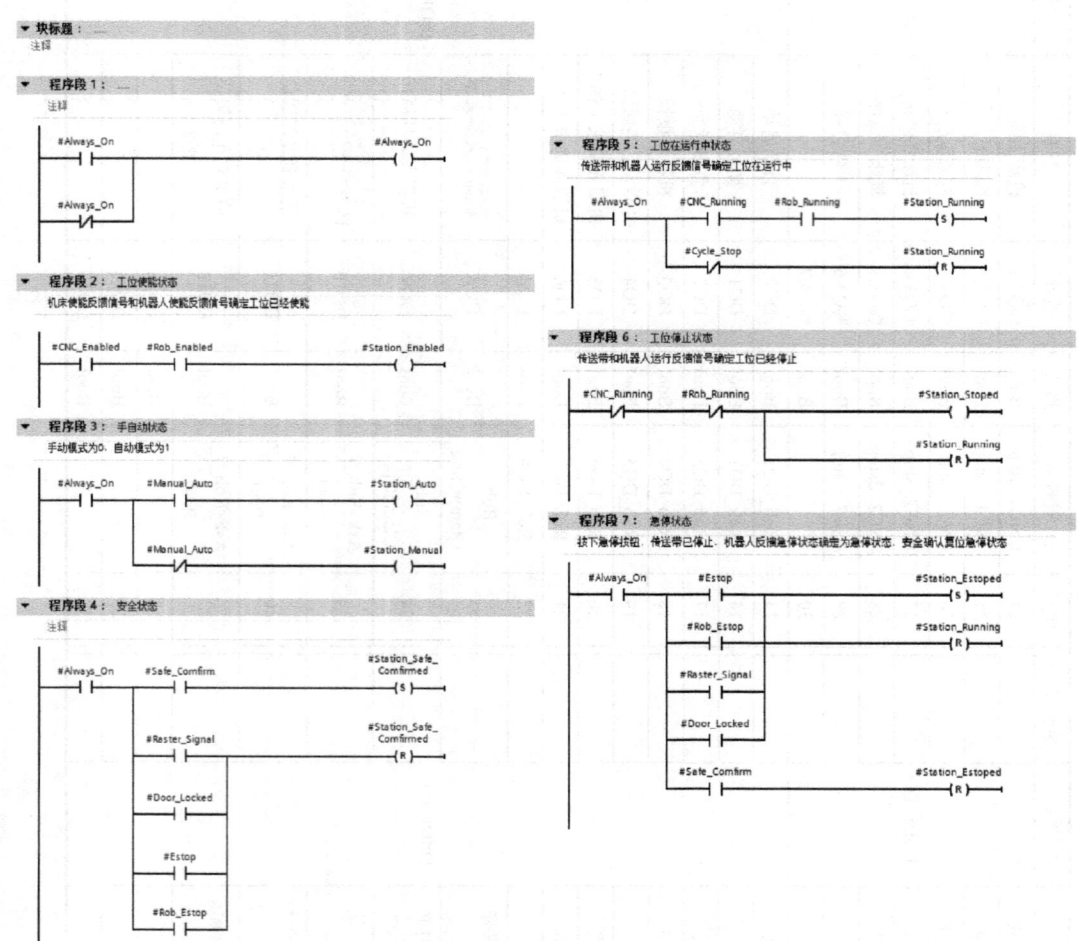

图 2-43　工位状态处理程序
Figure 2-43　Station Status Processing Program

3. 交互界面的设计

交互界面采用的是触摸屏的形式，触摸屏型号为西门子 TP900（精智面板）6AV2124-0JC01-0AX0。根据项目要求，并结合工程实际情况，完成触摸屏工程设计，如图 2-44 和图 2-45 所示。

3. Design of the interactive interface

The interactive interface adopts the touch screen whose model is Siemens TP900 (smart panel) 6AV2124-0JC01-0AX0. The engineering design of the touch screen is completed according to the requirements and actual situations of the program, as shown in Figure 2-44 and 2-45.

项目2　典型工业机器人机床上下料工作站系统的设计及应用

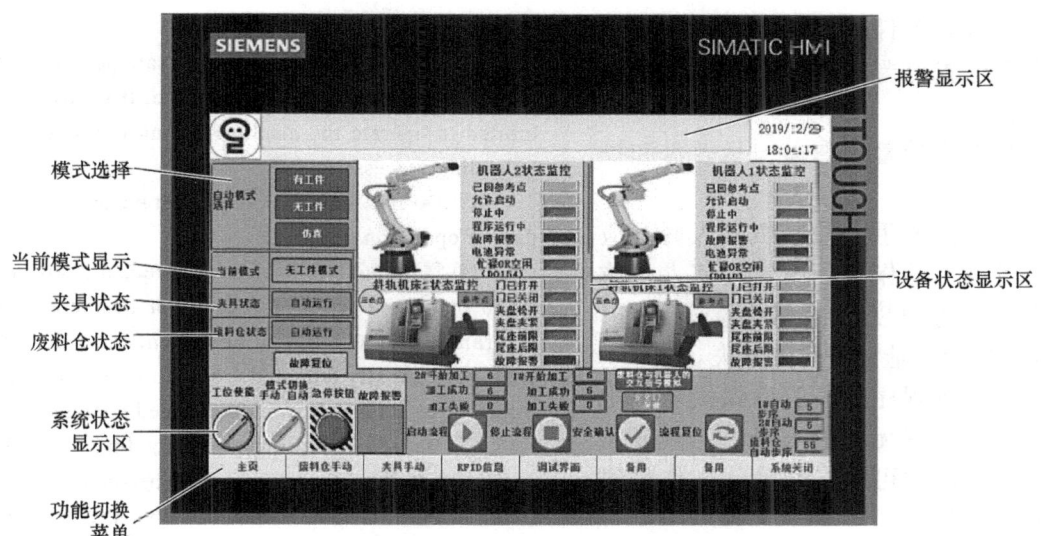

图 2-44　触摸屏主界面

Figure 2-44　Main Interface of Touch Screen

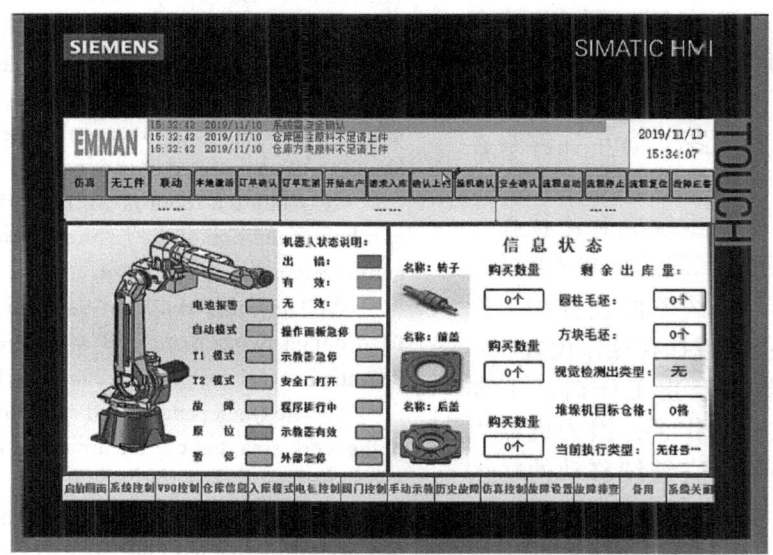

图 2-45　调试界面设计

Figure 2-45　Commissioning of Interface Design

2.5.3　工作站调试

完成工作站的安装与程序编写后，在正式投入使用前，需对工作站进行调试。调试的基本步骤如下：

1. 熟悉操作规范

①机器人周围区域必须清洁（无油、水及杂质等）。

2.5.3　Workstation Commissioning

When completing the installation and programming of the workstation, it is necessary to conduct commissioning of the workstation before it is officially put into operation. The basic steps for commissioning are as follows:

1. Be familiar with operating regulations

① The area around the robot has to be clean (free

②装卸工件前，先将机械手运动至安全位置，严禁装卸工件过程中操作机器。

③不要带着手套操作示教盘和操作盘。

④如需要手动控制机器人时，应确保机器人动作范围内无任何人员或障碍物，将速度由慢到快逐渐调整，避免速度突变造成伤害或损失。

⑤执行程序前，应确保机器人工作区内不得有无关的人员、工具或物品，工件摆放可靠，并确认工作程序与工件对应。

⑥机器人动作速度较快，存在危险性，操作人员应负责维护工作站正常运转秩序，严禁非工作人员进入工作区域。

⑦机器人运行过程中，严禁操作者离开现场，以确保意外情况的及时处理。

⑧机器人工作时，操作人员应注意查看夹具引线状况，防止其缠绕在机器人上。

⑨线缆不能严重绕成麻花状和与硬物摩擦，以防内部线芯折断或裸露。

⑩示教器和线缆不能放置在工作区域，应随手携带或挂在操作位置。

⑪当机器人停止工作时，不要认为其已经完成工作了，因为机器人很可能是在等待让它继续移动的输入信号。

⑫因故离开设备工作区域前应按下停止开关，避免突然断电或者关机零位丢失，并将示教器放置在安全位置。

⑬工作结束时，应使机械手置于零位位置或安全位置。

⑭严禁在控制柜内随便放置配件、工具、杂物、安全帽等，以免影响到部分线路，造成设备的异常。

2. 操作前准备

①上电前短路与电压检查。

of oil, water, impurities, etc).

② Move the robotic arm to a safe position before loading and unloading the workpiece. It is strictly prohibited to operate the machine during the loading and unloading process.

③ Do not wear gloves to operate the teach pendant or operation panel.

④ If manual control is required, it shall be ensured that there are no personnel or obstacles within the range of the robot's movement, and the speed shall be gradually adjusted from slow to fast to avoid injury or loss caused by sudden speed changes.

⑤ Before executing the procedure, it shall be ensured that there are no irrelevant personnel, tools, or items in the robot work area, and that the workpieces are placed reliably, and confirmed that the work procedure corresponds to the workpieces.

⑥ The robot moves relatively fast and poses risks. The operator shall be responsible for maintaining the order for normal operation of the workstation. No one except the relevant staff shall be allowed to enter the work area.

⑦ During operation of the robot, it is strictly prohibited for the operator to leave the site to ensure timely handling of unexpected situations.

⑧ When the robot is working, the operator shall check the condition of the fixture lead to prevent it from getting entangled on the robot.

⑨ The cable shall not be seriously wound into fried dough twists or rubbed with hard objects to prevent the internal core from breaking or being exposed.

⑩ The teach pendant and cable shall not be placed in the work area, but carried around or hung in the operating position.

⑪ Do not assume that the robot has completed the task when it stops working, as it is very likely waiting for an input signal to continue moving.

⑫ Before leaving the equipment working area for any reason, press the stop switch to avoid sudden power outage or loss of zero position, and place the teach pendant in a safe position.

⑬ At the end of the work, the robotic arm shall be placed in the zero position or safe position.

⑭ It is strictly prohibited to place accessories, tools, debris, safety helmets, etc. casually inside the control cabinet to avoid affecting some circuits that further cause equipment abnormalities.

利用万用表检查主回路是否存在短路故障，检查主回路与各设备供电电压等级是否正常。

②气路与气压检查。

启动工位前检查夹具气路情况，检查各连接器是否存在漏气，检查气管是否有磨损漏气，检查压力表观察压力是否足够。

③工位内安全检查。

检查机器人第 7 轴附近是否有物体阻挡其运动，有则清理；检查机器人工作区域是否有阻挡物体，检查机床内是否有遗留未加工完工件，有则进行清理。

3. 工作站启动流程

①合闸给机床、空压机和操作台上电。

②斜轨机床分别合闸，系统上电，选择手动模式—按下液压启动按钮—按 Reset 按钮—选择自动模式，运行 O0010 号程序进行回零。

③按工位上电按钮，给操作台和机器人上电。检查机器人是否在初始位置，如不在初始位置需要用机器人示教器手动回零。

④根据需要运行的模式在传送带上放置相应托盘，注意托盘方向不能放反。

⑤确认触摸屏上各信号点正常后，工位使能旋钮拨至右边，模式切换旋钮拨至右边，自动模式选择窗选有工件模式或无工件模式，按下安全确认按钮，按下启动流程按钮。

【课后巩固】

1. 请描述 PLC 程序中用于进行安全防护的程序功能。

2. 机器人上下料工作站中的机器人、PLC 和机床是如何通信的？

2. Preparation before operation

① Short circuit and voltage check before power on.

Use a multimeter to check if there is a short circuit fault in the main circuit, and check if the voltage level of the main circuit and the equipment is normal.

② Gas circuit and pressure check.

Before starting the station, check the gas circuit of the fixture to see whether there is any air leakage in the connector, the gas circuit is worn or leaking, and check the pressure gauge to see if the pressure is sufficient.

③ Safety inspection inside the station.

Check if there are any objects blocking the movement of the robot near the seventh axis, and clean them if there are any; check if there are any blocking objects in the working area of the robot, and any unfinished workpieces left in the machine tool, and clean them if there are any.

3. The workstation startup process

① Close the switch to power on the machine tool, air compressor and operating console.

② Switch on the inclined rail machines separately, power on the system, select the manual mode—press the hydraulic start button—press the Reset button—select the automatic mode, and run the program O0010 to return to zero.

③ Press the power on button on the station to power on the operating console and robot. Check if the robot is in its initial position, and if it is not, reset it to zero manually with the robot's teach pendant.

④ Place the corresponding tray on the conveyor belt according to the mode of operation set, and be careful not to place it in the opposite direction.

⑤ When all signal points on the touch screen are confirmed normal, turn the enable knob of the station to the right, the mode switch knob to the right, and select either the mode with workpiece or mode without workpiece from the automatic mode selection window. Press the confirm button for safety confirmation, and the start button to start the process.

[Consolidation after Class]

1. Please describe the program functions used for safety protection in the PLC program.

2. How do the robot, PLC and machine tool communicate in the robot loading and unloading workstation?

任务 2.6 技术交底材料的整理和编写
Task 2.6 Organization and Compilation of Technical Disclosure Materials

【知识目标】

1. 掌握工作站技术交底材料的整理方法。
2. 掌握工作站使用说明书的编写方法。

【技能目标】

1. 能够根据任务要求完成工业机器人机床上下料工作站技术交底材料的整理。
2. 能够编写工业机器人机床上下料工作站的使用说明书。

【素质目标】

1. 培养学生一丝不苟,具有严谨、全面、规范、标准、熟练的工作态度和工作作风。
2. 培养学生善于观察、总结归纳的能力。

【任务情景】

某企业需要将一数控加工工作站进行设计和改造,将机床上下料的工作全部设计由机器人自动完成,目前已完成工业机器人机床上下料工作站的设计方案、仿真项目、安装调试,请你根据现有的材料,完成机器人机床上下料工作站技术交底材料的整理和编写。

【任务分析】

1. 整理有关机器人机床上下料工

[Knowledge Objectives]

1. To master the method of organizing technical disclosure materials of workstation.
2. To master the method of compiling user manual of workstation.

[Skill Objectives]

1. To be able to organize technical disclosure materials for industrial robot machine tool loading and unloading workstations according to task requirements.
2. To be able to compile user manuals for industrial robot machine tool loading and unloading workstations.

[Competence Objectives]

1. To cultivate a work attitude and style of being rigorous, comprehensive, normative, standardized, and proficient of the student.
2. To cultivate students' ability to observe and summarize.

[Task Scenario]

An enterprise needs to design and transform a CNC machining workstation, with all the loading and unloading work of the machine tool to be automatically completed by robots. Currently, the design scheme, simulation program, installation and commissioning of the industrial robot machine tool loading and unloading workstation have been completed. Please complete the organization and compilation of the technical disclosure materials for the robot machine tool loading and unloading workstation based on the existing materials.

[Task Analysis]

1. Organize information related to the robot ma-

作站有关的资料。

2. 了解工作站的操作说明书中包含的内容、目录。

3. 编写机器人机床上下料工作站的操作说明书。

【知识准备】

完成工作站调试后，应开展技术资料的整理和编写工作，使工作站的使用、维护人员，明确本工作站的技术特点、系统构成、操作流程，注意具体技术要求和有针对性的关键技术措施，系统掌握工作站操作过程全貌和操作中的关键步骤。使参与工作站操作的每一个人，通过技术资料，能够了解工作站及其主要组成设备的情况，提高工作站的使用效率。

2.6.1 主要技术资料

工作站常见的技术资料主要有以下几点(见图2-46)：

1. 工作站操作说明书

1) 工作站概况和基本软硬件组成。

2) 基本操作流程，关键性的技术及操作中可能会存在的问题。

3) 操作的细节及其操作须知，特殊设备的处理。

4) 工作站开关机流程及注意事项。

5) 工作站常见故障的现象描述及处理方法。

6) 如果可以最好能将工作站的程序也列出，别做出注解。

2. 工作站全套图样

1) 工作站布局图及网络拓扑图。

2) 电气原理图及接线图。

3) 工作站系统安装图。

4) 非标件零件图及装配图。

chine tool loading and unloading workstation.

2. Understand the content contained in the workstation operating manual and and the table of contents.

3. Compile the operating manual for the robot machine tool loading and unloading workstation.

[Assumed Knowledge]

When the commissioning of the workstation is completed, technical documents shall be organized and compiled to enable the operation and maintenance personnel of the workstation to clearly understand the technical characteristics, system composition, operation process of the workstation, pay attention to specific technical requirements and targeted key technical measures, and systematically grasp the overall operation process and key steps of the workstation. Everyone involved in workstation operation can understand the situation of the workstation and its main equipment through technical information, which improves the efficiency of workstation.

2.6.1 Main Technical Documents

The common technical documents for workstations mainly include the following (as shown in Figure 2-46):

1. Workstation operating manual

1) Overview of the workstation and composition of basic software and hardware.

2) Basic operating procedures, key technologies, and potential problems in operation.

3) Details and instructions for the operation and handling of special equipment.

4) The startup and shutdown processes of the workstation and their precautions.

5) Description and handling methods of common faults in the workstation.

6) If possible, it would be better to list the programs of the workstation without making annotations.

2. Complete set of drawings of the workstation

1) Workstation layout and network topology diagram.

2) Electrical schematic diagram and wiring diagram.

3) Installation drawing of the workstation system.

4) Drawings of non-standard parts and their assembly.

3. 工作站设备程序

应将调试后的设备程序如工业机器人程序、PLC 程序、HMI 工程文件、变频及伺服设置文件以及数控程序等全部整理并标注好后交付使用方。

4. 工作站设备说明书

应将工作站中使用的成品设备说明书整理好后交付给使用方，以确保使用方在使用过程中可以方便地查阅资料。

5. 工作站仿真项目

将工作站仿真项目进行打包并交代客服工作人员仿真工作站的使用方法，如图 2-46 所示。

3. Programs of the workstation equipment

All equipment programs whose commissioning has been done such as industrial robot programs, PLC programs, HMI engineering files, frequency conversion and servo settings files, and CNC programs shall be organized and labeled before being delivered to the user.

4. Workstation equipment manual

The manuals for the finished equipment used in the workstation shall be organized and delivered to the user to ensure easy access to the information during use.

5. Workstation simulation program

Pack up the workstation simulation programs and explain to the customer service personnel the usage method of the simulation workstation, as shown in Figure 2-46.

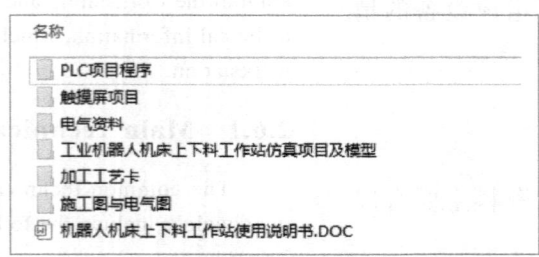

图 2-46　工业机器人机床上下料工作站技术交底材料整理

Figure 2-46　Organization of Technical Disclosure Materials of a Machine Tool Loading and Unloading Workstation of the Industrial Robot

2.6.2　使用说明书的编写

1. 目录编写

典型工业机器人机床上下料工作站使用说明书的目录样式如图 2-47 所示。

2. 正文编写

这里以"工位简介""设备参数""系统操作规程"的内容编写为例，为大家演示如何进行操作说明书的编写。操作说明书的其他内容请同学们按要求自行编写。

2.6.2　Compilation of User Manual

1. Compilation of the table of contents

The style of the table of contents of the user manual for typical industrial robot machine tool loading and unloading workstations is shown in Figure 2-47.

2. Compilation of the text

Here, the compilation of the content of " Station Introduction ", " Equipment Parameters " and " the System Operation Procedure " is taken as an example to demonstrate how to compile the operating manual. Other content of the operating manual shall be compiled by yourself according to the requirements.

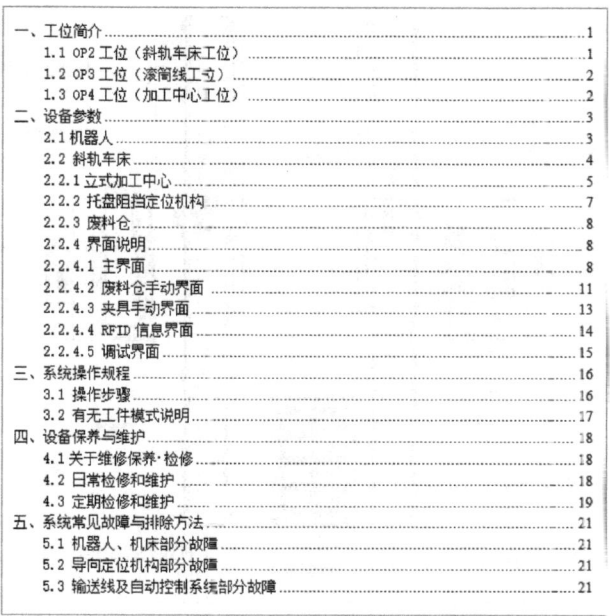

图 2-47 机器人机床上下料工作站使用说明书的目录样式

Figure 2-47 Style of the Table of Contents of the User Manual for the Robot Machine Tool Loading and Unloading Workstation

(1) 工位简介

OP2 工位由 2 台 FANUC 机器人、2 台斜轨车床、1 套废料仓库、1 个操作台、1 个显示看板组成，如图 2-48 所示。该工位主要完成轴类工件的自动上下料和加工过程。

OP3 工位由 6 条 8m 直线滚筒线、2 个 180° 转弯滚筒线、4 套托盘阻挡定位机构、1 个显示看板组成，如图 2-49 所示。电动机由变频器控制。

OP4 工位由 1 台 FANUC 机器人、2 台立式加工中心、1 套工件翻转机构、1 套废料仓库、1 个操作台、1 个显示看板组成，如图 2-50 所示。该工位主要完成法兰类工件的自动上下料和加工过程。

(2) 设备参数

OP2 和 OP4 工位为 6 轴 FANUC M-20iA 机器人，负载为 20kg。机器人 M-20iA 如图 2-51 所示。机器人参数见表 2-12。

(1) Station introduction

The OP2 station consists of 2 FANUC robots, 2 inclined rail lathes, 1 waste store, 1 operation console, and 1 display board, as shown in Figure 2-48. This station mainly completes the automated loading, unloading, and machining of workpieces of shaft.

The OP3 station consists of 6 straight drum lines of 8m, 2 drum lines of 180 degree turning, 4 sets of positioning mechanism with tray blocking, and 1 display board, as shown in Figure 2-49. The motor is controlled by a frequency converter.

The OP4 station consists of 1 FANUC robot, 2 vertical machining centers, 1 set of workpiece turnover mechanism, 1 set of waste store, 1 operation console, and 1 display board, as shown in Figure 2-50. This station mainly completes the automated loading, unloading, and machining of workpieces of flange.

(2) Equipment parameters

The OP2 and OP4 stations are equipped with 6-axis FANUC M-20iA robots with a load of 20kg. The robot M-20iA is shown in Figure 2-51, and its various parameters in Table 2-12.

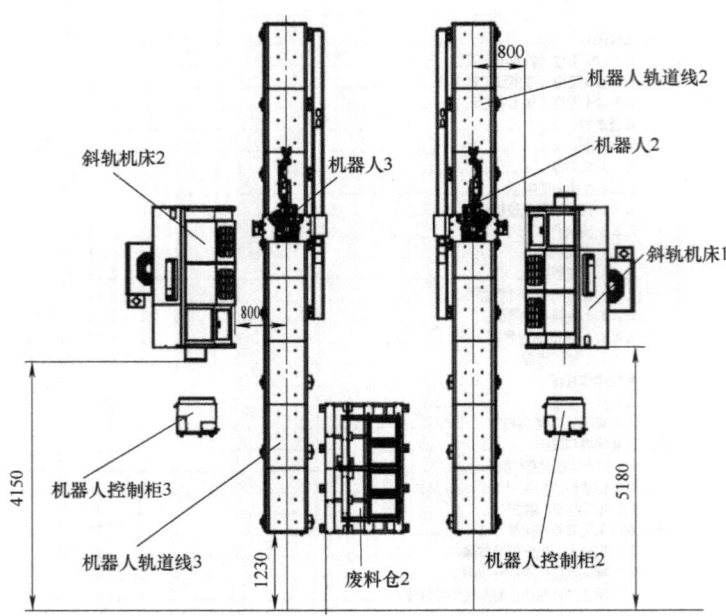

图 2-48　OP2 斜轨车床工位示意图
Figure 2-48　Schematic Diagram of Inclined Rail Lathe Station OP2

图 2-49　OP3 滚筒线工位示意图
Figure 2-49　Schematic Diagram of Station OP3 of the Drum Line

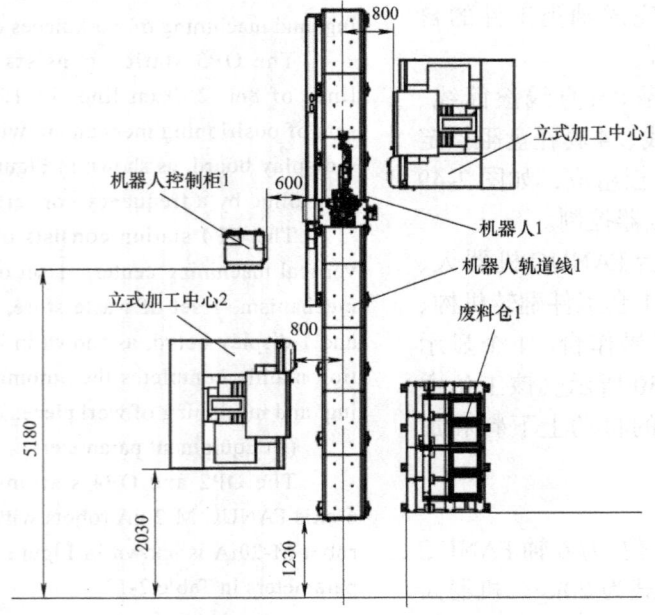

图 2-50　OP4 加工中心工位示意图
Figure 2-50　Schematic Diagram of Station OP4 of the Machining Center

项目2 典型工业机器人机床上下料工作站系统的设计及应用

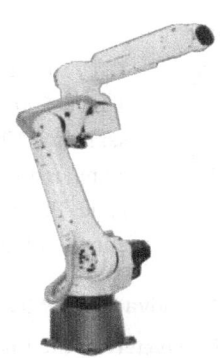

图 2-51　M-20iA
Figure 2-51　M-20iA

表 2-12　机器人参数
Table 2-12　Robot Parameters

机器人基本参数 Basic parameters of the robot	型号 Model	M-20iA
	轴数 Number of axis	6 轴 6-axis
	负载 Load	20kg
	重复定位精度 Repetitive positioning precision	±0.04mm
	本体重量 Weight of the body	250kg
	能耗 Energy consumption	3kW
	可达半径 Reachable radius	1811mm
	本体防护等级 Protection level of the body	IP65
	电柜防护等级 Protection level of electrical cabinet	IP43
机器人活动范围 Motion range of the robot	1 轴 1-axis	340°/s
	2 轴 2-axis	260°/s
	3 轴 3-axis	458°/s
	4 轴 4-axis	400°/s
	5 轴 5-axis	360°/s
	6 轴 6-axis	900°/s

1#斜轨车床型号为 Viva T2C/500，2#斜轨车床型号为 Viva T2Cm/500。参数基本一致，Viva T2Cm/500 比 Viva T2C/500 增加了 2 个动力头，是铣削复合式车床。具体斜轨车床参数见表 2-13。

The model of 1 # inclined rail lathe is Viva T2C/500, and that of 2 # inclined rail lathe is Viva T2Cm/500. While their parameters are basically the same, the Viva T2Cm/500 has additional 2 power heads compared to Viva T2C/500, and is a compound milling lathe. Please refer to Table 2-13 in the manual of the inclined rail lathe for details.

表 2-13 Viva T2Cm/500 参数
Table 2-13 Parameters of the Viva T2Cm/500

项目 Item		单位 Unit	规格 Specifications	备注 Remarks
床身上最大回转直径 Maximum swing diameter on the lathe		mm	560	
最大切削长度 Maximum cutting length		mm	500	标准配置 Standard configuration
最大切削直径 Maximum cutting diameter		mm	280	标准配置 Standard configuration
滑板上最大回转直径 Maximum turning diameter on the sliding plate		mm	350	
主轴端部型式及代号 Type and code of the main axle end			A2-6	
主轴孔直径 Hole diameter of the main axle		mm	65	
最大通过棒料直径 Maximum diameter of bar passing through		mm	50	标准配置 Standard configuration
单主轴主轴箱 Main axle box for single main axle	主轴转速范围， Speed range of the main axle, 主轴最大输出扭矩 Maximum output torque of the main axle	r/min, N·m	50～4500, 235	FANUC 0i -TF
主电动机输出功率 Main motor output power	30min/连续 30 minutes/continuous	kW	15/11	β iIP22/6000
标准卡盘 Standard chuck	卡盘直径 Chuck diameter	in	8	
X 轴快移速度 X-axis quick movement speed		m/min	30	滚动导轨 Rolling guide rail
Z 轴快移速度 Z-axis quick movement speed		m/min	30	滚动导轨 Rolling guide rail

（续）

项目 Item		单位 Unit	规格 Specifications	备注 Remarks
X轴行程 X-axis stroke		mm	200	
Z轴行程 Z-axis stroke		mm	560	
尾座行程 Tailstock stroke		mm	450	
尾座主轴锥孔锥度 Tailstock main axle taper		莫氏 Morse	5号 fifth	
标准刀架形式 Standard tool holder form			卧式12工位伺服动力刀架 Horizontal 12-station servo dynamic tool holder	
刀具尺寸 Tool size	外圆刀 Outer cylindrical knife	mm	25×25	
	镗刀杆直径 Boring bar diameter	mm	$\phi 40, \phi 32, \phi 25, \phi 20$	
刀盘可否就近选刀 Can the cutterhead be selected nearby			可 can	
机床质量 Machine tool weight	总质量 Total weight	kg	4050	
最大承重 Max bearing capacity	盘类件 Disk parts	kg	200（含卡盘等机床附件） 200 (including machine tool accessories, e. g. chuck)	
	轴类件 Shaft parts	kg	500（含卡盘等机床附件） 500 (including machine tool accessories, e. g. chuck)	
机床外形 External dimension of machine tool	长×宽×高 length × width × height	mm	2950×1860×1850	不含排屑器 Without chip conveyor

立式加工中心型号为VMC850B，其中1#加工中心最高转速为10000r/min，使用BT40刀具，2#加工中心最高转速为18000r/min，使用HSK63A刀具。其余参数基本一致。加工中心具体参数见表2-14，设备使用环境见表2-15。

The model of vertical machining centers is VMC850B, with 1 # machining center having a maximum speed of 10000 rpm using BT40 tools, and 2 # machining center having a maximum speed of 18000 rpm using HSK63A tools. The other parameters are basically the same. Please refer to Table 2-14 for the specific instructions of the machining center, and Table 2-15 for the equipment operating environment.

表 2-14　18000r/min VMC850B 参数
Table 2-14　Parameters of VMC850B with 18000 rpm

名称 Name			规格 Specifications	单位 Unit
工作台 workbench	工作台尺寸 Workbench size		1000 × 500	mm
	允许最大荷重 Maximum allowable load		600	kg
	T 形槽尺寸 T-groove size		18 × 5	mm × 个 mm × pc
加工 范围 Machining range	工作台最大行程 –X 轴 Maximum stroke of workbench – X-axis		850	mm
	滑座最大行程 –Y 轴 Maximum stroke of sliding seat – Y-axis		560	mm
	主轴最大行程 –Z 轴 Maximum stroke of main axle – Z-axis		650	mm
	主轴端面至 工作台面距离 Distance from the main axle end face to the workbench face	最大 maximum	800	mm
		最小 minimum	150	mm
	主轴中心到导轨基面距离 Distance from the main axle center to the guide rail base face		665	mm
主轴 Main axle	锥孔 Conical hole		HSK–A63	
	转数范围 Revolution range		50～18000	r/min
	最大输出扭矩 Maximum output torque		29.4/43.6	N·m
	主轴电动机功率 Main axle motor power		18.5/26	kW
	主轴传动方式 Main axle transmission mode		电主轴形式 Motorized main axle	
刀具 Tools	刀柄型号 Knife handle model		HSK63A	

（续）

名称 Name			规格 Specifications		单位 Unit
进给 Feed	快速移动 Quick movement	X轴 X-axis	32		m/min
		Y轴 Y-axis	32		
		Z轴 Z-axis	30		
	三轴拖动电动机功率 (X/Y/Z) Three axis drag motor power (X/Y/Z)		2.5/2.5/3		kW
	三轴拖动电动机扭矩 (X/Y/Z) Three axis drag motor torque (X/Y/Z)		20/20/27		N·m
	进给速度 Feed rate		1-20000		mm/min
刀库 Tool magazine	刀库形式 Tool magazine form		机械手 Robotic arm		
	选刀方式 Tool selection method		双向就近选刀 Bidirectionally nearby tool to be selected		
	刀库容量 Magazine capacity		24		把 tool
	最大刀具长度 Maximum tool length		300		mm
	最大刀具质量 Maximum tool weight		7		kg
	最大刀盘直径 Maximum cutter diameter	满刀 Full tool	φ80		mm
		相邻空刀 Adjacent empty tool	φ150		mm
	换刀时间 Tool changing time		2.5		s
定位精度 Positioning precision			JISB6336-4：2000	GB/T 18400.4—2010	
	X轴 X-axis		0.016	0.016	mm
	Y轴 Y-axis		0.012	0.012	mm
	Z轴 Z-axis		0.012	0.012	mm
重复定位精度 Repetitive positioning precision	X轴 X-axis		0.010	0.010	mm
	Y轴 Y-axis		0.008	0.008	mm
	Z轴 Z-axis		0.008	0.008	mm

(续)

名称 Name	规格 Specifications	单位 Unit
机床质量 Machine tool weight	6800	kg
电气总容量 Total electrical capacity	25	kV·A
机床轮廓尺寸（长×宽×高） Outline dimension (length × width × height) of machine tool	4355 × 2275 × 2920	mm

表 2-15 设备使用环境
Table 2-15 Equipment Operation Environment

序号 S/N	项目 Item	参数 Parameter
1	环境温度 Ambient temperature	2~40℃
2	相对湿度 Relative humidity	≤85%
3	空气粉尘含量 Dust content in the air	≤10mg/m³
4	工作环境 Working environment	周围无腐蚀性气体、导电粉尘及爆炸气体 There is no corrosive gas, conductive dust, or explosive gas around
5	压缩气体 Compressed air	≥0.55MPa

托盘阻挡定位机构部件包括前阻挡气缸、定位气缸、后阻挡气缸及其电磁阀、光电开关、激光传感器、导向条等，如图 2-52 所示。

废料仓包括 XYZ 轴滑台、三行四列共 12 个仓位、1 个机器人放置废料取料位、料框光电传感器、XYZ 轴回零传感器等。废料仓伺服电机为增量式编码器，自动运行时会自动进行回零动作，也可以手动进行回零，如图 2-53 所示。

The components of the positioning mechanism with tray blocking include front blocking cylinder, positioning cylinder, rear blocking cylinder and its solenoid valve, photoelectric switch, laser sensor, guide bar, as shown in Figure 2-52.

The waste store includes an X-axis, Y-axis, and Z-axis sliding platform, a total of 12 bins in three rows and four columns, a waste placing and taking position for the robot, photoelectric sensors for the waste bins, and X-axis, Y-axis, and Z-axis return-to-zero sensors. The servo motor of the waste bin is an incremental encoder, which automatically performs the action of return-to-zero under automated operation or can be manually operated to return to zero, as shown in Figure 2-53.

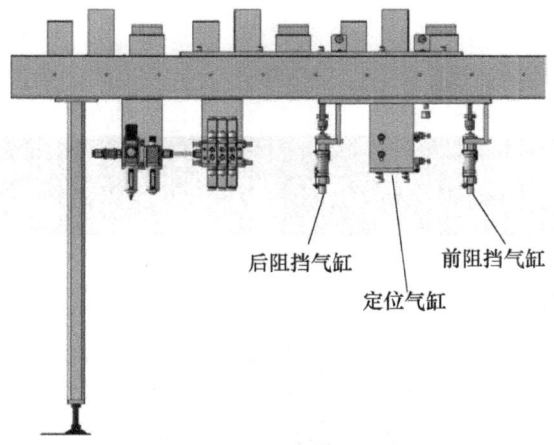

图 2-52 托盘阻挡定位机构
Figure 2-52 Positioning Mechanism with Tray Blocking

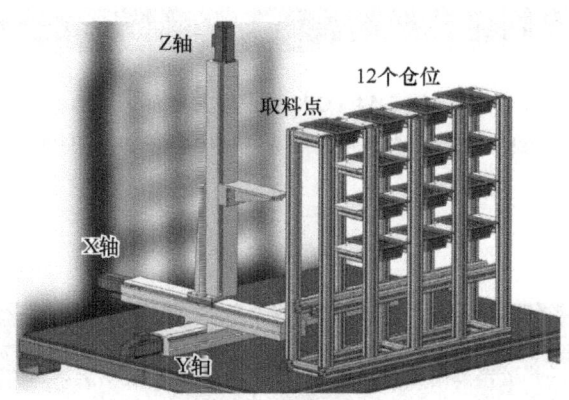

图 2-53 废料仓
Figure 2-53 Waste Store

触摸屏主界面上部为报警信息显示窗，中间部分为设备运行状态及操作区域，下部为功能切换按钮，如图 2-54～图 2-56 所示。

界面说明：

模式选择栏："有工件模式"为上料真实加工的模式，"无工件模式"为模拟加工的演示模式。需人工进行选择操作。

当前模式栏：选择模式后自动显示当前的模式状态。

The upper part of the main interface of the touch screen is the alarm information display window, the middle part is the equipment running status and operation area, and the lower part is the function switch button, as shown in Figure 2-54 to 2-56.

Interface Description:

Mode selection bar: " with workpiece mode " refers to the mode of actual machining with workpieces, and " without workpiece mode " refers to the demonstration mode that simulates machining. Manual selection is required.

Current mode bar: automatically displays the current mode status after selecting a mode.

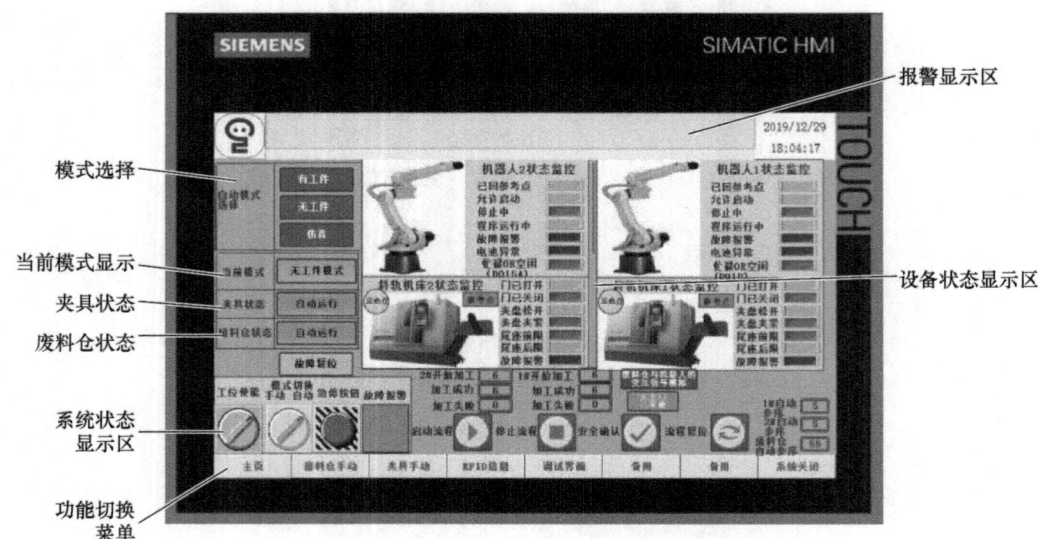

图 2-54　OP2 触摸屏主界面
Figure 2-54　Main Interface of the OP2 Touch Screen

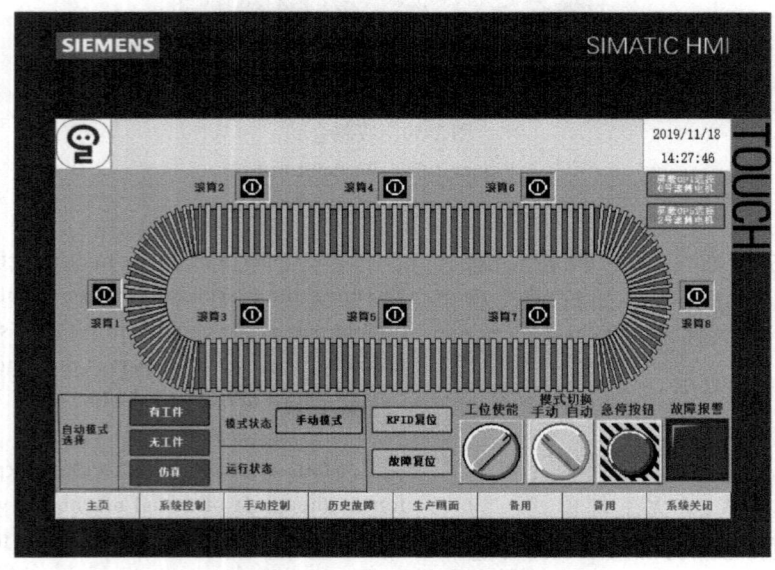

图 2-55　OP3 触摸屏主界面
Figure 2-55　Main Interface of the OP3 Touch Screen

项目2　典型工业机器人机床上下料工作站系统的设计及应用

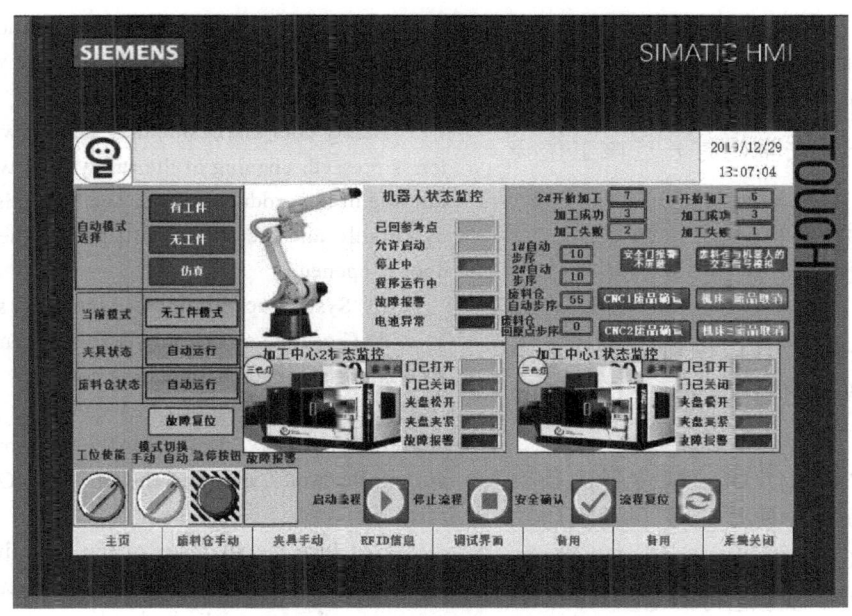

图 2-56　OP4 触摸屏主界面

Figure 2-56　Main Interface of the OP4 Touch Screen

夹具状态栏：显示托盘阻挡定位机构是否处于初始位置，如显示"未初始化"，需要人工操作使定位机构处于初始位置，否则工位不能运行。

废料仓状态栏：本工位自动运行后废料仓自动进入自动运行状态。

机器人状态监控栏：显示机器人的状态。

机床状态监控栏：显示机床的状态。

机床三色灯：显示当前机床信号灯颜色。

机床参考点：当机床各轴处于初始位置并且机床门打开时，机床参考点图标显示绿色。

RFID 复位按钮：当 RFID 读写头显示红灯报警时单击此按钮进行复位。

故障复位按钮：当本工位有故障或警告发生时单击此按钮进行复位。

废品确认按钮：当此按钮出现时，可以单击确认按钮定义当前工件加工

Fixture status bar: displays whether the positioning mechanism with tray blocking is in the initial position. If it displays "not initialized", manual operation is required to make the positioning mechanism to be in the initial position, otherwise the station cannot run.

Waste bin status bar: when the station runs automatically, the waste bin will enter the state of automatic running automatically.

Robot status monitoring bar: displays the status of the robot.

Machine tool status monitoring bar: displays the status of the machine tool.

Machine tool tricolor light: displays the current color of the signal light of the machine tool.

Machine tool reference point: when all axis of the machine tool are in their initial position and the the door of the machine tool door open, the icon of the machine tool reference point shows green.

RFID reset button: when the RFID read-write head displays a red light alarm, click this button to reset.

Fault reset button: Click this button to reset when there is a fault or warning in this station.

Waste confirm button: when this button appears,

完毕后作为废品进入废料仓。

安全门报警屏蔽按钮：此按钮按下后安全门打开不会导致生产线停机，默认未按下，安全门打开后该工位会暂停。

(3) 系统操作规程及步骤

1) 合闸给机床、空压机和操作台上电。

2) 2 台加工中心分别合闸，系统上电，选择自动模式，运行 O0010 号程序进行回零。

3) 2 台斜轨机床分别合闸，系统上电，选择手动模式—按下液压启动按钮—按 Reset 按钮—选择自动模式，运行 O0010 号程序进行回零。

4) OP2—OP4 分别合闸，按工位上电按钮，给操作台和机器人上电。检查机器人是否在初始位置，如不在初始位置，需要用机器人示教器手动回零。

5) 根据需要运行的模式放置相应托盘，注意托盘方向不能放反。

6) OP4 工位确认触摸屏上各信号点正常后，工位使能旋钮拨至右边，模式切换旋钮拨至右边，在自动模式选择窗口选择有工件模式或无工件模式。按下安全确认按钮，按下启动流程按钮。

7) OP3 工位确认触摸屏上各信号点正常后，工位使能旋钮拨至右边，模式切换旋钮拨至右边，在自动模式选择窗口选择有工件模式或无工件模式。按下安全确认按钮，按下启动按钮。

OP3 电视使用遥控器进入应用界面，选择"我的应用"，打开极速浏览器，插上鼠标，双击"CNC"图标。

8) OP2 工位确认触摸屏上各信号点正常后，工位使能旋钮拨至右边，

one can click the confirm button to define the current workpiece as a waste which will enter the waste bin after machining.

Safety door alarm shielding button: when this button is pressed, opening of the safety door will not cause the stop of the production line. The button is not pressed by default, and the station will pause when the safety door is opened.

(3) System operating procedures and steps

1) Close the switch to power on the machine tool, air compressor, and the operating console.

2) Close the switches of the 2 machining centers separately, power on the system, select the automatic mode, and run the program O0010 to return to zero.

3) Close the switches of the 2 inclined rail machine tools separately, power on the system, select the manual mode—press the hydraulic start button—press the Reset button—select the automatic mode, and run the program O0010 to return to zero.

4) Close the switches of OP2—OP4 respectively, press the power on button at the station to power on the operating console and robot. Check if the robot is in its initial position, and if it is not, reset it to zero manually with the robot's teach pendant.

5) Place the corresponding tray according to the mode of operation set, and be careful not to place it in the opposite direction.

6) When all signal points on the touch screen of the OP4 station are confirmed normal, turn the enable knob of the station to the right, the mode switch knob to the right, and select either the mode with workpiece or mode without workpiece from the automatic mode selection window. Press the confirm button for safety confirmation, and the start button to start the process.

7) When all signal points on the touch screen of the OP3 station are confirmed normal, turn the enable knob of the station to the right, the mode switch knob to the right, and select either the mode with workpiece or mode without workpiece from the automatic mode selection window. Press the confirm button for safety confirmation, and the start button to start the process.

Make the OP3 TV enter the application interface with the remote, select "My Applications", open the

模式切换旋钮拨至右边，在自动模式选择窗口选择有工件模式或无工件模式。按下安全确认按钮，按下启动流程按钮。

OP2电视使用遥控器进入应用界面，选择"我的应用"，打开极速浏览器，插上鼠标 双击"CNC"图标。

(4) 设备保养与维护
- 进行检修作业的人员，必须是由接受过特殊指导教育或规定时间的教育、熟知相关内容的人员担任。
- 检修作业必须在确认周围的安全、确保躲避危险所必需的通道和场所的前提下安全地进行作业。
- 进行机器人或其他带电设备的日常检修和部件更换作业时，请务必先切断电源然后再进行。另外，为了防止其他作业者不小心接通电源，请在一级电源等位置挂上"禁止接通电源"的警示牌。
- 进行检修作业时，可能会发生由于机器人机械臂落下、移动等而导致危险的场合，请务必进行机械臂的固定后再进行作业。

日常维护为每次开机前进行的维护和保养，具体维护和保养项目见表2-16。

speed browser, plug in the mouse, and double-click the "CNC" icon.

8) When all signal points on the touch screen of the OP2 station are confirmed normal, turn the enable knob of the station to the right, the mode switch knob to the right, and select either the mode with workpiece or mode without workpiece from the automatic mode selection window. Press the confirm button for safety confirmation, and the start button to start the process.

Make the OP2 TV enter the application interface with the remote, select "My Applications", open the speed browser, plug in the mouse, and double-click the "CNC" icon.

(4) Equipment upkeep and maintenance
- The personnel conducting maintenance and repair operations must be those who have received special guidance education or the education for a specified time and are familiar with relevant content.
- Maintenance and repair operations shall only be carried when confirming the safety of the surrounding area and ensuring the availability of necessary passages and places to avoid danger.
- When carrying out operations of daily maintenance, repair, and component replacement on robots or other electrically-charged equipment, it shall be sure that the power supply is cut off before proceeding. In addition, to prevent other operators from accidentally hooking up the power supply, please hang a warning sign of "Do not hook up the power supply" at positions such as the primary power supply.
- During maintenance and repair, dangerous events may occur due to the falling or moving of the robotic arm. Please be sure to fix the robotic arm before proceeding with the said operation.

Daily maintenance refers to the maintenance and upkeep carried out before starting up the machine every time, and the specific maintenance content is shown in Table 2-16.

表 2-16 维护和保养项目
Table 2-16 Maintenance and Upkeep Items

设备 Equipment	部件 Parts	项目 Item	维修 Repair
整体 The whole	所有部件 All components	清理铝屑、污垢，保持设备整体干净 Clean up aluminum shavings and dirt, and maintain cleanliness of the overall equipment	
导向定位机构 Guiding and positioning mechanism	传感器及电磁阀接线 Sensor and solenoid valve wiring	是否松动或断开 Whether they are loose or disconnected	紧固或更换 Tighten or replace
	气缸 Cylinder	安装螺栓是否有松动 Whether the installation bolts are loose	拧紧 Tighten
	导向条 Guide bar	导向条是否松动 Whether the guide bar is loose	紧固或更换 Tighten or replace
传感器 Sensor	传感器 Sensor	定期检查各传感器和开关是否工作正常 Regularly check if all sensors and switches are working normally	紧固或更换 Tighten or replace
机床 Machine tool	切削液 Cutting fluid	检查液位高度是否不足 Check if the liquid level is insufficient	添加切削液 Adding cutting fluid
	加工平台和导轨护板 Machining platform and guide rail guard board	检查是否生锈 Check for rust	清理锈迹 Clean the rust
	润滑油 Lubricating oil	检查润滑油液位高度是否不足 Check if the lubricating oil level is insufficient	添加46#机油 Add 46 # engine oil
	液压站液压油 Hydraulic oil of the hydraulic station	检查液压油液位高度是否不足 Check if the hydraulic oil level is insufficient	添加46#机油 Add 46 # engine oil
机器人 Robot	机器人手爪 Robot gripper	检查是否有损坏 Check for damage	修复或更换 Repair or replace
	油水过滤器 Oil-water filter	检查油水过滤器中杂物 Check for impurities in the oil-water filter	及时清理 Timely cleaning
	接近开关及电磁阀接线 Proximity switch and solenoid valve wiring	是否松动或断开 Whether they are loose or disconnected	紧固或更换 Tighten or replace

【课后巩固】

1. 请编写一份使用工业机器人机床上下料仿真工作站的使用说明书,并将仿真工作站的运行流程录制视频。

2. 请将工业机器人机床上下料仿真工作站的实际接线图拍照,并存档于技术交底材料中。

[Consolidation after Class]

1. Please compile a user manual for the use of an industrial robot machine tool loading and unloading simulation workstation, and record a video of the running process of the simulation workstation.

2. Please take photos of the actual wiring diagram of the industrial robot machine tool loading and unloading simulation workstation and archive them in the technical disclosure materials.

项目 3 典型工业机器人搬运工作站系统的设计及应用

Program 3 Design and Application of a Typical Industrial Robot Handling Workstation System

【项目场景】

某汽车企业在汽车门饰板生产过程中需要对门饰板进行焊接,将焊接后的门饰板通过 2D 视觉检测焊接的质量是否合格。因为汽车门饰板体积大,重量重,使用人工完成工作需要多人合作进行搬运,增加了人工成本,降低了生产效率。为了提高生产效率,降低生产成本,需要设计和改造检测、搬运工艺,将汽车门饰板的检测、搬运工作实现自动化,如图 3-1 所示。

检测与搬运的产品是汽车的门饰板,如图 3-2 所示。

【项目描述】

工业机器人搬运工作站由 2 台 6 自由度的 FANUC 工业机器人组成。现需要完成机器人搬运工作站系统设计,项目设计从仓库搬运汽车门饰板到工作台模拟焊接,并通过视觉检测车门饰板焊接点的数量是否满足要求,合格产品放入仓库中,不合格产品放到传送带传送到下一个工位。通过两台机器人协作,实现汽车门饰板的自动化生产控制。

[Program Scenario]

An automobile enterprise needs to weld the car door trim panel during its production process, and detect the quality of welding with 2D vision to see if it is qualified. Due to the large volume and heavy weight of the car door trim panel, multiple people are required for its manual handling, which increases labor costs and reduces production efficiency. In order to improve production efficiency and reduce production costs, it is necessary to design and modify the detection and handling process of the car door trim panel to realize automation of the detection and handling, as shown in Figure 3-1.

The product to be detected and handled is the car door trim panel, as shown in Figure 3-2.

[Program Description]

An industrial robot handling workstation is composed of 2 FANUC industrial robots with 6 degrees of freedom. Now it is necessary to complete the design of the robot handling workstation system. In this program, it is designed to carry the car door trim panel from the warehouse to the workbench for simulated welding, and detect visually whether the number of welding points of the door trim panel meets the requirements. Qualified products will be put into the warehouse, while the unqualified ones be put on the conveyor belt and sent to the next station. The collaboration of two robots realizes the automated control of the car door trim panel production.

项目3 典型工业机器人搬运工作站系统的设计及应用

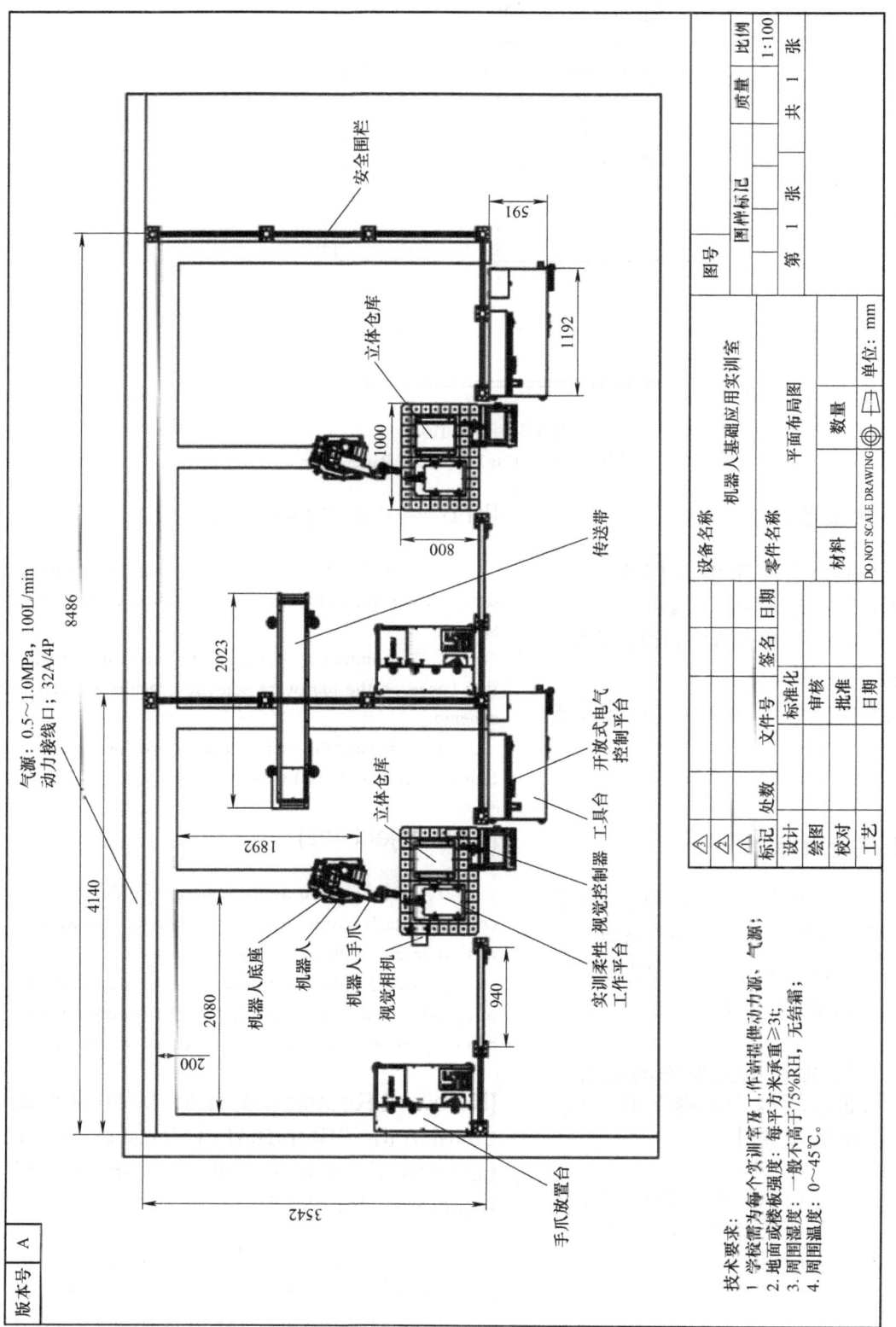

图 3-1 Overall Layout of a Handling Workstation

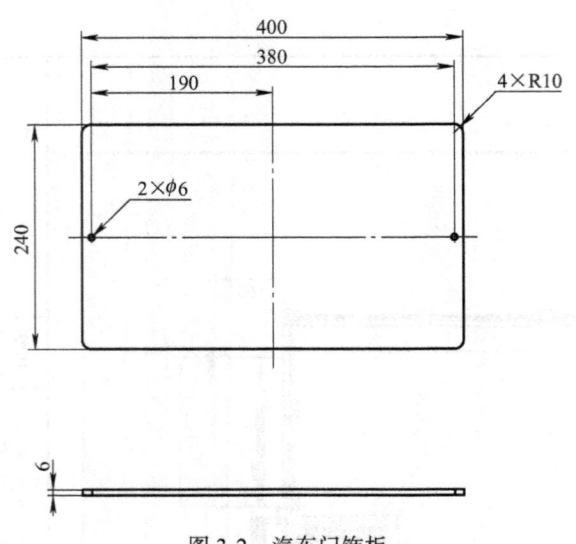

图 3-2 汽车门饰板
Figure 3-2　Car Door Trim Panel

【知识目标】

1. 了解设计方案的结构和要素，学习设计方案的基本体例编制。

2. 依据硬件选型结果和设计方案，进行建模和仿真验证。

3. 掌握编写技术交底材料的方法和步骤。

【技能目标】

1. 学会根据任务要求分析硬件需求，并根据硬件需求进行硬件选型。

2. 能够运用所学知识，综合完成工业机器人搬运工作站的集成与调试。

【"工业机器人操作与运维职业技能等级标准"对中级的相关要求】

1.1.1　能根据操作手册的安全规范要求，对工业机器人工作站物理环境进行安全检查。

[Knowledge Objectives]

1. To understand the structures and elements of design schemes, and learn the preparation of their basic styles.

2. To conduct modeling and simulation verification based on the hardware selection results and design scheme.

3. To master the methods and steps for the compilation of technical disclosure materials.

[Skill Objectives]

1. To be able to analyze hardware demands according to task requirements, and conduct hardware selection based on the hardware demands.

2. To be able to complete comprehensively the integration and commissioning of the industrial robot handling workstation based on the knowledge learned.

[Relevant Requirements for Intermediate Level in the "Standard of Vocational Skill Level for Operation and Maintenance of Industrial Robot"]

1.1.1　Be able to conduct safety checks on the physical environment of industrial robot workstations in accordance with the safety specifications in the operating manual.

1.2.2 能安装工业机器人应用系统液压、气动控制回路。

1.2.4 依据技术文件要求，能选用和安装光电开关、磁性开关、视觉相机等常用传感器。

1.3.2 能对工业机器人进行信号处理调试。

1.3.3 能对工业机器人及周边辅助设备（液压、气动、电气、夹具等）进行联调。

2.1.7 能根据工业机器人典型应用（搬运码垛、装配）的任务要求，编写工业机器人程序。

2.2.2 能根据工作站应用系统（搬运码垛、装配）的通信要求，配置和调试工业机器人和 PLC 控制设备的通信。

2.3.3 能完成视觉系统的具体参数配置（像素格式、触发方式、通信协议等）。

2.3.5 能在视觉系统中编程实现物料形状、颜色、尺寸、位置等信息数据的识别与输出。

3.2.3 能根据维护保养的要求，进行工业机器人周边电气设备程序、参数的设置与备份。

4.2.4 能根据工业机器人故障现象查询故障码，并排除。

1.2.2 Be able to install hydraulic and pneumatic control circuits for industrial robot application systems.

1.2.4 Be able to select and install commonly used sensors such as photoelectric switches, magnetic switches, and visual cameras according to the requirements of technical documents.

1.3.2 Be able to conduct the commissioning of signal processing for industrial robots.

1.3.3 Be able to conduct joint commissioning of industrial robots and peripheral auxiliary equipment (hydraulic, pneumatic and electrical equipment, fixtures, etc.).

2.1.7 Be able to write industrial robot programs based on the task requirements of typical applications (handling, stacking, and assembly) of industrial robots.

2.2.2 Be able to conduct configuration and commissioning of communication between industrial robots and PLC controlled equipment according to the communication requirements of application systems (handling, stacking, assembly) of the workstation.

2.3.3 Be able to configure specific parameters (pixel format, triggering method, communication protocol, etc.) of the visual system.

2.3.5 Be able to write programs in the visual system to recognize and output information data such as material shape, color, size, and position.

3.2.3 Be able to set and backup the programs and parameters of the electrical equipment around industrial robots according to requirements of maintenance and upkeep.

4.2.4 Be able to to query fault codes based on the industrial robot faults and have them troubleshooted.

任务 3.1　搬运工作站系统分析及硬件选型
Task 3.1　Handing Workstation System Analysis and Hardware Selection

【知识目标】

[Knowledge Objectives]

1.掌握典型工业机器人搬运工作站硬件选型方法和步骤。

1. To master the hardware selection methods and steps of typical industrial robot handling workstations.

2. 掌握搬运工作站工艺要求分析方法。

【技能目标】

1. 根据工艺要求完成标准部件和非标准部件的选型。
2. 能编写搬运工作站硬件选型方案。

【素质目标】

1. 树立学生乐观积极、务实进取的人生态度。
2. 强化学生的专业技术应用能力、沟通协调能力和再学习能力等职业能力的培养。

【任务情景】

某汽车企业在汽车门饰板生产中使用人工完成搬运任务效率低，无法满足当前生产要求，需要便捷、高效、高性价比的生产设备，实现生产自动化。自动化搬运系统需要分析生产工艺，对搬运系统的硬件进行选型，达到经济高效生产。

【任务分析】

根据对工业机器人搬运工作站的生产工艺系统分析，选择搬运工作站所需要的主要硬件设备，包括工业机器人、工作站夹具、对应的控制软件、PLC等其他电气控制系统。

2. To master the analysis methods of process requirement for handling workstations.

[Skill Objectives]

1. To complete the selection of standard and non-standard components according to process requirements.
2. To be able to compile hardware selection schemes of handling workstations.

[Competence Objectives]

1. Cultivate in the student an optimistic, positive, pragmatic, and enterprising attitude towards life.
2. Improve the student's vocational abilities such as the abilities of professional technology application, communication and coordination, and relearning.

[Task Scenario]

Workers are used to complete the handling task in the production of car door trim panels of an automobile enterprise, which is inefficient and cannot meet the current production requirements. Convenient, efficient, and cost-effective production equipment are required to achieve production automation. Production process analysis and hardware selection need to be conducted for the automated handling system to achieve economical and efficient production.

[Task Analysis]

Select the main hardware equipment required for the handling workstation, including industrial robots, workstation fixtures, corresponding control software, PLC and other electrical control systems, according to the analysis of the production process system of the industrial robot handling workstation.

【知识准备】

3.1.1 搬运工作站系统分析

用触摸屏配合按钮及转换开关实现全线自动运行功能及单站运行功能的切换,运行流程图如图3-3所示。

搬运工作站以长方形汽车门饰板搬运为例,如图3-2所示,利用工业机器人在生产线搬运汽车门饰板,将其从仓库中搬运到工作台进行模拟焊接,完成模拟焊接后再将汽车门饰板搬运到传送带流转到下一个工位;下一个工位同样利用工业机器人将汽车门饰板搬运到视觉相机进行检测,合格的汽车门饰板将放入仓库存储,不合格的汽车门饰板放至下一级传送带进行其他处理。

搬运工作站可以独立运行,可以实现单站的搬运及模拟焊接,也可以实现单站的视觉检测;又可以进行联机运行,组成一套自动化运行系统,完成搬运、模拟焊接、视觉检测流程,如图3-3所示。

搬运工作站的控制可以通过触摸屏实现启动、暂停、停止功能,也可以通过按钮实现这部分功能。

系统设计要求如下:

1)需要对整个工作站进行合理的布局和规划。

2)需要在控制期间对整个现场设备进行控制。

3)需要配置工业机器人代替人工进行汽车门饰板的搬运。

4)需要设计出能够安装在机器人第六轴法兰盘上的吸盘及焊枪,用于搬运和焊接汽车门饰板。

5)需要合理选择传送带,用于机器人之间运送汽车门饰板。

[Assumed Knowledge]

3.1.1 Handling Workstation System Analysis

Conduct the switch between the function of the automated operation of the entire line and the function of only the workstation operation with a touch screen, buttons and transfer switches, and the flow is as shown in Figure 3-3.

The handling of rectangular aluminum plates at the workstation is taken as an example, as shown in Figure 3-2. Industrial robots are used for the handling of aluminum plates in the production line, handling them from the warehouse to the workbench for simulated welding and, upon its completion, handling them to the conveyor belt which sends them to the next station, where industrial robots are used to handle the plates to the visual camera for detection. Qualified aluminum plates will be stored in the warehouse, while unqualified ones be placed on the next conveyor belt for machining of other kinds.

The handling workstation can operate independently, with the possibility of realizing handling, simulated welding, and visual detection respectively of the single workstation; it can also perform a joint operation to form an automated operation system, completing the flow of handling, simulated welding, and visual detection, as shown in Figure 3-3.

The control of the handling workstation can realize the start, pause and stop functions through both the touch screen and the buttons.

The system design requirements are as follows:

1) It is necessary to have a reasonable layout and planning for the entire workstation.

2) It is necessary to control all the on-site equipment during the control period.

3) It is necessary to equip industrial robots replacing humans for the handling of car door trim panels.

4) It is necessary to design suction cups and welding guns that can be installed on the flange of the sixth axis of the robot for the handling and welding of car door trim panels.

5) It is necessary to select reasonably conveyor belts for the handling of car door trim panels between robots.

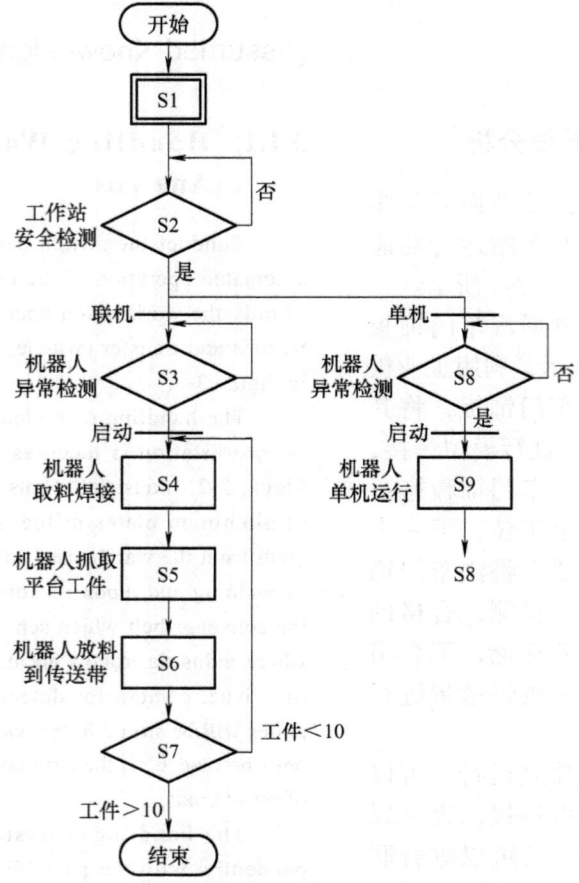

图 3-3 搬运工作站系统运行流程图
Figure 3-3　Flow Chart of the Handling Workstation System

6) 需要选择合适的网络，并合理进行配置，使得智能设备间能够实现有效通信。

7) 需要编写智能控制器的程序，使得整个控制流程达到工艺要求。

8) 需要整理出相关资料，完成材料的技术交底。

3.1.2　主要硬件选型

1. 工业机器人选型

搬运工作站中工业机器人是系统的主要执行机构，是硬件选型的重要设备。当前市场中工业机器人的品牌较多，国内常见的工业机器人品牌主

6) It is necessary to select a suitable network with reasonable configurations to enable effective communication between intelligent equipment.

7) It is necessary to write programs for the intelligent controller to ensure that the entire control flow meets the process requirements.

8) It is necessary to organize relevant materials and complete their technical disclosure.

3.1.2　Main Hardware Selection

1. Selection of industrial robots

The industrial robot in the handling workstation is the main executive mechanism of the system and the important equipment for hardware selection. There are many brands of industrial robot in the market nowadays, and those common brands in China mainly include

要有 KUKA、ABB、YASKAWA、FANUC、STUBCI、OTC、COMAU、华数等，其中 KUKA、ABB、FANUC 和华数的应用较广。

首先，选择工业机器人应考虑所选工业机器人的功能必须满足生产工艺要求，是否通过了可靠性测试，是否解释了无故障时间。生产工艺不同，对工业机器人的动作类型、承载能力、运动范围、速度和机器人的重复精度也有不同的要求。

其次，工业机器人除了满足生产工艺要求外，还要保证稳定可靠的质量，而且要经济合理。关注减速器、伺服系统、控制器的品牌和质量水平；考虑机械结构是否优化设计，结构是否合理，是否有足够的刚性和稳定性。

最后，工业机器人的操作、教学和编程应该简单易学。评估机器人本身质量，评估机器人制造商质量保证体系的完善和可信度，对工业机器人整体生产企业进行评估 ISO 认证相关标准。

本项目是对汽车门饰板进行点焊及搬运，汽车门饰板最大重量不超过 2kg，工具重量不超过 2kg，工作范围应不小于 1100mm，根据工艺要求和布局要求，给出两个方案进行选择：

1) FANUC 垂直关节多轴机器人，如图 3-4 所示。图 3-5 给出了 FANUC 机器人 M-10iA 工作范围及外围尺寸。

2) ABB 工业机器人 IRB 2600-12/1.65，如图 3-6 所示。图 3-7 给出了 ABB 机器人 IRB 2600-12/1.65 工作范围及外围尺寸。

KUKA, ABB, YASKAWA, FANUC, STUBCI, OTC, COMAU, and Huashu. Among them, KUKA, ABB, FANUC, and Huashu are more popular.

Firstly, the functions of the industrial robots to be selected shall meet the production process requirements, and such issues shall be considered as whether the robots have passed reliability testing, and whether the fault free time has been explained when selecting the industrial robots. Different production processes require different robots in terms of their action type, bearing capacity, range of motion, speed, and repetitive precision.

Secondly, in addition to meeting the production process requirements, industrial robots shall have a good quality, stability and reliability with a reasonable price. Attention shall be paid to the brand and quality of reducers, servo systems, and controllers; such issues shall be taken into account as whether optimized design has been done to the mechanical structure, whether the structure is reasonable, rigid and stable as required.

Finally, the operation, teaching, and programming of industrial robots shall be simple and easy to learn. Evaluate the quality of robots, and the completeness and credibility of the quality assurance system of the robot manufacturers, as well as ISO certification standards related to the overall production enterprises of industrial robot.

Spot welding and handling of aluminum plates are performed in this program. The maximum weight of the aluminum plate do not exceed 2kg, that of the tool do not exceed 2kg, and the motion range shall not be less than 1100mm. Two options are provided for selection according to process and layout requirements:

1) FANUC vertical joint multi-axis robot, as shown in Figure 3-4. The motion range and external dimensions of the FANUC robot M-10iA are shown in Figure 3-5.

2) The ABB Industrial Robot IRB 2600-12/1.65 is as shown in Figure 3-6, and its motion range and external dimensions are shown in Figure 3-7.

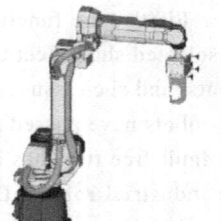

M-10iA/12L
标准轴数：6轴
Standard number of axis: 6
手臂负载：12kg
Arm load: 12kg
工作范围：1632mm
Motion range: 1632mm

图 3-4　M-10iA 参数
Figure 3-4　Parameters of the M-10iA

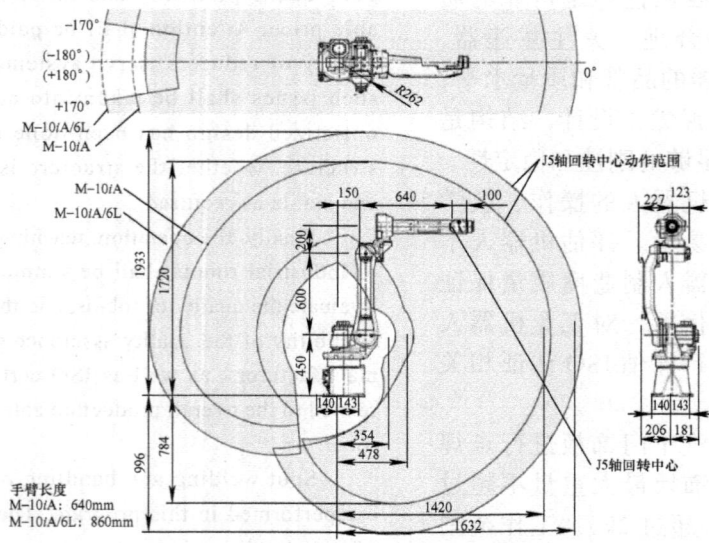

图 3-5　M-10iA 工作范围及外围尺寸
Figure 3-5　Motion Range and External Dimensions of the M-10iA

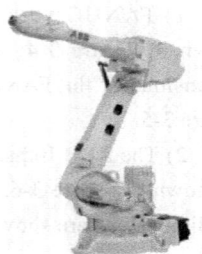

IRB 2600-12/1.65
标准轴数：6轴
Standard number of axis: 6
手臂负载：12kg
Arm load: 12kg
工作范围：1650mm
Motion range: 1650mm

图 3-6　IRB 2600-12/1.65 参数
Figure 3-6　Parameters of the IRB 2600-12/1.65

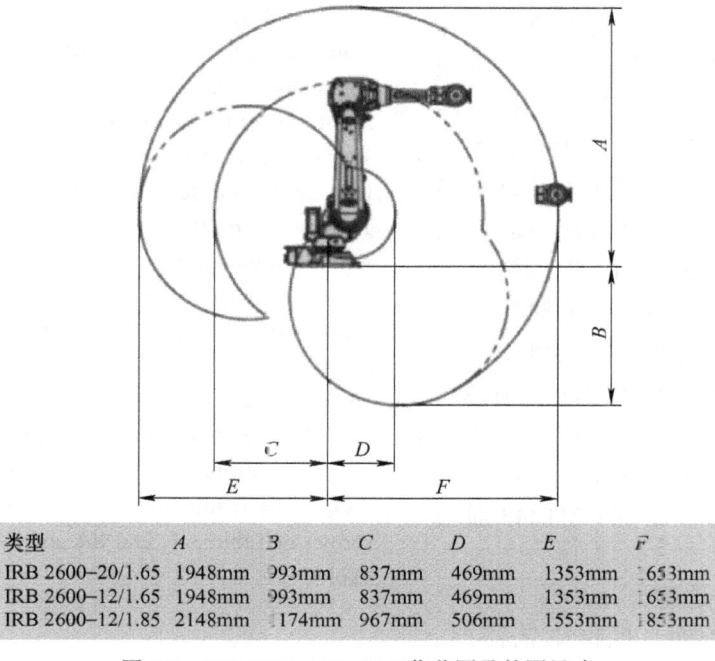

类型	A	B	C	D	E	F
IRB 2600-20/1.65	1948mm	993mm	837mm	469mm	1353mm	1653mm
IRB 2600-12/1.65	1948mm	993mm	837mm	469mm	1353mm	1653mm
IRB 2600-12/1.85	2148mm	1174mm	967mm	506mm	1553mm	1853mm

图 3-7　IRB 2600-12/1.65 工作范围及外围尺寸

Figure 3-7　Motion Range and External Dimensions of the IRB 2600-12/1.65

考虑到搬运工作站对搬运的精度要求不高，可以选用国产性价比较高的工业机器人来实现，选用国产工业机器人也必须要考虑到上述所说的满足工艺要求原则、可靠性原则、易操作原则、优化配置原则、质量保证原则。例如：华数 HSR-JR612，自由度 6，额定负载 12kg，最大工作半径 1555mm。

2. PLC 选型

PLC 的主要功能是控制外部系统。该系统可以是单机、集群或生产过程。不同类型的 PLC 有不同的应用范围。根据生产过程的要求，分析被控对象的复杂性，对 I/O 点和 I/O 点的类型进行计数，本项目输入点数为 38 个点，输出点数为 29 个点。在不浪费资源的前提下，合理估计内存容量，确定合适的模型。结合市场情况，对 PLC 生

As the required handling precision is not high for the handling workstation, the cost-effective Chinese industrial robots can be selected based on the above-mentioned principles of meeting process requirements, being reliable, and easy to operate with optimized configuration and assured quality. For instance, the Huashu HSR-JR612 with 6 degrees of freedom, rated load of 12kg, and maximum working radius of 1555mm can be selected.

2. Selection of PLCs

The main function of PLCs is to control external systems, which can be single machines, clusters, or production processes. PLCs of different types have different application ranges. There are 38 input points and 29 output ones in this program according to the requirements of the production process, the analysis of the complexity of the controlled object, and the calculation based on the number and type of the I/O points. Estimate reasonably the memory capacity and determine a suitable model without wasting resources. Select PLC models with higher cost-effectiveness based on the market situation, and the comprehensive survey results on the products, after-sales service, technical support,

产厂家的产品及售后服务、技术支持、网络通信等综合情况进行调查，选择性价比较高的 PLC 机型。通过对搬运工作站控制系统的分析，工业机器人与 PLC 可以实现 PROFINET 通信和 PROFIBUS 通信。因此，可以选择带有 2 个 PROFINET 通信接口的西门子 S7-1215C DC/DC 型 PLC。该类型 PLC 只有 14 个输入和 10 个输出，必须增加 2 个含有 16 输入的 SM1221 输入模块和 2 个含有 16 输出的 SM1222 输出模块增加输入/输出点数。如果使用 PROFIBUS 通信，除了已经选择的 PLC 外，还必须增加 CM1243 通信模块。

3. 其余设备选型

除主要设备外，根据工艺需要，项目其余设备清单见表 3-1，软件配置见表 3-2。

and network communication of the PLC manufacturers. It is found by the analysis of the control system of the handling workstation that PROFINET and PROFIBUS communications can be achieved between the industrial robot and PLC. Therefore, two PLCs of Siemens S7-1215C DC/DC with PROFINET communication interface can be selected. Since PLCs of this type have only 14 inputs and 10 outputs, two SM1221 input modules with 16 inputs and two SM1222 output modules with 16 outputs have to be added to increase the number of input/output points. If PROFIBUS communication is adopted, a CM1243 communication module has to be added in addition to the selected PLC.

3. Selection of other equipment

In addition to the main equipment, and based on the process requirements, the list of other equipment is shown in Table 3-1, and the software configuration in Table 3-2.

表 3-1 项目其余设备清单
Table 3-1 List of Other Equipment for the Program

序号 S/N	名称 Name	数量 Qty	单位 Unit	备注 Remarks
1	机器人工夹具 Robot manual fixture	1	套 set	
2	立体仓库 Stereoscopic store	1	套 set	设 10 层 1 列共 10 个仓位 Set up a total of 10 bins on 10 layers and 1 column
3	变频器 Frequency converter	1	套 set	6SL3210–5BB11–2UV0
4	触摸屏 Touch screen	1	台 piece	TP1200 精致面板，TFT 显示屏，PROFINET/工业以太网接口（2 个端口） TP1200 exquisite panel, TFT display screen, PROFINET/industrial Ethernet interface (2 ports)
5	安全防护系统 Safety protection system	1	套 set	防止意外闯入、保护人员安全 Prevent accidental entry and protect personnel safety
6	电气控制柜 Electrical control cabinet	1	套 set	用于放置电气元件和电气设备 For placement of electrical components and equipment

表 3-2　项目软件配制清单

Table 3-2　List of the Software Configuration for the Program

序号 S/N	软件名称 Software name	数量 Qty	单位 Unit	基本功能 Basic function
1	TIA 软件 TIA software	1	套 set	负责周边设备及机器人控制，实现智能制造单元的流程和逻辑总控 Being responsible for controlling peripheral equipment and robots to achieve the flow and overall logic control of intelligent manufacturing units
2	机器人仿真软件 Robot simulation software	1	套 set	单元设备模拟，虚拟安装调试，布局优化 Unit device simulation, virtual installation and commissioning, layout optimization
3	SOLIDWORKS 或 NX 软件 SOLIDWORKS or NX software	1	套 set	三维模型设计和编程，编制零件加工工艺 3D model design and programming, and parts machining process compiling
4	EPLAN 电气设计软件 EPLAN electrical design software	1	套 set	电气线路绘制 Electrical circuit drawing

任务 3.2　设计方案的编写
Task 3.2　Compilation of Design Schemes

【知识目标】

1. 掌握工业机器人搬运工作站简介及布局。

2. 掌握工业机器人搬运工作站工作流程及控制要求。

3. 了解工业机器人搬运工作站主要设备清单。

4. 掌握整体设计方案的编写方法。

[Knowledge Objectives]

1. To master the introduction and layout of industrial robot handling workstations.

2. To master the workflow and control requirements of industrial robot handling workstations.

3. To understand the list of main equipment of industrial robot handling workstations.

4. To master the compiling methods of overall design schemes.

【技能目标】

1. 根据工业机器人搬运工作站的要求选择设备清单。

2. 根据工艺要求说明工作流程及控制要求。

3. 能够设计工业机器人搬运工作站的整体方案。

[Skill Objectives]

1. To select the list of equipment according to the requirements of the industrial robot handling workstations.

2. To explain the workflow and control requirements according to the process requirements.

3. To be able to compile over design schemes of industrial robot handling workstations.

【素质目标】

1. 养成良好的自主学习习惯。
2. 培养团队协作精神。

【任务情景】

工业机器人搬运工作站系统是高效实现机器人对汽车门饰板的搬运，一个好的方案能加快实现搬运工作站的生产。某汽车企业需要一份工业机器人搬运工作站的设计方案实现自动化搬运工作，该方案包括工作站的简介、搬运工件说明、搬运工作站设备清单、工作流程等内容。

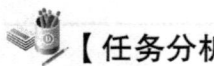【任务分析】

在项目实施的过程中，必须要编写出设计方案交给客户。设计方案必须详细地叙述出项目实施的优势、项目能够给企业带来的利益和生产效率的提升、项目实施过程中的设备选型和布局、项目的工程预算等。一个好的设计方案对于项目的推进和实施有着重要的意义和作用。

【知识准备】

3.2.1 设计方案的结构和要素

1. 工作站简介及布局

说明编写设计方案的主要目的和设备布局。

2. 加工工件说明

说明加工工件毛坯尺寸、加工图纸以及零件加工工艺。

[Competence Objectives]

1. To develop good self-directed learning habits.
2. To cultivate the teamwork spirit.

[Task Scenario]

The industrial robot handling workstation system is to efficiently realize the handling of the car door trim panel with robots, and a good scheme can speed up its production. For realizing automated handling, an automobile enterprise needs the design scheme of an industrial robot handling workstation which includes the introduction of the workstation, description of handling workpieces, list of equipment for the handling workstation, and workflow, etc.

[Task Analysis]

During the program implementation process, it is necessary to compile and submit to the client a design scheme which shall include the detailed description of the advantages of the program, the benefits that the program can bring to the enterprise, the improvement in the production efficiency, the selection and layout of equipment during the program implementation process, the engineering budget of the program, etc. A good design scheme is of great significance for the promotion and implementation of the program.

[Assumed Knowledge]

3.2.1 Structure and Elements of Design Schemes

1. Introduction and layout of the workstation

Explain the main purpose and equipment layout of the design scheme.

2. Description of workpieces to be machined

Explain the dimensions of the workpiece blanks, machining drawings, and parts machining process.

3. 工艺动作流程

说明工作站工作流程及控制要求。

4. 工业机器人搬运工作站主要设备清单

说明工作站主要设备，以及这些设备的技术配置及参数。

3.2.2 设计方案编写示例

1. 目录

典型工业机器人搬运工作站系统设计方案的目录样式如图 3-8 所示。

3. Process action flow

Explain the workflow and control requirements of the workstation.

4. List of main equipment of the industrial robot handling workstation

Explain the main equipment of the workstation, and their technical configuration and parameters.

3.2.2 An Example of Design Scheme Compilation

1. Table of contents

The style of the table of content of the design scheme of a typical industrial robot handling workstation system is shown in Figure 3-8.

图 3-8　典型工业机器人搬运工作站系统设计方案的目录样式

Figure 3-8　Style of the Table of Contents of the Design Scheme of a Typical Industrial Robot Handing Workstation System

2. 正文

这里以"工业机器人搬运工作站系统设计方案"的内容编写为例,为大家演示如何进行方案文档编写,方案的其他内容按要求自己编写。

示例:

搬运作业是指用一种设备握持工件,从一个加工位置移动到另外一个加工位置的过程。如果采用工业机器人来完成这个任务,整个搬运系统则构成了工业机器人搬运工作站。工业机器人工作站以典型工业机器人搬运工作站系统实际与工艺要求为设计依据,将一台或多台工业机器人、PLC控制系统、输送线系统、立体仓库、视觉检测设备、末端执行器等设备,集成为完成某一特定工序作业的独立生产系统。

工业机器人搬运工作站主视图和俯视图如图3-9和图3-10所示,包含6轴多关节机器人、机器人安装底座、在线检测单元、输送线系统、立体仓库、PLC控制系统等。

2. Text

The compilation of the content of the "Design Scheme of the Industrial Robot Handling Workstation System" is taken as an example to demonstrate how to compile the scheme documents. Other content of the scheme shall be compiled by yourself according to the requirements.

Example:

Handling operation refers to the process of using a device to hold a workpiece and move it from one machining position to another. If industrial robots are used to complete this task, the entire handling system constitutes an industrial robot handling workstation. Based on the actual situation of typical industrial robot handling workstation systems and process requirements, the industrial robot workstation integrates one or more industrial robots, PLC control system, conveying line system, warehouse, visual detection equipment, end effectors and other equipment into an independent production system to complete a specific process operation.

The structure of an industrial robot handling workstation is shown in Figure 3-9 and Figure 3-10, including six-axis multi-joint robots, robot mounting bases, online testing units, conveying line system, warehouses, PLC control system, etc.

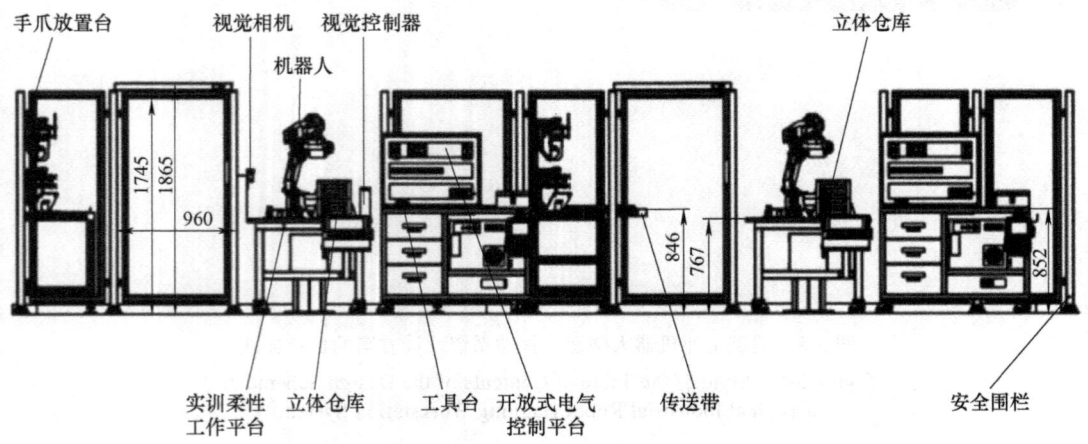

图 3-9 工业机器人搬运工作站主视图

Figure 3-9 Front View of an Industrial Robot Handling Workstation

项目 3 典型工业机器人搬运工作站系统的设计及应用

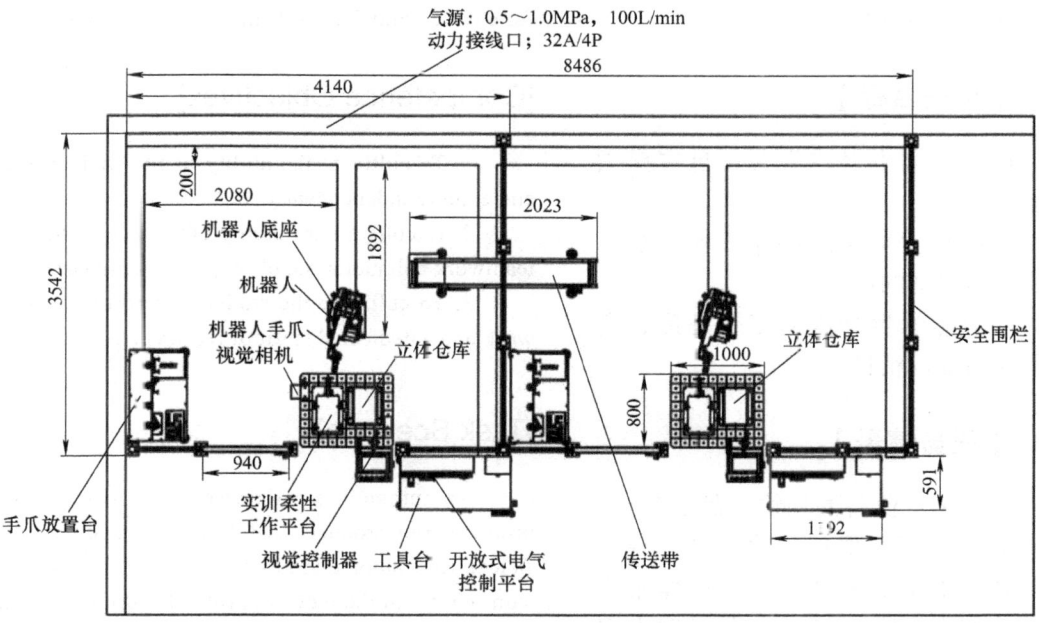

图 3-10 工业机器人搬运工作站俯视图
Figure 3-10 Top View of an Industrial Robot Handling Workstation

任务 3.3 施工图的设计及建模
Task 3.3 Design and Modeling of Detailed Drawings

【知识目标】

1. 掌握利用 SOLIDWORKS 或者 NX 软件为典型工业机器人搬运工作站系统施工图建模。
2. 了解工业机器人搬运工作站系统施工工艺流程。
3. 掌握电气原理图方案设计知识。
4. 掌握非标准零件工程图的设计与绘制。

[Knowledge Objectives]

1. To master the modeling of detailed drawings of typical industrial robot handling workstation systems by the SOLIDWORKS or NX software.
2. To understand the construction process of industrial robot handling workstation systems.
3. To master the knowledge of scheme design of electrical schematic diagram.
4. To master the design and preparation of non-standard parts engineering drawings.

【技能目标】

1. 能够根据工业要求选择正确的软件完成建模。
2. 根据工业要求完成标准部件和非标准部件的选型。
3. 能编写搬运工作站系统施工

[Skill Objectives]

1. To be able to select the correct software to complete modeling according to industrial requirements.
2. To be able to complete the selection of standard and non-standard components according to industrial requirements.
3. To be able to compile design selection schemes

图的设计选型方案。

【素质目标】

1. 培养学生自主学习、自主探索的能力。

2. 培养学生团队协作、善于观察以及总结归纳的能力。

3. 培养学生严谨的工作态度、潜心研究的敬业精神。

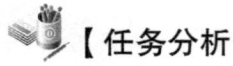
【任务情景】

某汽车企业需要加快产品的生产，提高生产效率实现自动化生产，现在已经做出了工业机器人搬运工作站系统设计方案，需要设计工业机器人搬运工作站系统的施工图，及对非标准零部件进行建模。

【任务分析】

在项目实施的过程中，当完成方案的设计和设备选型以后，在具体生产、安装和调试阶段，往往需要一个团队来完成。负责设计和施工的人员并不一定是同一人或同一小组，因此必须先设计出相关非标准产品的图纸和具体系统的施工图，这样才能便于项目实施的更好开展。需设计的图纸有设备布局图、系统框图、电气原理图和非标准零件工程图。

【知识准备】

3.3.1 设备布局图

工业机器人搬运工作站布局图如图3-11所示，包含FANUC工业机器人、传送带、手爪放置台、柔性工作平台、开放式电气控制平台和安全围栏等。

for detailed drawings of handling workstation system.

[Competence Objectives]

1. To cultivate the ability of the student to learn and explore independently.

2. To cultivate the ability of the student to perform teamwork collaboration, observe, and summarize.

3. To cultivate the student a rigorous attitude to work and a dedicated spirit of research.

[Task Scenario]

An automobile enterprise needs to speed up the production of products, and desires to have improve efficiency and automation in production. It has made a design scheme of the industrial robot handling workstation system now, and needs the design of detailed drawings of the system and the modeling of non-standard components and parts.

[Task Analysis]

In the process of program implementation, when the design scheme and and equipment selection have been completed, a team is often required to complete production, installation, and commissioning at the corresponding stages. As the person or team responsible for design may not be the same as that of for construction, it is necessary to first complete the design drawings of relevant non-standard products and the detailed drawings of the specific system in order to have better implementation of the program. The drawings to be designed include: equipment layout, system chart, electrical schematic diagram, and non-standard parts engineering diagram.

[Assumed Knowledge]

3.3.1 Equipment Layout Diagram

The layout of the industrial robot handling workstation is shown in Figure 3-11, including the FANUC industrial robot, conveyor belt, gripper placement platform, flexible working platform, open electrical console, and safety fence.

项目3 典型工业机器人搬运工作站系统的设计及应用

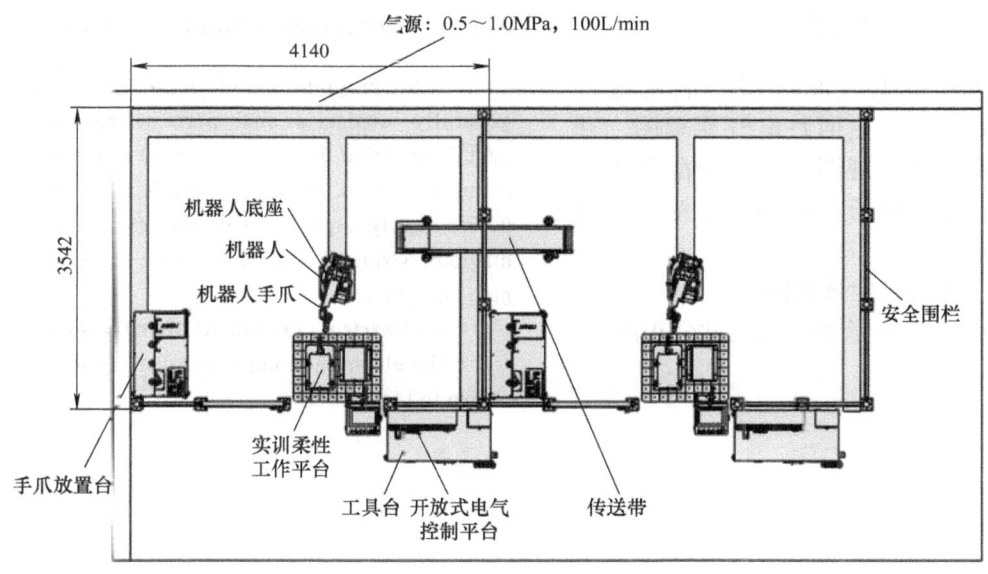

图 3-11 工业机器人搬运工作站布局图
Figure 3-11 Layout of the Industrial Robot Handling Workstation

3.3.2 系统框图

典型搬运工作站各主要设备之间的连接关系及控制关系的系统框图如图 3-12 所示。系统框图根据给定的系统功能要求，进行相应的搬运工作站系统设计。在设计之初，需要设计系统框图，为接下来的电路和程序设计提供一个基础。

3.3.2 System Chart

The system chart of the connection and control relationship between the main equipment of the typical handling workstation is shown in Figure 3-12. In the system chart, the handling workstation is designed according to the given functional requirements of the system. At the beginning of design, it is necessary to design a system chart which provides a foundation for the subsequent design of circuits and procedures.

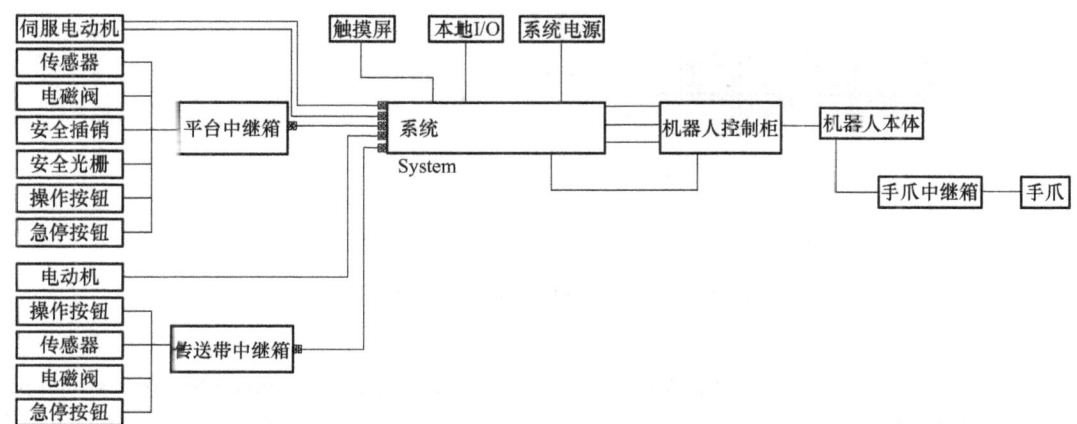

图 3-12 典型搬运工作站各主要设备之间的连接关系及控制关系系统框图
Figure 3-12 System Chart of the Connection and Control Relationship between the Main Equipment of the Typical Handing Workstation

3.3.3 电气原理图

项目中电气原理图一般由电气元件分布图、主回路电气原理图、开关电源电路原理图、信号分配电气原理图、接线端子与现场信号接线原理图等部分组成。

1. 电气元件分布图

图3-13所示为电气元件分布图。

3.3.3 Electrical Schematic Diagram

In the program, the electrical schematic diagram generally consists of such parts as electrical component distribution diagram, electrical schematic diagram of main circuits, circuit schematic diagram of switch power supply, signal distribution electrical schematic diagram, wiring terminal and on-site signal wiring schematic diagram.

1. Electrical component distribution diagram

The electrical component distribution diagram is shown in Figure 3-13.

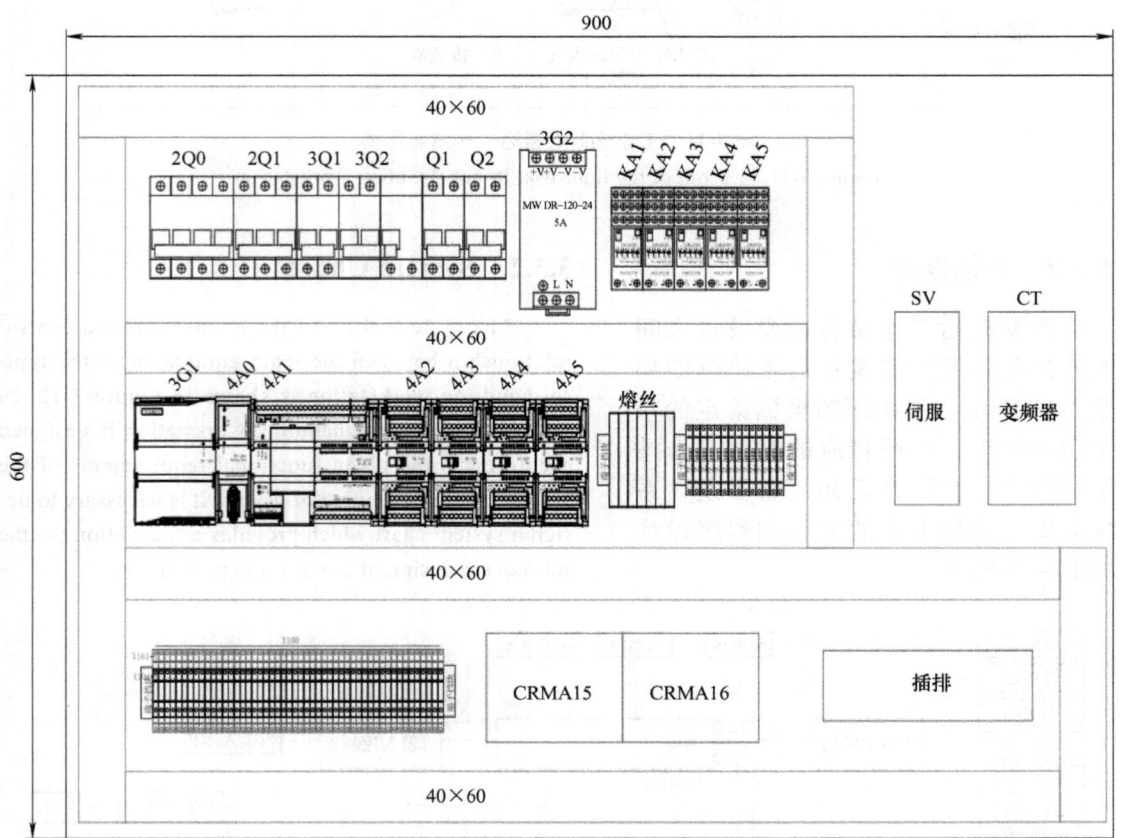

图 3-13　电气元件分布图
Figure 3-13　Electrical Component Distribution Diagram

第一层为断路器、开关电源和继电器，第二层为PLC模块、熔断器、伺服控制器和变频器，第三层为信号接线端子、信号配置板和排插。

The first layer is for circuit breakers, switch power supplies, and relays, while the second layer for PLC modules, fuses, servo controllers, and frequency converters, and the third layer for signal wiring terminals, signal configuration boards, and sockets.

2. 主回路电气原理图

电源进线经过开放式电气控制平台的总断路器分配给工业机器人、PLC 24V 电源、触摸屏及传感器 24V 电源、伺服驱动、变频器断路器，形成了整个电柜的主电源进线，如图 3-14 所示。

2. Electrical schematic diagram of the main circuit

The incoming lines of power supply are distributed to industrial robots, 24V power supply to the PLC, 24V power supply to the touch screen and sensors, servo drive, and the circuit breaker of frequency converter through the main circuit breaker of the open electrical console, forming the incoming lines of the main power supply of the entire electrical cabinet, as shown in Figure 3-14.

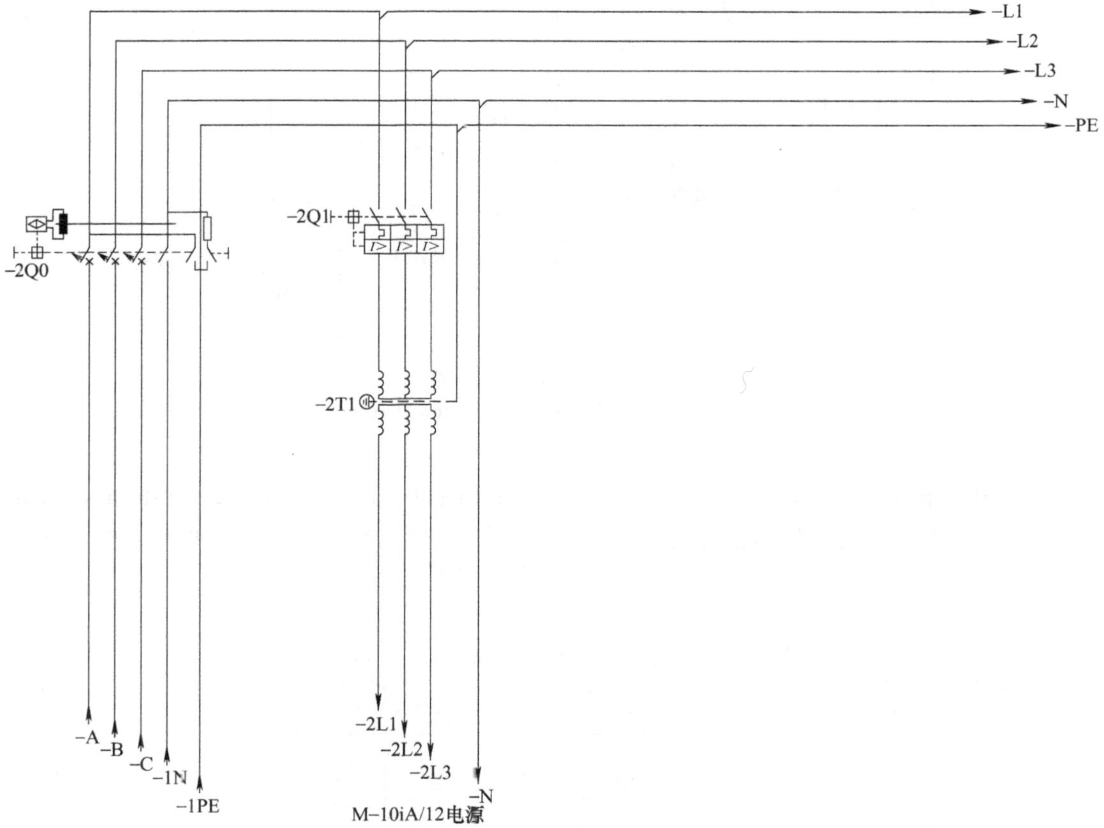

图 3-14 主电源进线图

Figure 3-14　Diagram of the Incoming Lines of the Main Power Supply

3. 直流电源电气原理图

主电源断路器分别对开关电源等分配控制电源，电气控制平台断路器直流电源分配如图 3-15 所示。

3. Electrical schematic diagram of DC power supply

The circuit breaker of the main power supply distributes control power to the switch power supply, etc. The distribution of DC power to the circuit breaker of the electrical console is shown in Figure 3-15.

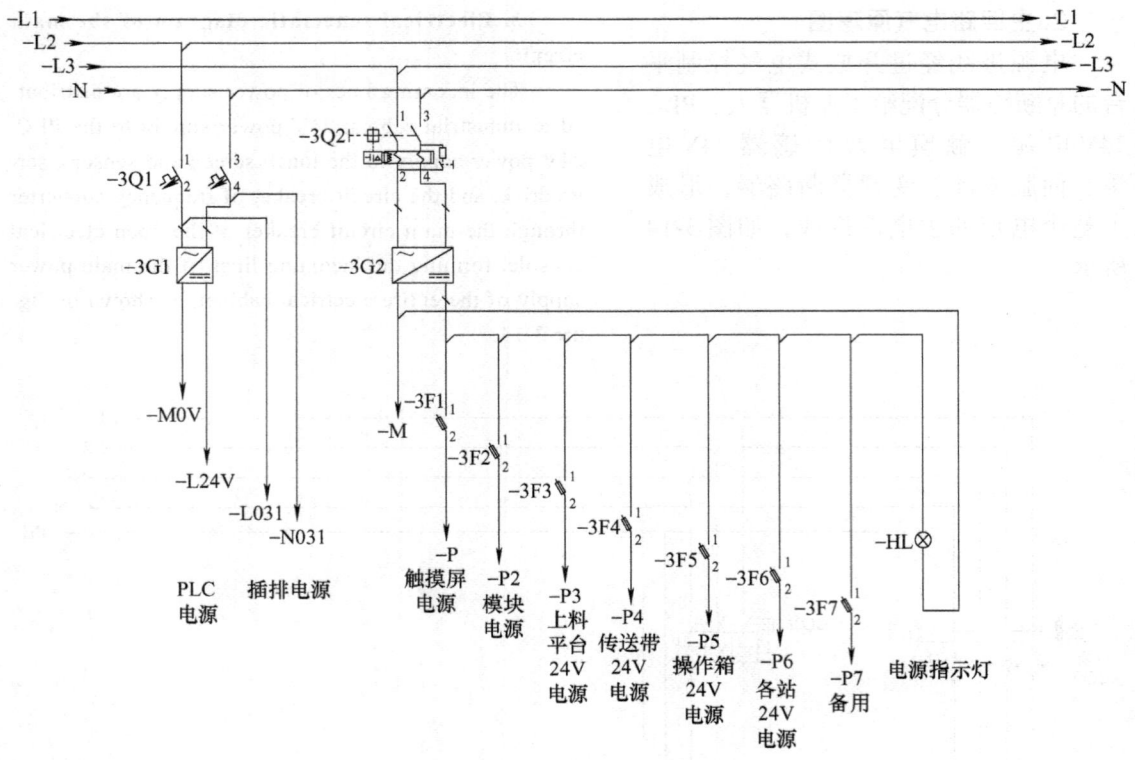

图 3-15　直流电源分配图

Figure 3-15　DC Power Distribution Diagram

4. 主机架电路原理图

主机架电路原理如图 3-16 所示。

4. Circuit schematic diagram of the host frame

The circuit philosophy of the host frame is shown in Figure 3-16.

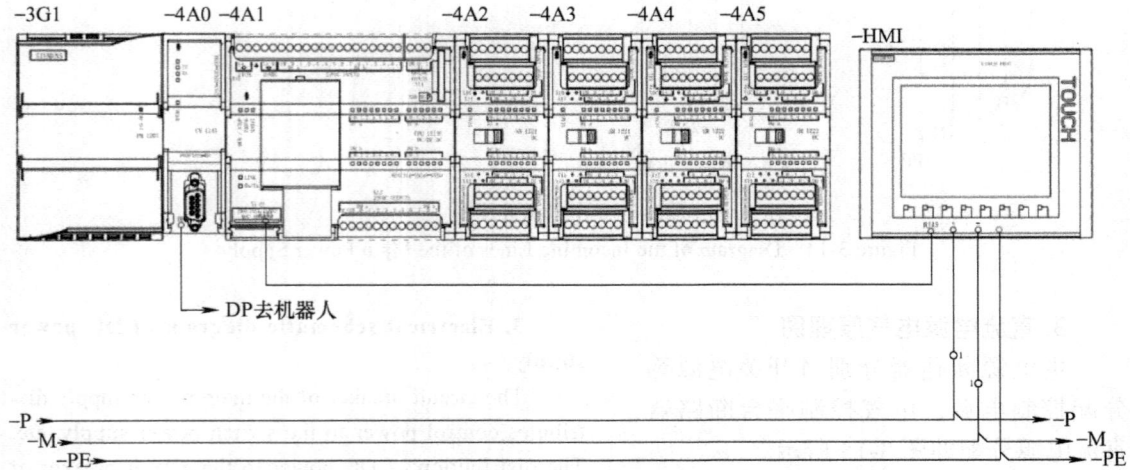

图 3-16　主机架电路图

Figure 3-16　Circuit Diagram of the Host Frame

5. 模块总览图

(1) CPU 模块 4A1 总览图

PLC 的基本单元主要做伺服控制的脉冲和方向输出、上料台夹具的到位信号输入及安全信号信号输入，信号输入输出接线图如图 3-17 所示。

5. Module overview

(1) Overview of the CPU module 4A1

The basic unit of PLC mainly performs pulse and directional output for servo control, in-place signal input for the loading table fixture, and safety signal signal input. The signal input wiring is shown in Figure 3-17.

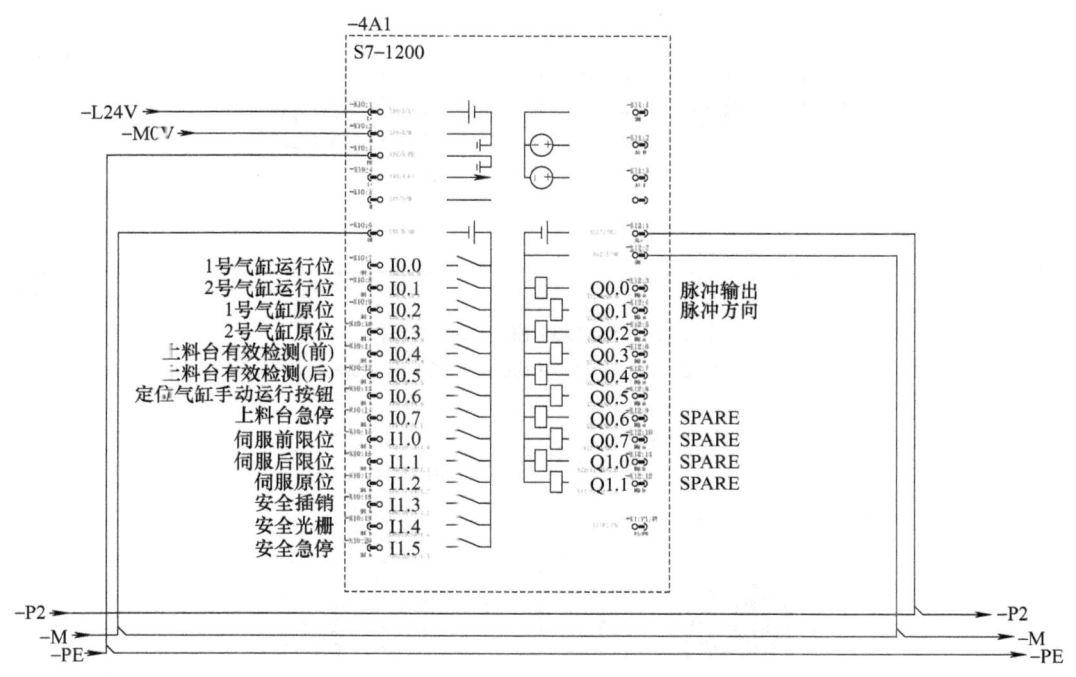

图 3-17 信号输入输出接线图

Figure 3-17 Signal Input/Output Wiring Diagram

(2) SM1221 模块 4A2 总览图

SM1221 模块分为四个部分，分别是传送带信号运行信号输入、启停控制输入、伺服准备就绪及报警输入和气压检测及操作面板急停输入，如图 3-18 所示。

(3) SM1221 模块 4A3 总览图

SM1221 模块是读取 FANUC 工业机器人信号的模块，根据读取的机器人信号进行相应的输出控制，信号通过网络进行数据的读取，如图 3-19 所示。

(2) Overview of the SM1221 module 4A2

The SM1221 module is divided into four parts, namely conveyor belt signal operation signal input, start stop control input, servo readiness and alarm input, air pressure detection and operation panel emergency stop input. The signal input wiring is shown in Figure 3-18.

(3) Overview of the SM1221 module 4A3

The SM1221 module is a module that reads signals from FANUC industrial robots, and performs corresponding output control based on the read robot signals. The signals are read through the network for data, as shown in Figure 3-19.

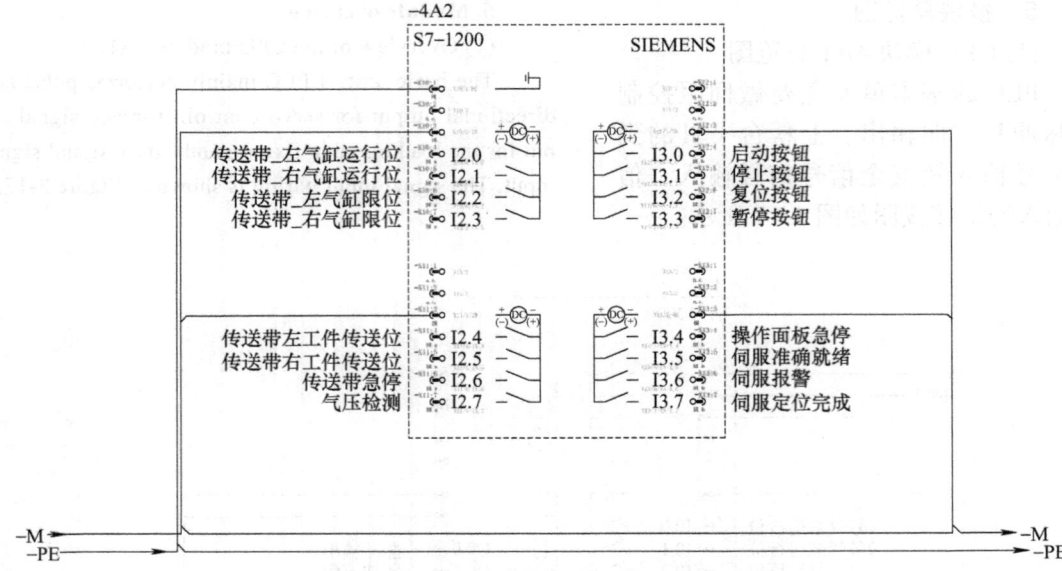

图 3-18　SM1221 模块 4A2 总览图
Figure 3-18　Overview of the SM1221 Module 4A2

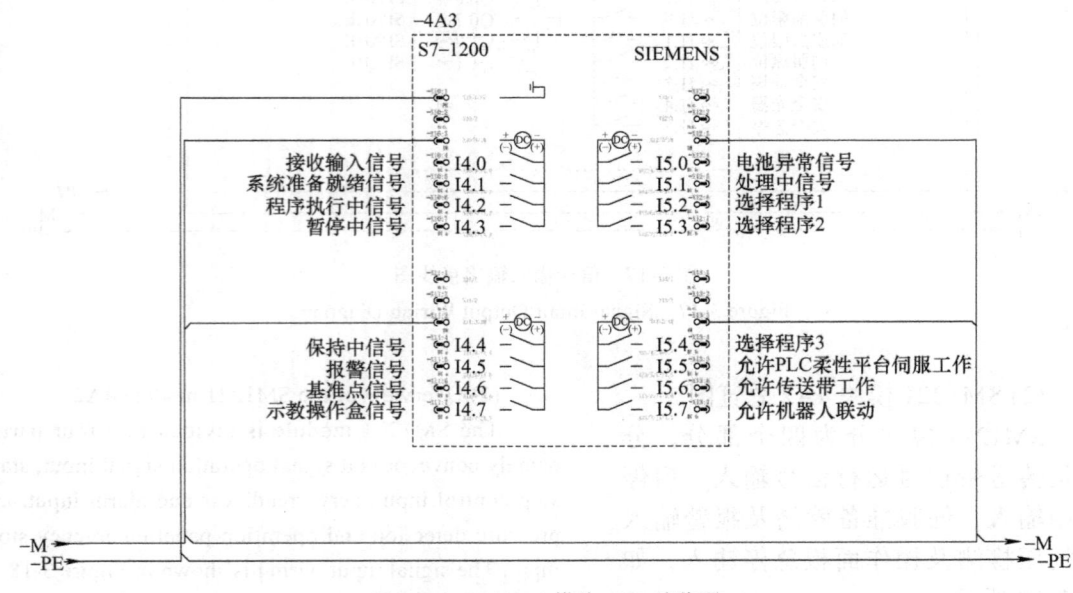

图 3-19　SM1221 模块 4A3 总览图
Figure 3-19　Overview of the SM1221 Module 4A3

(4) SM1222 模块 4A4 总览图

SM1222 模块是控制 FANUC 工业机器人搬运系统的指示灯及警示灯控制、变频器速度控制、传送带控制、联机控制的模块，如图 3-20 所示。

(4) Overview of the SM1222 module 4A4

The SM1222 module is for the control of indicators and warning lights, the speed of frequency converter, conveyor belt, and the online control of the FANUC industrial robot handling system, as shown in Figure 3-20.

(5) SM1222 模块 4A5 总览图

SM1222 模块是控制 FANUC 工业机器人搬运系统的 PLC 信号、机器人使能信号、机器人控制信号、机器人程序选择信号和机器人联机控制信号数据交换的模块,如图 3-21 所示。

(5) Overview of the SM1222 module 4A5

The SM1222 module is for the control of the PLC signal, robot enable signal, robot control signal, and the robot program selection signal, as well as the data exchange of the online robot control signals of the FANUC industrial robot handling system, as shown in Figure 3-21.

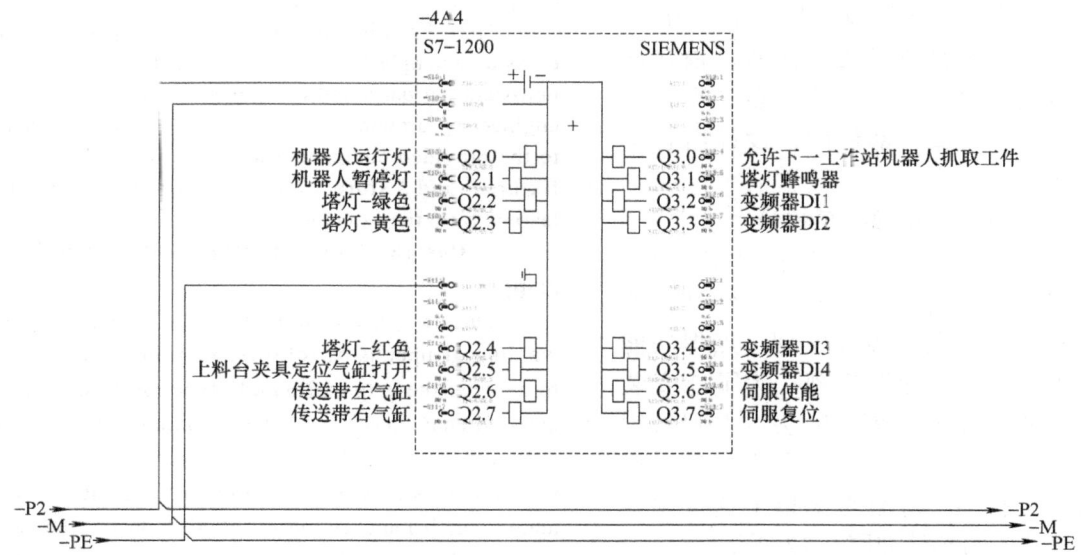

图 3-20　SM1222 模块 4A4 总览图

Figure 3-20　Overview of the SM1222 Module 4A4

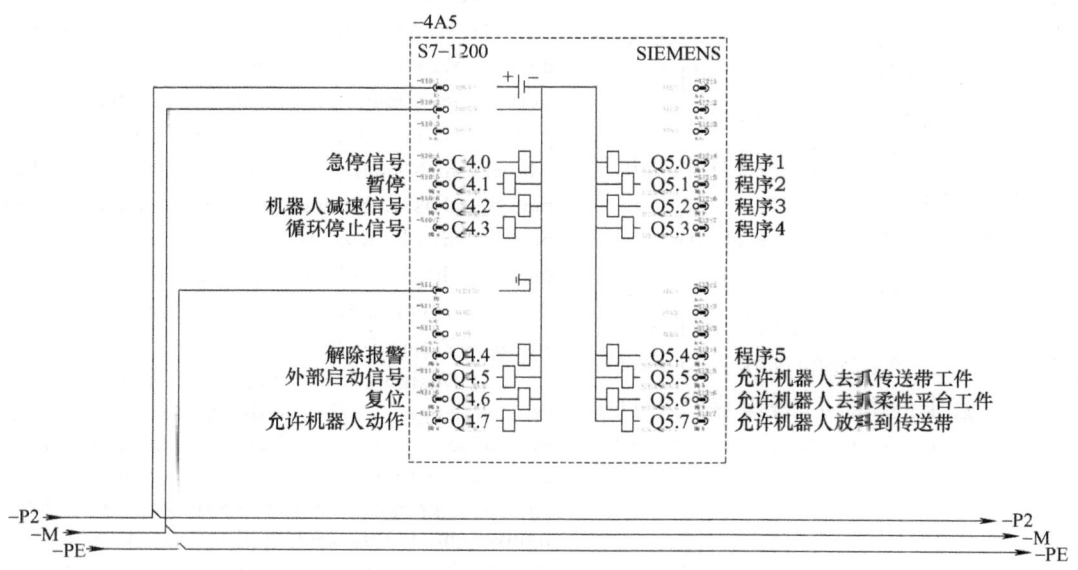

图 3-21　SM1222 模块 4A5 总览图

Figure 3-21　Overview of the SM1222 Module 4A5

3.3.4 非标件工程图

在本项目中,需要利用工业机器人将工件"汽车门饰板"从"仓库"取出放置在工作台上,然后对"汽车门饰板"进行点焊,点焊完成后再将"汽车门饰板"抓取搬运放到输送带上。

下面以"搬运及点焊手爪"为例,说明非标零部件的设计过程。请在此学习的基础上,根据提供的"平台夹具"和"仓库"部件图,设计出各个零件的工程图。

1. 搬运及点焊手爪设计

搬运及点焊手爪的设计思路是:根据"汽车门饰板"的外形特征,抓取搬运功能利用吸盘实现,点焊功能则采用"点焊钳"实现;为了能安装在工业机器人的第6轴法兰上则采用法兰结构;为了减轻重量则尽量采用铝合金材料。综合考虑以上这些因素后设计出的部件如图3-22所示。

1)"复合手爪法兰连接板"部件的设计。"复合手爪法兰连接板"的作用是将其他零件连接、组合成为一体,因此它的结构设计要与其他零件能够配合、组装;在重量上为了轻量化则选用铝合金型材。综合考虑以上这些因素后,设计出"复合手爪法兰连接板"如图3-23所示。

2)"搬运手爪法兰"部件的设计。设计"搬运手爪法兰"时主要考虑它与"复合手爪法兰连接板"的连接、与工业机器人第6轴的连接,以及轻量化、气管走线等。综合考虑以上这些因素后,设计出"搬运手爪法兰"如图3-24所示。

3.3.4 Engineering Drawings of Non-standard Parts

In this program, it is necessary to use industrial robots to take out the "loading and unloading products" of workpieces from the "store" and place them on the workbench, where spot welding is performed on the "loading and unloading products", which will be grabbed and handled onto the conveyor belt upon the completion of spot welding.

The "handling and spot welding gripper" is taken as an example here to explain the design process of non-standard parts and components. Please design the engineering drawing of each component in accordance to the provided component drawings of the "platform fixture" and "store" and based on what you have learned from the example.

1. Design of the handling and spot welding gripper

The design concept of the handling and spot welding gripper is as follows: suction cups are used to achieve the grip-and-handle function based on the external characteristics of the "loading and unloading products", while "spot welding pliers" are used to realize the spot welding function; a flange structure is adopted as it matches the installation on the flange of the 6th axis of the robot; aluminum alloy materials shall be used as far as possible in order to reduce weight. The designed components based on the above factors are shown in Figure 3-22.

1) Design of the component of the "flange connection plate of the composite gripper". The function of the "flange connection plate of the composite gripper" is to connect and combine other parts into a whole, which entails its structural design to be able to match and assemble with other parts; aluminum alloy profiles shall be used as far as possible in order to reduce weight. The designed "flange connection plate of the composite gripper" based on the above factors is as shown in Figure 3-23.

2) Design of the component of the "handling gripper flange". When designing the "handling gripper flange", such main factors are taken into account as its connection with the "flange connection plate of composite gripper" and 6th axis of the industrial robot, lightweight, and the routing of the gas circuits. The designed "handling gripper flange" based on the above factors is as shown in Figure 3-24.

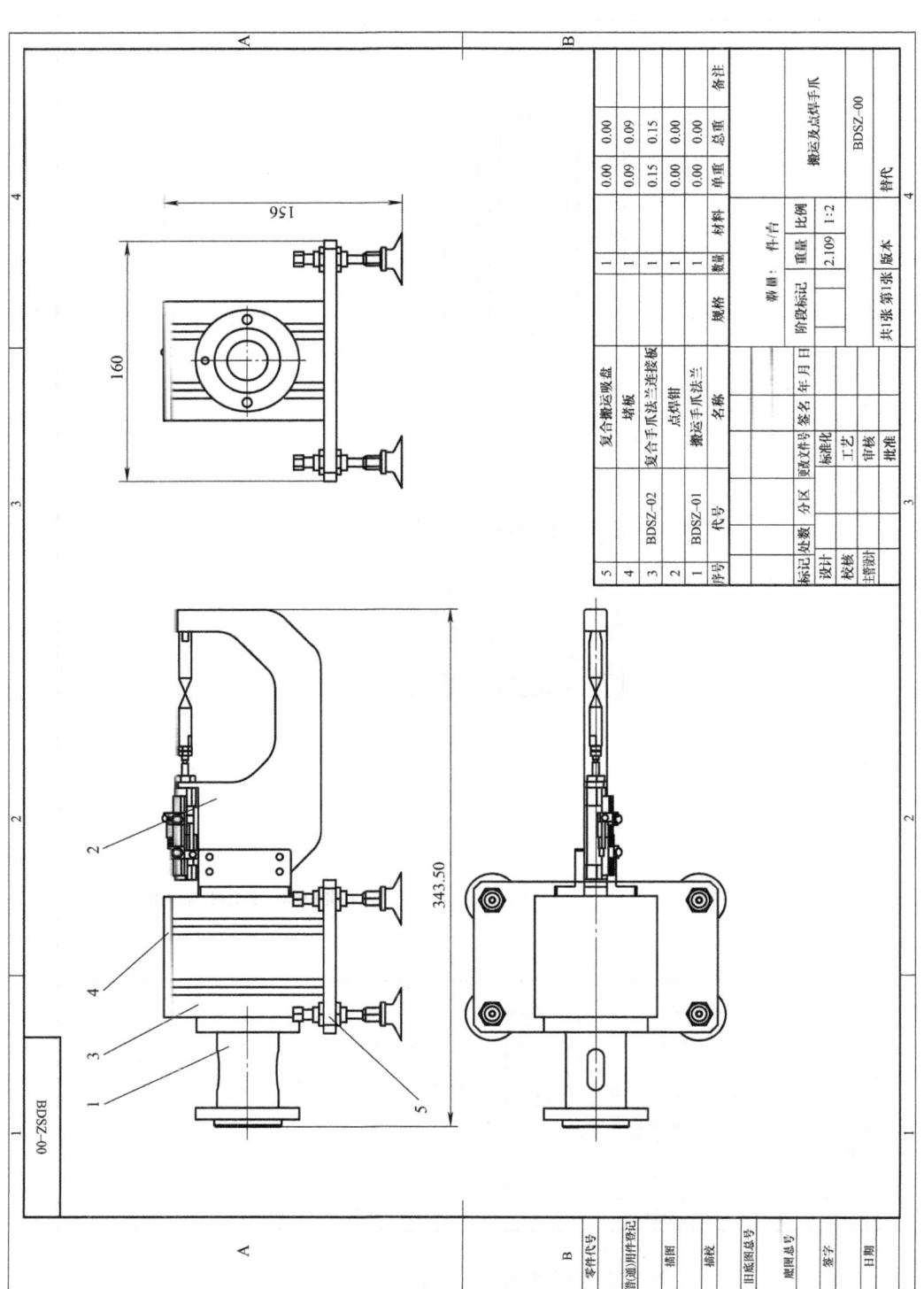

图 3-22 搬运及点焊手爪部件图

Figure 3-22 Component Drawing of a Handling and Spot Welding Gripper

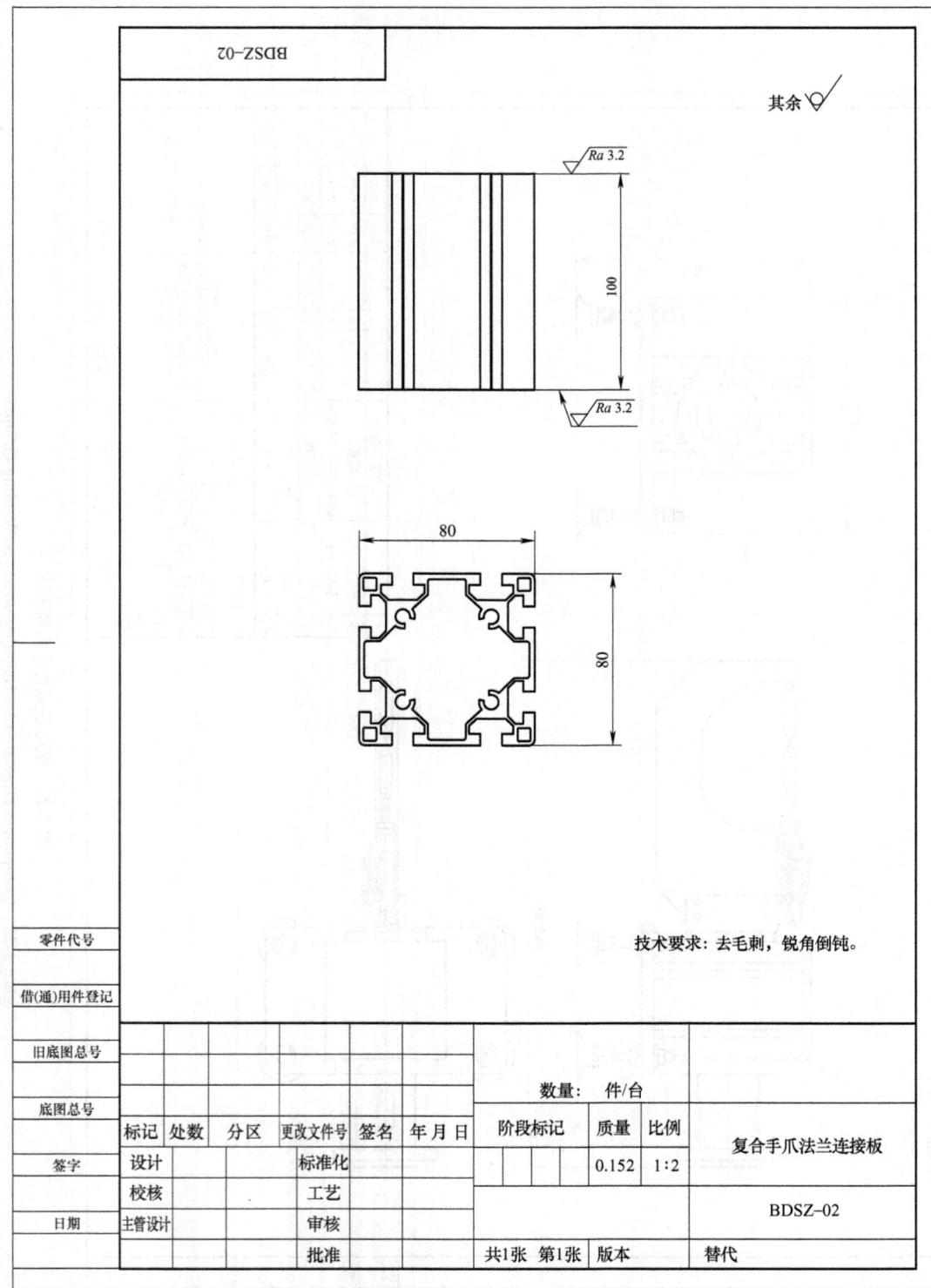

图 3-23 复合手爪法兰连接板工程图

Figure 3-23 Engineering Drawing of a Flange Connection Plate of Composite Gripper

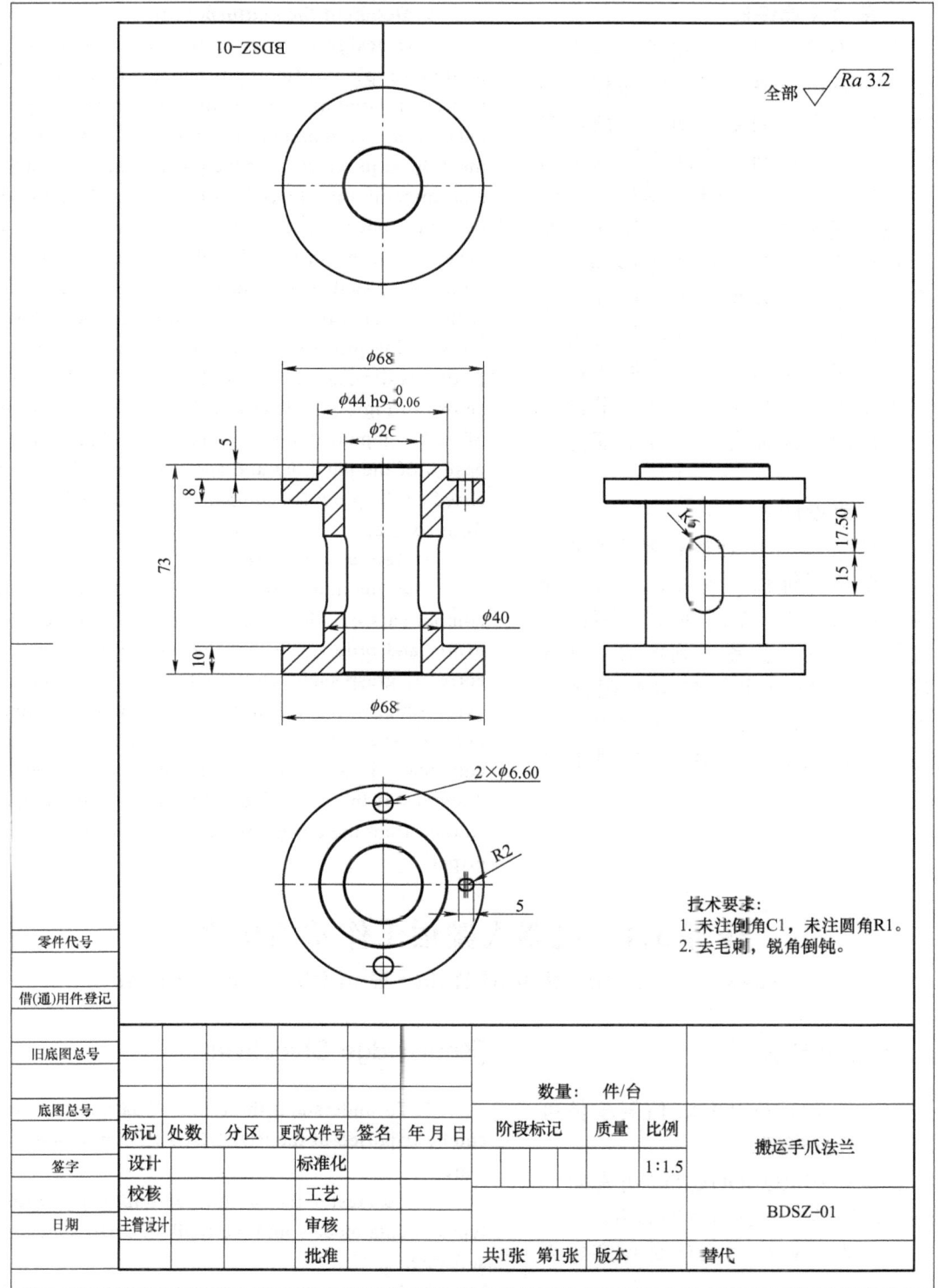

图 3-24 搬运手爪法兰工程图

Figure 3-24 Engineering Drawing of a Handling Gripper Flange

2. 平台夹具设计

平台夹具的设计思路是：采用销钉定位，可以实现"汽车门饰板"的正确定位；采用伺服电动机与销钉组合的形式，可以满足不同长度规格的"汽车门饰板"定位要求；采用气缸可以实现快速装夹；各部件的尺寸和安装位置要确保工作过程中不能与"搬运及点焊手爪"发生干涉和碰撞。综合考虑以上因素后，设计出平台夹具部件图，如图 3-25 所示。参照搬运及点焊手爪设计，根据提供的"平台夹具"部件图，详细设计出各个部件的工程图。

3. 仓库设计

"仓库"的作用是整齐规整地放置数个"汽车门饰板"零件，以便工业机器人在抓取"汽车门饰板"零件时，每次都能按同一位置正确抓取。其部件图如图 3-26 所示。在学习搬运及点焊手爪设计的基础上，根据提供的"仓库"部件图，设计出各个零件的工程图。

2. Design of the platform fixture

The design concept of the platform fixture is as follows: pin positioning is adopted to achieve correct positioning of the " loading and unloading products " ; the combination of servo motors and pins can meet the requirements for the positioning of " loading and unloading products " with different lengths; the use of cylinder can achieve rapid clamping; it shall be ensured that the size and installation position of each component shall cause no interference or collision with the " handling and spot welding gripper " during the working process. The component of the platform fixture is designed based on the above factors and as shown in Figure 3-25. Detailed engineering drawings of each component are designed with reference to the design of the handling and spot welding gripper, and based on the provided component drawing of the platform fixture.

3. Design of the store

The function of the " store " is to place several components of " loading and unloading products " neatly and orderly, so that the industrial robots can correctly grasp such components at the same position every time. The components are as shown in Figure 3-26. Please design the engineering drawing of each component in accordance to the provided component drawings of the " store " and based on what you have learned from the design of handling and spot welding grippers.

任务 3.4　机器人搬运工作站的仿真
Task 3.4　Simulation of Robot Handling Workstation

【知识目标】

1. 了解 ROBOGUIDE 仿真软件可以导入的数据模型类型。

2. 掌握在 ROBOGUIDE 仿真软件中导入各种类型的数据模型的方法。

3. 掌握工业机器人仿真程序导入方法。

4. 掌握程序仿真方法。

[Knowledge Objectives]

1. To understand the types of data models that can be imported to the ROBOGUIDE simulation software.

2. To master the method of importing various types of data models into the ROBOGUIDE simulation software.

3. To master the methods of importing industrial robot simulation programs.

4. To master the methods of program simulation.

项目3 典型工业机器人搬运工作站系统的设计及应用

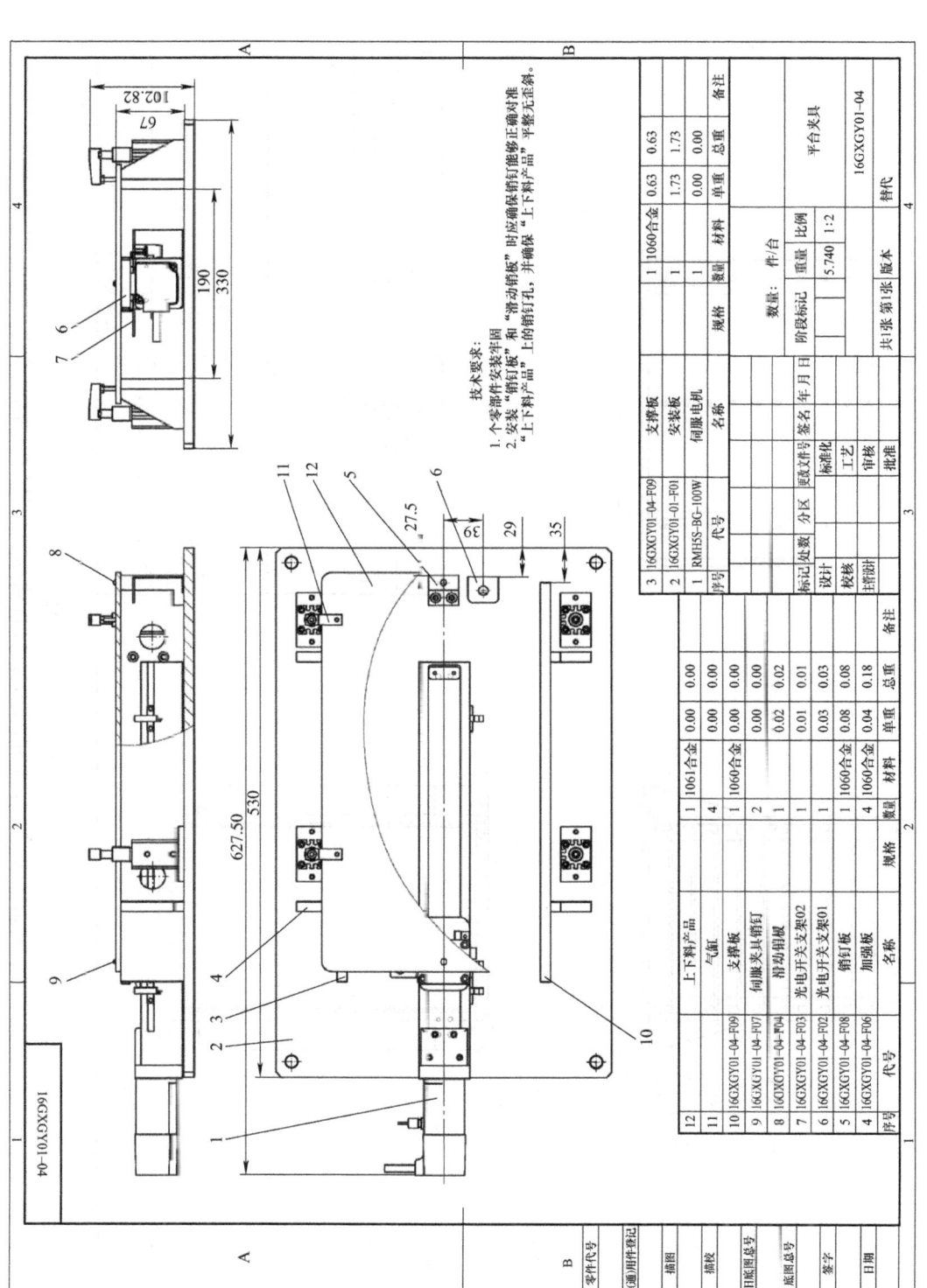

图 3-25 平台夹具部件图

Figure 3-25 Component Drawing of a Platform Fixture

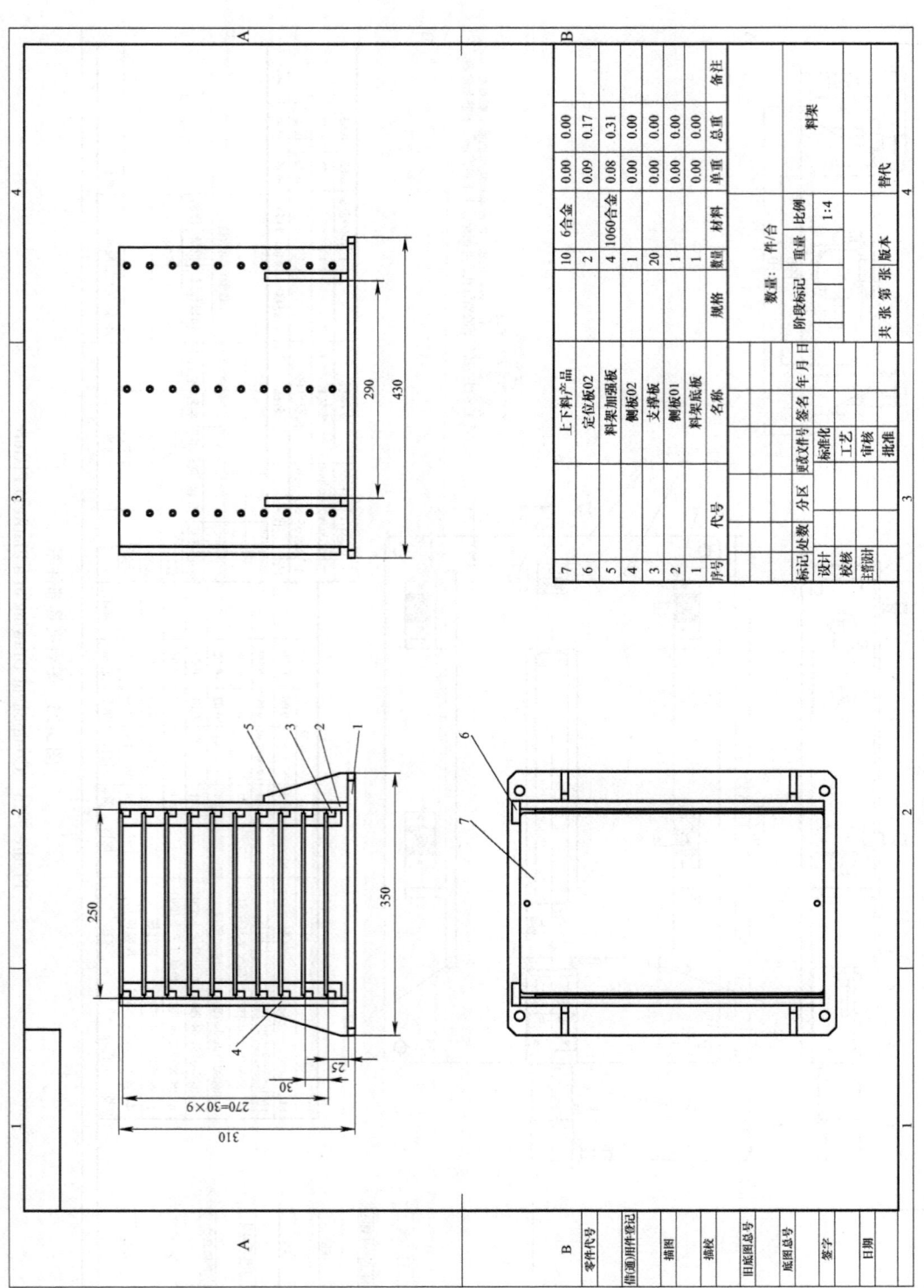

图 3-26 仓库部件图

Figure 3-26 Component Drawing of a Loading and Unloading Rack

【技能目标】

1. 能按照工艺要求和规范进行机器人工作站程序编写。
2. 能正确导入/导出工业机器人仿真程序。
3. 能够导入工业机器人搬运工作站仿真布局。

【素质目标】

1. 培养学生沟通、协作能力。
2. 培养学生自主探索、善于观察的能力。

【任务情景】

某汽车生产企业为了加快生产速度，提高生产效率，准备建立一条工业机器人搬运工作站生产线，前期已完成了生产线的设计方案编写，现在需要在软件上对生产工艺进行模拟仿真，软件使用 FANUC 仿真软件 ROBOGUIDE，请写出程序并进行仿真调试。

【任务分析】

当所研究的系统造价昂贵、实验的风险性大或需要很长的时间才能了解系统参数变化所引起的后果时，仿真是一种特别有效的研究手段。利用计算机实现对于系统的仿真研究不仅方便、灵活，而且也是经济的。在本项目中，通过利用 FANUC 机器人配套的软件 ROBOGUIDE 进行仿真，也可以用 VISIUALONE 和 PROCESS 软件进行仿真，具体选用哪款软件，可根据自身条件进行选用，这里以 ROBOGUIDE 作为案例进行讲解。

[Skill Objectives]

1. To be able to write programs for robot workstation according to process requirements and codes.
2. To be able to correctly import/export industrial robot simulation programs.
3. To be able to import the simulation layout of industrial robot handling workstations.

[Competence Objectives]

1. To cultivate the students' abilities of communication and collaboration.
2. To cultivate the students' abilities of exploration and observation.

[Task Scenario]

In order to speed up production and improve its efficiency, an automobile manufacturer plans to establish a production line of industrial robot handling workstation. As the design scheme of the production line has been completed in the early stage, it is now necessary to simulate the production process with software, for which purpose the ROBOGUIDE software of FANUC is supposed to be adopted. Please write the program and conduct commissioning of the simulation.

[Task Analysis]

Simulation is a particularly effective research tool for the studied systems which are expensive, or with high experimental risks, or whose parameter changes result in consequences that takes a long time to be understood. Conducting simulation on such systems with computer is a research method that is not only convenient, flexible, but also economical. In this program, simulation can be carried out with the ROBOGUIDE software that matches up with the FANUC robots, or the VISIUALONE or PROCESS software, which can be selected based on one's actual conditions. The ROBOGUIDE software is taken as an example for explanation in this task.

【知识准备】

3.4.1 仿真环境搭建

参照工作站布局图,在仿真软件中导入设备及已设计好的部件模型,仿真环境如图 3-27 所示。

[Assumed Knowledge]

3.4.1 Establishment of the Simulation Environment

Import the equipment and designed component models into the simulation software with reference to the workstation layout, and the simulation environment is shown in Figure 3-27.

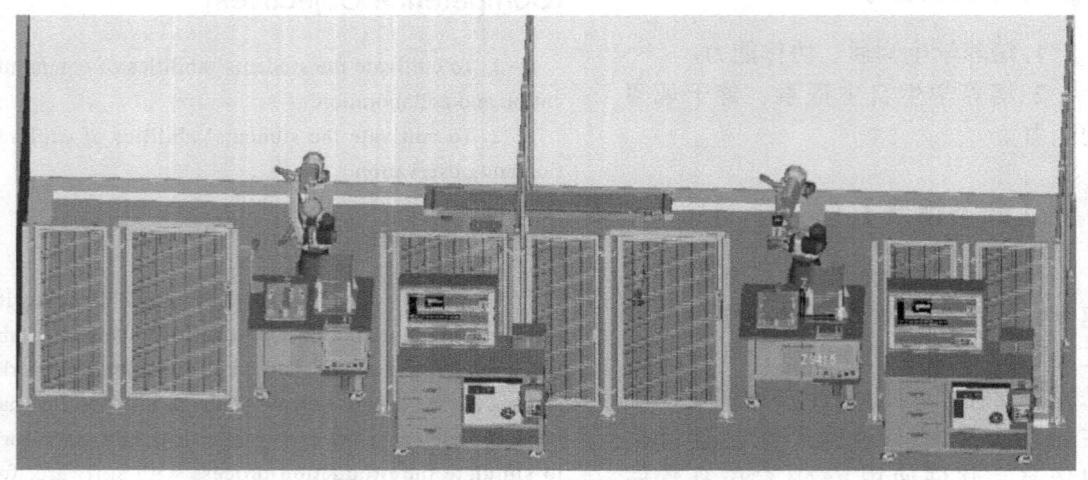

图 3-27 工业机器人搬运工作站仿真环境
Figure 3-27 Simulation Environment of an Industrial Robot Handling Workstation

3.4.2 工作站仿真

1. 添加运动部件

焊接平台转头的作用是在汽车门饰板进行焊接时,将汽车门饰板夹紧固定,方便机器人按照工艺要求准确点焊作业。在仿真布局中,需要在搬运焊接的机器人焊接平台中添加运动的【Link】,焊接平台在【Machines】中添加(在视觉检测的工业机器人工作站中,焊接平台在【Fixtures】中添加),由机器人的 DO[1] 实现控制,转头到位信号给 DI[1],设置方法见表 3-3。

3.4.2 Workstation Simulation

1. Add the parts in motion

The function of the welding platform swivel is to clamp and fix the car door trim panel during welding, making it convenient for the robot to accurately perform spot welding according to the process requirements. In the simulation layout, it is required to add the moving [Link] to the robot welding platform which is for handling and welding. The welding platform is added from the [Machines] (but from the [Fixtures] in the industrial robot workstation for visual detection), which is controlled by the DO [1] of the robot. The swivel-in-place signal is sent to the DI [1], and the setting method is shown in Table 3-3.

项目 3　典型工业机器人搬运工作站系统的设计及应用

表 3-3　焊接平台转头和输送线设置步骤
Table 3-3　Steps for Setting the Welding Platform Swivel and Conveying Line

序号 S/N	步骤 Step	图示 Illustration
1	打开 Cell Browser 面板 Open the Cell Browser panel	—
2	右键单击【Machines】，选择【Add Machine】，出现 7 个选项，选择第三个选项【CAD File】 Right click on the [Machines] and select the [Add Machine]. Among the seven options appearing, select the third option of [CAD File]	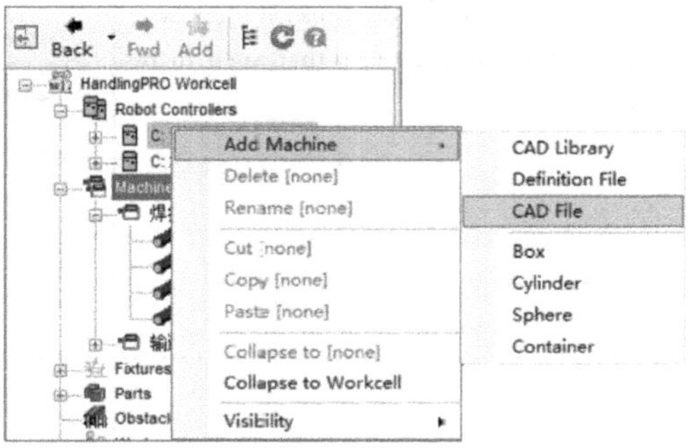
3	在弹出来的对话框中选择需要添加的【Machine】中的焊接平台.CSB 文件 In the pop-up dialog box, select the welding platform.CSB file that needs to be added from the [Machine]	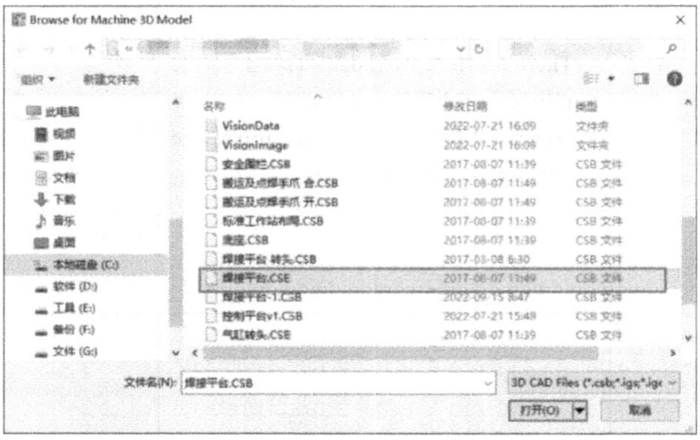
4	选择焊接平台 1，单击右键【Add Link】 Select welding platform 1, right-click on the [Add Link]	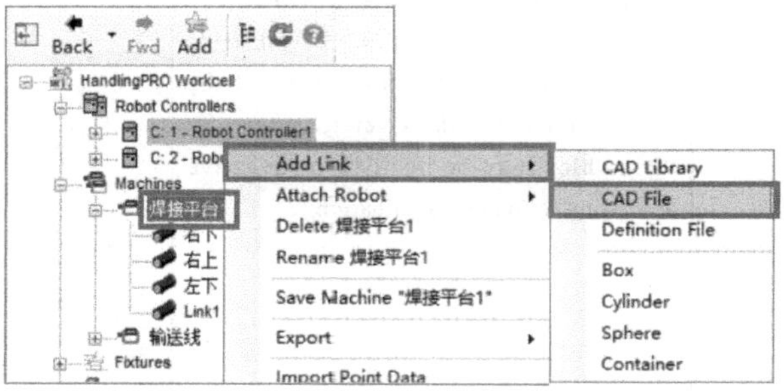

（续）

序号 S/N	步骤 Step	图示 Illustration
5	在弹出的【Add Link】对话框中，选择需要添加的运动部件 In the pop-up [Add Link] dialog box, select the moving parts that need to be added	
6	在焊接工作台上有4个气缸转头，所以第5个步骤需要重复4次，完成4个气缸转头的添加 There are four cylinder swivels on the welding workbench, so step 5 needs to be repeated by 4 times to complete the addition of 4 cylinder swivels	
7	气缸转头的运动设置，4个气缸转头各旋转90°，需要根据实际情况设置正负值 For the setting of motion, each of the 4 cylinder swivels is set to rotate by 90°, which may be positive or negative according to the actual situation	

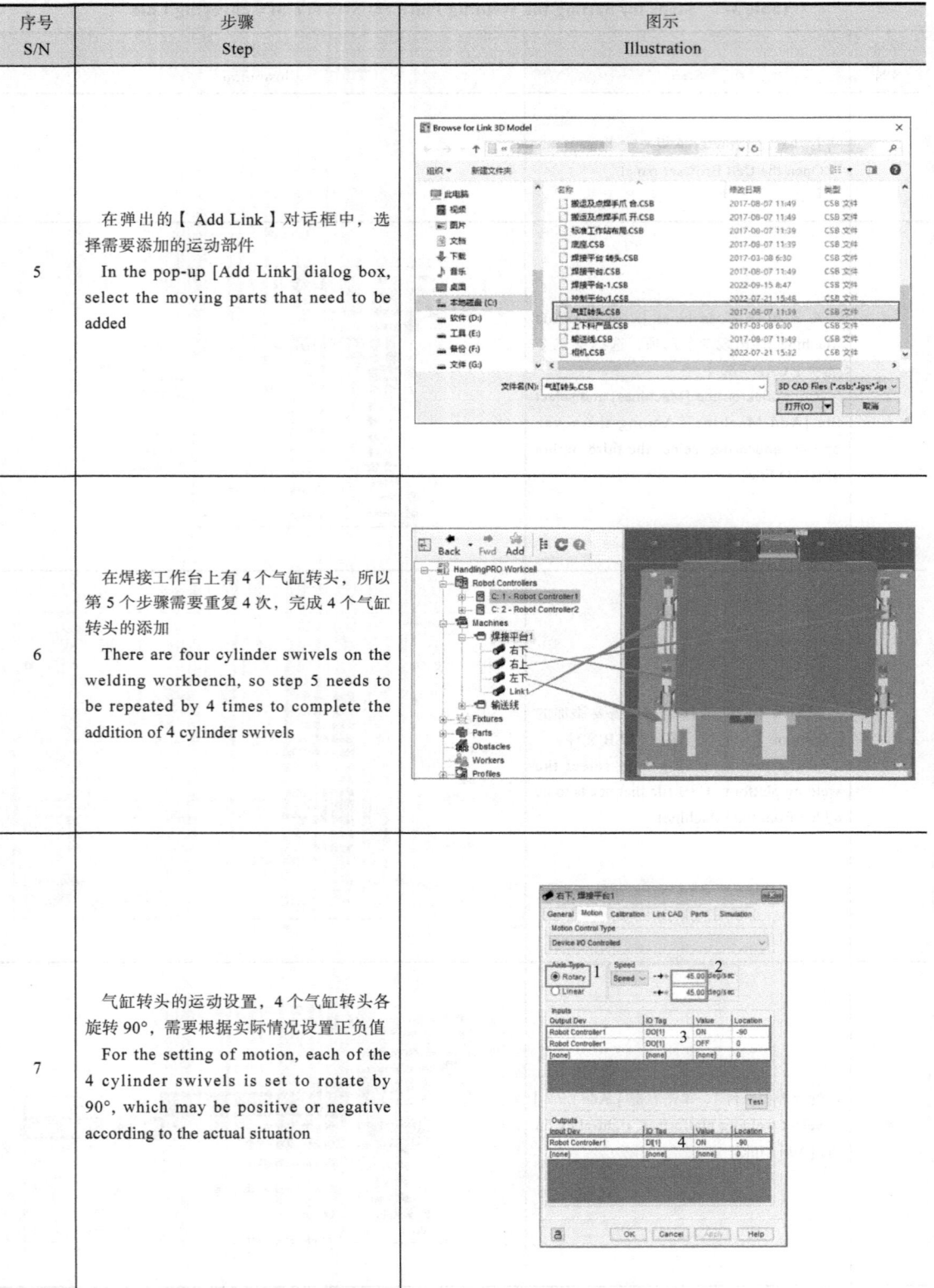

（续）

序号 S/N	步骤 Step	图示 Illustration
8	添加输送线的方法与添加焊接平台转头的方法相同，不同的地方是焊接平台转头在参数设置中是旋转运动，输送线的运动是直线运动；仿真中使用了焊接机器人的DO[2]、DI[2]对输送线进行信号交换控制，而搬运检测机器人与输送线没有信号的交换 The method of adding a conveying line is the same as that of adding a welding platform swivel, with the differences being as follows: on the parameter settings, the welding platform swivel is set with rotational motion while the conveying line is with rectilinear motion; in the simulation, the DO [2] and DI [2] of the welding robot are used for signal exchange control of the conveying line, while there is no signal exchange between the handling detection robot and the conveying line	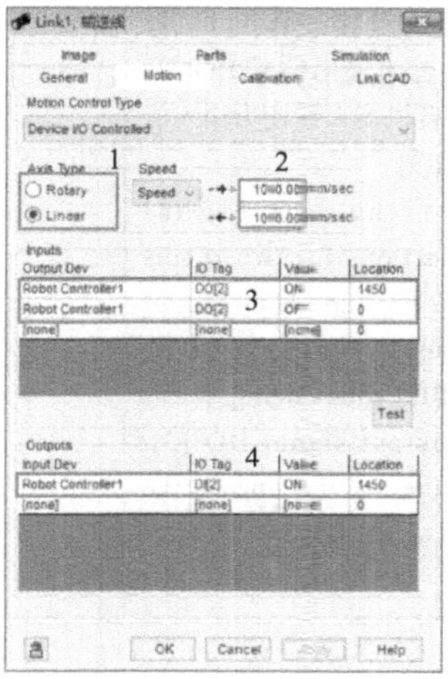

2. 添加 I/O 信号连接
设置方法见表 3-4。

2. Add I/O signal connections
The setting method is shown in Table 3-4.

表 3-4 I/O 信号连接设置方法

Table 3-4 Setting Method of I/O Signal Connections

序号 S/N	步骤 Step	图示 Illustration
1	两台机器人在联机仿真过程中需要进行信号交换，仿真前需要做好信号交换的设置 As two robots need to exchange signals during the online simulation process, signal exchange needs to be set before simulation	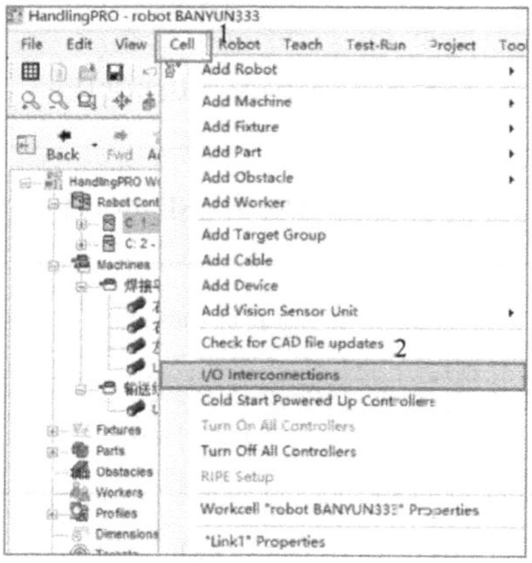

序号 S/N	步骤 Step	图示 Illustration
2	在弹出的 I/O 信号连接对话框中设置两台机器人的 I/O 信号连接,第一台机器人设置为输出信号 DO[3],第二台机器人设置为输入信号 DI[3] Set the I/O signal connections for two robots in the pop-up dialog box of I/O signal connection, with the first robot being set as the output signal DO [3], and the second one as the input signal DI [3]	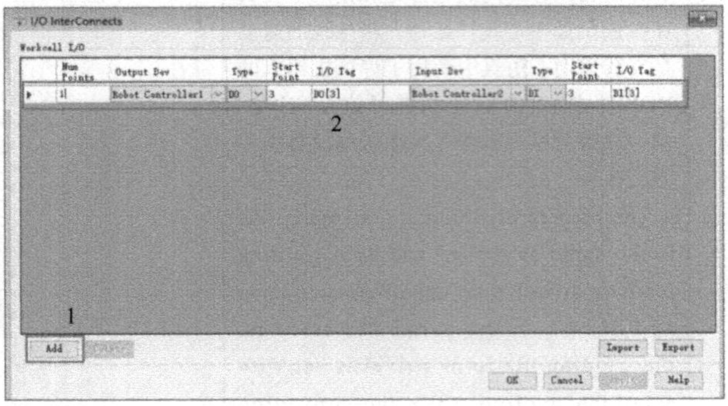

3. 仿真工作站机器人程序编写

仿真环境建立后,必须在仿真软件中按照工作站工艺流程编写相应的机器人程序才能够进行运动仿真。

1) 搬运焊接仿真主程序见表 3-5。

3. Robot programming for the simulation workstation

When the simulation environment is established, the simulation software must be programmed with corresponding robot program according to the process flow of the workstation for the simulation of motion.

1) The main program for the handling and welding simulation is shown in Table 3-5.

表 3-5 搬运焊接仿真主程序
Table 3-5 Main Program for Handling and Welding Simulation

程序号 Program number	程序 Program	注释 Notes
1	UTOOL_NUM[GP1]=1	使用工具坐标 1 Use tool coordinate 1
2	UFRAME_NUM[GP1]=0	使用用户坐标 0 Use user coordinate 0
3	J P[1] 100% FINE	回到 HOME 点 Return to the HOME point
4	DO[2]=OFF	传送带回 0 conveyor belt returns to 0
5	LBL[1]	循环标签 Loop label

（续）

程序号 Program number	程序 Program	注释 Notes
6	DO[1]=OFF	气缸转头松开 Release of cylinder swivel
7	DO[3]=OFF	与第2台机器人发信号（工作未完成信号） Signal with the second robot (signal of unfinished work)
8	CALL subprogram1	调用出库子程序 Call the subprogram for store-exit
9	DO[1]=ON	气缸转头夹紧 Clamping of cylinder swivel
10	WAIT DI[1]=ON	等待气缸转头夹紧到位 Wait for the clamp-in-place of cylinder swivel
11	CALL subprogram2	调用点焊子程序 Call the subprogram for spot welding
12	DO[1]=OFF	气缸转头松开 Release of cylinder swivel
13	WAIT2.00 (sec)	等待松开2s Wait 2 seconds for release
14	CALL subprogram3	调用搬运传送带子程序 Call the subprogram for the handling conveyor belt
15	DO[2]=ON	传送工件到传送带末端 Transfer the workpiece to the end of the conveyor belt
16	WAIT DI[2]=ON	等待传送带到位信号 Wait for the signal of conveyor belt-in-place
17	DO[3]=ON	给第2台机器人发送工作完成信号 Send the signal of work completion to the second robot
18	J P[58] 100% FINE	回到HOME点 Return to the HOME point
19	WAIT5.00 (sec)	等待5s Wait 5 seconds
20	DO[2]=OFF	传送带回到起点 The conveyor belt returns to the starting point
21	JMP LBL[1]	跳转到标签处 Jump to label

2) 出库子程序见表 3-6。
3) 点焊子程序见表 3-7。
4) 搬运传送带子程序见表 3-8。
5) 搬运检测仿真程序见表 3-9。

2) The subprogram for store-exit is shown in Table 3-6.
3) The subprogram for spot welding is shown in Table 3-7.
4) The subprogram for handling conveyor belt is shown in Table 3-8.
5) The program for handling and detection simulation is shown in Table 3-9.

表 3-6 出库子程序
Table 3-6　Subprogram for Store-exit

程序号 Program number	程序 Program	注释 Notes
1	J P[2] 100% FINE	出库接近点 1 Store-exit approach point 1
2	L P[3] 2000mm/sec FINE	出库接近点 2 Store-exit approach point 2
3	L P[4] 2000mm/sec FINE	抓取点 Pickup point
4	WAIT.50 (sec)	等待 0.5s Wait 0.5s
5	! Pickup	抓取仿真动画 Pickup simulation animation
6	WAIT.50 (sec)	等待 0.5s Wait 0.5s
7	L P[5] 2000mm/sec FINE	出库逃离点 1 Store-exit escape point 1
8	J P[6] 100% FINE	出库逃离点 2 Store-exit escape point 2
9	J P[7] 100% FINE	夹台接近点 1 Clamping platform approach point 1
10	L P[8] 2000mm/sec FINE	夹台接近点 2 Clamping platform approach point 2
11	L P[9] 2000mm/sec FINE	放置点 Dropping point
12	WAIT.50 (sec)	等待 0.5s Wait 0.5s
13	! Drop	放置仿真动画 Dropping simulation animation
14	WAIT.50 (sec)	等待 0.5s Wait 0.5s
15	L P[10] 2000mm/sec FINE	夹台逃离点 Clamping platform escape point

表 3-7　点焊子程序
Table 3-7　Subprogram for Spot Welding

程序号 Program number	程序 Program	注释 Notes
1	J P[11] 100% FINE	
2	L P[12] 2000mm/sec FINE	第 1 个点焊位置 1st spot welding position
3	L P[13] 2000mm/sec FINE	
4	L P[14] 2000mm/sec FINE	
5	WAIT1.00 (sec)	
6	L P[15] 2000mm/sec FINE	第 2 个点焊位置 2nd spot welding position
7	L P[16] 2000mm/sec FINE	
8	L P[17] 2000mm/sec FINE	
9	L P[18] 2000mm/sec FINE	
10	WAIT1.00 (sec)	
11	L P[19] 2000mm/sec FINE	第 3 个点焊位置 3rd spot welding position
12	L P[20] 2000mm/sec FINE	
13	L P[21] 2000mm/sec FINE	
14	WAIT1.00 (sec)	
15	L P[22] 2000mm/sec FINE	第 4 个点焊位置 4th spot welding position
16	L P[23] 2000mm/sec FINE	
17	L P[24] 2000mm/sec FINE	
18	WAIT1.00 (sec)	
19	L P[25] 2000mm/sec FINE	第 5 个点焊位置 5th spot welding position
20	L P[26] 2000mm/sec FINE	
21	L P[27] 2000mm/sec FINE	
22	L P[28] 2000mm/sec FINE	
23	WAIT1.00 (sec)	
24	L P[29] 2000mm/sec FINE	第 6 个点焊位置 6th spot welding position
25	L P[30] 2000mm/sec FINE	
26	WAIT1.00 (sec)	
27	L P[31] 2000mm/sec FINE	第 7 个点焊位置 7th spot welding position
28	L P[32] 2000mm/sec FINE	
29	L P[33] 2000mm/sec FINE	
30	L P[34] 2000mm/sec FINE	
31	L P[35] 2000mm/sec FINE	
32	WAIT1.00 (sec)	

（续）

程序号 Program number	程序 Program	注释 Notes
33	L P[36] 2000mm/sec FINE	
34	L P[37] 2000mm/sec FINE	
35	L P[38] 2000mm/sec FINE	第8个点焊位置 8th spot welding position
36	L P[39] 2000mm/sec FINE	
37	L P[40] 2000mm/sec FINE	
38	WAIT1.00 (sec)	
39	L P[41] 2000mm/sec FINE	逃离点 Escape point
40	L P[42] 2000mm/sec FINE	安全位置 Safe location

表 3-8　搬运传送带子程序

Table 3-8　Subprogram for Handling Conveyor Belt

程序号 Program number	程序 Program	注释 Notes
1	J P[43] 100% FINE	夹台接近点 Clamping platform approach point
2	L P[44] 2000mm/sec FINE	夹台抓取点 Clamping platform pickup point
3	WAIT.50 (sec)	等待 0.5s Wait 0.5s
4	! Pickup	抓取仿真程序 Pickup simulation program
5	WAIT.50 (sec)	等待 0.5s Wait 0.5s
6	L P[45] 2000mm/sec FINE	夹台逃离点 Clamping platform escape point
7	J P[46] 100% FINE	传送带接近点 Conveyor belt approach point
8	L P[47] 2000mm/sec FINE	传送带放置点 Conveyor belt dropping point
9	WAIT.50 (sec)	等待 0.5s Wait 0.5s
10	! Drop	放置仿真程序 Dropping simulation program
11	WAIT.50 (sec)	等待 0.5s Wait 0.5s
12	L P[48] 2000mm/sec FINE	逃离点 Escape point
13	J P[49] 100% FINE	HOME 点 HOME point

表 3-9 搬运检测仿真程序
Table 3-9 Program for Handling and Detection Simulation

程序号 Program number	程序 Program	注释 Notes
1	UTOOL_NUM[GP1]=1	使用工具坐标 1 Use tool coordinate 1
2	UFRAME_NUM[GP1]=0	使用用户坐标 0 Use user coordinate 0
3	J P[1] 100% FINE	机器人回 HOME 点 Robot returns to the HOME point
4	LBL[1]	跳转标签 Jump label
5	WAIT DI[3]=ON	等待第 1 台机器人发送工作完成信号 Waiting for the first robot to send a work completion signal
6	J P[2] 100% FINE	传送带接近点 1 Conveyor belt approach point 1
7	J P[3] 100% CNT100	传送带接近点 2 Conveyor belt approach point 2
8	L P[4] 2000mm/sec FINE	传送带接近点 3 Conveyor belt approach point 3
9	L P[5] 2000mm/sec FINE	抓取点 Pickup point
10	WAIT.50 (sec)	等待 0.5s Wait 0.5s
11	! Pickup	抓取仿真程序 Pickup simulation program
12	WAIT.50 (sec)	等待 0.5s Wait 0.5s
13	L P[6] 2000mm/sec FINE	传送带逃离点 1 Conveyor belt escape point 1
14	L P[7] 2000mm/sec FINE	传送带逃离点 2 Conveyor belt escape point 2
15	J P[8] 100% FINE	视觉接近点 1 Visual approach point 1
16	J P[9] 100% FINE	视觉接近点 2 Visual approach point 2
17	L P[10] 2000mm/sec CNT50	视觉接近点 3 Visual approach point 3
18	L P[11] 2000mm/sec FINE	视觉检测点 1 Visual detection point 1

（续）

程序号 Program number	程序 Program	注释 Notes
19	L P[12] 2000mm/sec FINE	视觉检测点 2 Visual detection point 2
20	L P[13] 2000mm/sec FINE	视觉检测点 3 Visual detection point 3
21	L P[14] 2000mm/sec FINE	视觉检测点 4 Visual detection point 4
22	J P[15] 100% FINE	入库接近点 1 Store-entry approach point 1
23	J P[16] 100% FINE	入库接近点 2 Store-entry approach point 2
24	L P[17] 2000mm/sec FINE	入库点 Store-entry point
25	WAIT.50 (sec)	等待 0.5s Wait 0.5s
26	! Drop	放置仿真程序 Dropping simulation program
27	WAIT.50 (sec)	等待 0.5s Wait 0.5s
28	L P[18] 2000mm/sec FINE	入库逃离点 Store-entry escape point
29	J P[19] 100% FINE	HOME 点 HOME point
30	WAIT2.00 (sec)	等待 2s Wait 2 seconds
31	JMP LBL[1]	跳转到标签 1 Jump to label 1

4. 仿真验证

程序编制完毕后，单击仿真运行，仿真系统将会按照任务 1 所描述的功能运行，直到第 10 个汽车门饰板放置完毕后，两个机器人才回到初始位停止运行。运行效果图如图 3-28 所示。

4. Simulation verification

When the programming is completed, single click to operate the simulation, and the simulation system will operate with the functions described in Task 1, and won't stop until the 10th car door trim panel is placed as required and both robots return to the initial position. The operation is rendered as shown in Figure 3-28.

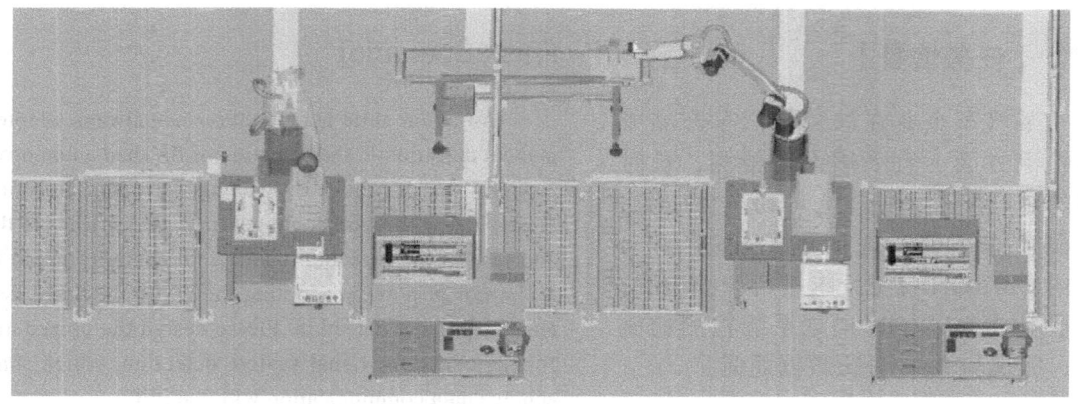

图 3-28 仿真运行效果图
Figure 3-28 Rendering of Simulation Operation

任务 3.5　视觉系统的调试
Task 3.5　Commissioning of the Visual System

【知识目标】

1. 了解机器视觉的发展。
2. 掌握机器视觉的典型应用。
3. 掌握 OMRON 视觉与 PLC 的通信方法。

【技能目标】

1. 能编写视觉检测流程。
2. 能正确设置 OMRON 视觉与 PLC 通信参数。
3. 能调试工业机器人搬运工作站视觉检测作业。

【素质目标】

1. 培养学生查询资料、解决问题能力。
2. 培养学生沟通协作、善于思考的能力。

[Knowledge Objectives]

1. To understand the development of machine vision.
2. To master typical applications of machine vision.
3. To master the communication methods between the OMRON vision and PLC.

[Skill Objectives]

1. To be able to write visual detection procedures.
2. To be able to correctly set communication parameters between the OMRON vision and PLC.
3. To be able to conduct visual detection operation of the industrial robot handling workstations.

[Competence Objectives]

1. To cultivate students' ability to search for information and solve problems.
2. To cultivate students' ability to communicate, collaborate, and think effectively.

【任务情景】

某汽车企业在加工的汽车产品成品检测方面一直采用人工检测，长时间的对产品检测使工人容易产生疲劳，检测质量下降。现在需要改进生产，采用OMRON视觉系统进行产品检测，请设计出OMRON视觉系统检测的流程，实现与PLC进行数据通信。

【任务分析】

工业机器人搬运工作站在汽车门饰板搬运入库之前，对加工的汽车门饰板进行检测。项目采用OMRON FH-L550系统，主要对点焊的汽车门饰板进行焊点数量及质量进行检测。项目需要完成OMRON视觉系统与PLC之间的通信数据传输和OMRON视觉检测流程的设计两部分内容。

【知识准备】

3.5.1 视觉概述

机器视觉是一门涉及人工智能、神经生物学、心理物理学、计算机科学、图像处理、模式识别等诸多领域的交叉学科。随着工业自动化技术的飞速发展和各领域消费者对产品品质要求的不断提高，零缺陷、高品质、高附加值的产品成为企业应对竞争的核心。为了赢得竞争，可靠的质量控制不可或缺。由于生产过程速度加快，产品工艺高度集成，体积缩小且制造精度提高，人眼已无法满足许多企业对外形质量控制的检测需要。机器视觉代替人类视觉自动检测产品外形特征，实现100%在线全检，已成为解决各行业制造商大批

[Task Scenario]

An automobile manufacturer has always adopted manual detection in the detection of finished automotive products that it produces. Prolonged product detection makes it easy for workers to get tired, which leads to the decrease in the detection quality. Productivity needs to be improved now by adopting the OMRON visual system for product detection. Please design the procedures for the OMRON visual system detection, which shall achieve data communication with the PLC.

[Task Analysis]

The processed car door trim panel needs to be detected at the industrial robot handling workstation before it is handled to the store. In this program, the OMRON FH-L550 system is adopted mainly for the detection of the quantity and quality of the welding points of the car door trim panel, on which spot welding has been performed. The two parts that need to be completed in the program are the communication data transmission between the OMRON visual system and PLC, and the design of the OMRON visual detection procedures.

[Assumed Knowledge]

3.5.1 Visual Overview

Machine vision is an interdisciplinary subject involving many fields such as artificial intelligence, neurobiology, psychophysics, computer science, image processing, and pattern recognition. With the rapid development of industrial automation technology and the continuous improvement of product quality required by consumers in various fields, products with zero defects, high quality, and high added value have become the core for enterprises to response to and win competitions. For this purpose, reliable quality control is indispensable. For many enterprises, manual detection by human eye is no longer able to meet the detection needs for appearance quality control, because of the accelerated production speed, highly integrated product processes, and improved manufacturing precision with smaller volume. Machine vision is adopted to replace human vision

量、高速度、高精度产品检测的主要趋势。简言之，机器视觉就是用机器代替人眼来做测量和判断，视觉系统框图如图 3-29 所示。

for automatic detection of appearance features of the product and realization of 100% online detection, which has become the main trend of detection with high speed and precision of mass-produced products for manufacturers in various industries. In short, machine vision is to use machine instead of human eyes for measurement and judgment, and the visual system chart is as shown in Figure 3-29.

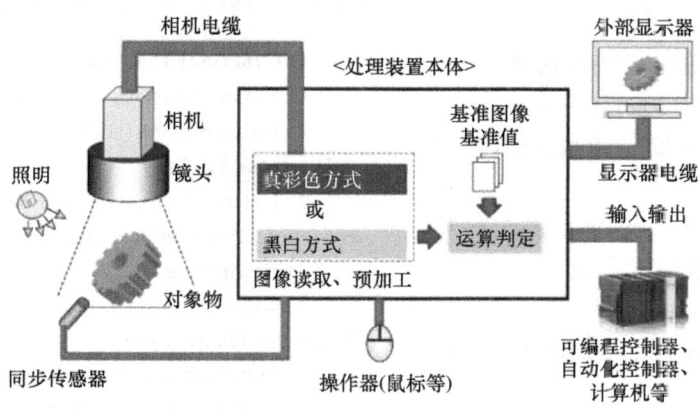

图 3-29 视觉系统框图
Figure 3-29 Visual System Chart

机器视觉由三部分组成，分别是光学系统、图像处理系统、执行机构及人机界面，三个部分缺一不可。选取合适的光学系统，采集适合处理的图像，是完成视觉检测的基本条件，而开发稳定可靠的图像处理软件是视觉检测的核心任务，可靠的执行机构和人性化的人机界面是实现最终功能的关键因素。

光学系统是机器视觉系统中不可缺少的部分，如果没有适合的光学系统采集适于处理的图片则难以有效完成图像检测，甚至导致检测失败。因此，我们认为适合的光学系统是成功完成机器视觉应用的前提条件。一个典型的光学系统包括光源、相机和镜头。

图像处理系统取得图像后，要对图像进行处理，分析计算并输出检测结果。图像处理部分包括软件和硬件。

Machine vision consists of three parts: optical system, image processing system, and actuator and human-machine interface, none of which is dispensable. Choosing a suitable optical system and collecting images suitable for processing are the basic conditions for the completion of visual detection, while developing a stable and reliable image processing software is the core task of visual detection, and a reliable actuator and user-friendly human-machine interface are the critical factors that determine the realization of the final functions.

The optical system is an indispensable part of the machine vision system. It is difficult, even impossible, to effectively conduct image detection without a suitable optical system that captures images suitable for processing. Therefore, we believe that a suitable optical system is a prerequisite for the successful application of machine vision. A typical optical system includes the light source, camera, and lens.

The image processing system needs to process, and analyze by calculation the image that has been obtained, and output the detection results. The image processing section includes software and hardware.

When all the image acquisition and processing

执行机构及人机界面在完成所有图像采集和图像处理工作后，需要输出图像处理结果，进行动作(报警、剔除、位移等)，并通过人机界面显示生产信息，以及在型号或参数发生改变时对系统进行切换和修改工作。

3.5.2 机器视觉应用

1. 机器视觉可以实现的功能

主要是下面4项功能：

1) 测量(如长度、角度、圆弧及半径测量)。

2) 检测(如有无检测、残次品检测、数字统计、瑕疵检测)。

3) 定位。

4) 识别(如读码、OCR/OCV、颜色识别)。

2. 机器视觉的主要应用领域

(1) 工业领域机器视觉

由于工业机器视觉可以克服人眼标准的不一致性，除了可以制定更高的行业品质管控的数字标准，还能在高速、高光谱、高分辨率、高灵敏度、高可靠性等方面超越人眼极限，因此工业机器视觉系统可以代替人眼完成检测、测量、识别和定位等工作。

目前工业机器视觉系统已广泛应用于电子制造、包装印刷、汽车制造、食品饮料等众多生产和服务行业。

(2) 安防监控领域机器视觉

安防监控领域机器视觉具有技术先进、防范能力极强、操作便利等优势，因为其可以对所监控范围内的情况进行实时监视和分析，并且可以将被监控范围内的场景全部记录下来，为以后处理某些意外情况和事件等提供有力的证据和支持。

近年来，随着成千上万台监控摄像机密布在国内城镇的大街小巷，使

work is completed, the actuator and human-machine interface will need to output the results of image processing, take actions (such as to give an alarm, remove, displace), and display production information on the human-machine interface, as well as switch and modify the system when the model or parameter changes.

3.5.2 Machine Vision Applications

1. Functions that machine vision can achieve

The main functions are as follows:

1) Measurement (such as measurement of length, angle, arc, and radius).

2) Detection (such as detection of presence or absence, defective products, digital statistics, and defects).

3) Positioning.

4) Identification (such as code reading, OCR/OCV, color identification).

2. The main application fields of machine vision

(1) Machine vision in the industrial field

Industrial machine vision systems can replace the human eye to complete tasks such as detection, measurement, identification, and positioning, as it can overcome the inconsistency of judgement by the human eye, and surpass the limits of the human eye in all aspects of high-speed, hyper-spectrum, high-resolution, high-sensitivity, and high-reliability, in addition to the higher digital standards that can be formulated for quality control in the industry.

At present, industrial machine vision systems have been widely used in many production and service industries such as electronic manufacturing, packaging and printing, automobile manufacturing, and food and beverage.

(2) Machine vision in the field of security monitoring

In the field of security monitoring, machine vision has the advantages of being technically advanced, strong in the prevention ability, and convenient in the operation, as it can monitor, analyze, and record all the scenes within the monitored range in real-time, which provides the strongest evidence and support for handling certain unexpected situations and events in the future.

In recent years, with thousands of surveillance cameras densely distributed in the streets and alleys of the cities and towns of China, the video surveillance systems at the city level have gradually transitioned

国内城市级的视频监控系统从数字化、网络化，逐步过渡到高清化、智能化，目前正进一步向包含视频云、AI 及视频大数据的深度智能时代发展。

(3) 体验交互领域机器视觉

体验交互领域机器视觉主要包含新兴技术如 VR/AR 等带来的用户交互及体验提升。VR/AR 产业是由硬件制造和组装开始，集成了操作系统与开发工具、应用、内容、销售分发等多种供应商的生态系统。产业价值链中的技术和设备环节对于行业发展有极大影响。

其中视觉感知与处理是 AR/VR 的一项关键技术，包括硬件设备与操作系统、内容制作等环节。甚至从零部件到输入设备、输出设备、芯片中都需要考虑图像视觉信息的传递与处理。

from the stage of digitization and network to a new level where they are of high-definition and intelligence. Currently, they are further advancing towards an era of advanced intelligence when they include video cloud, AI, and big data of video.

(3) Machine vision in the field of experiential interaction

Machine vision in the field of experiential interaction mainly includes interaction with and improved experience by the user brought about by emerging technologies such as VR/AR. The VR/AR industry starts with hardware manufacturing and assembly, and integrates an ecosystem of the operating system and development tools, applications, content, and sales distribution which are supplied by different suppliers. The links of technology and equipment in the industrial value chain have a great impact on the development of the industry.

Among them, visual perception and processing is a key technology in AR/VR, including hardware equipment and operating systems, and content production. The transmission and processing of image visual information need to be taken into account for the input equipment, output equipment, and chips even their parts and components.

3.5.3 PLC 与视觉软件通信

要实现 PLC 与视觉控制器之间的网络通信，首先应该将两种设备使用以太网线缆连接起来，使数据从硬件上能形成数据流。其次需要设置/知晓通信双方的通信方式及 IP 地址，使双方都能准确找到通信设备。除此之外，PLC 端还需要添加通信功能块，以激活/断开以太网连接。

1. 物理网络连接

物理网络连接方法见表 3-10。

2. 视觉通信设置

(1) 视觉通信设置

设置方法见表 3-11。

3.5.3 Communication between the PLC and Visual Software

To achieve network communication between the PLC and visual controller, they should firstly be connected with Ethernet cables, so that a data stream can be formed from the hardware. Secondly, it is necessary to set/know the communication method and IP addresses of both communication parties, so that they can accurately find the communication equipment. In addition, the PLC end needs to added with the communication function block to activate/disconnect the Ethernet connection.

1. Physical network connection

The method of physical network connection is shown in Table 3-10.

2. Visual communication settings

(1) The method of visual communication settings is shown in Table 3-11.

表 3-10 视觉与 PLC 物理网络连接方法
Table 3-10 Method of Physical Network Connection between the Vision and PLC

序号 S/N	步骤 Step	图示 Illustration	序号 S/N	步骤 Step	图示 Illustration
1	使用以太网线缆一头接入检测单元视觉控制器的外接端口 Connect one end of the Ethernet cable to the external port of the visual controller of the detection unit		2	以太网线缆另一端接入总控单元以太网交换机，总控单元接入 PLC 网络 Connect the other end of the Ethernet cable to the Ethernet switch of the main control unit which connects to the PLC network	

表 3-11 视觉通信设置方法
Table 3-11 Method of Visual Communication Settings

序号 S/N	步骤 Step	图示 Illustration
1	从菜单栏【工具】中打开【系统设置】窗口 Open the [System Settings] window from the menu bar [Tools]	
2	设置完成后，单击主界面的【保存】按钮，保存设置。然后单击菜单栏中【功能】，选择【系统重启】，重启系统 When the setting is finished, click the [Save] button on the main interface to save the settings. Then, click the [Function] in the menu bar, and select [System Restart] to restart the system	

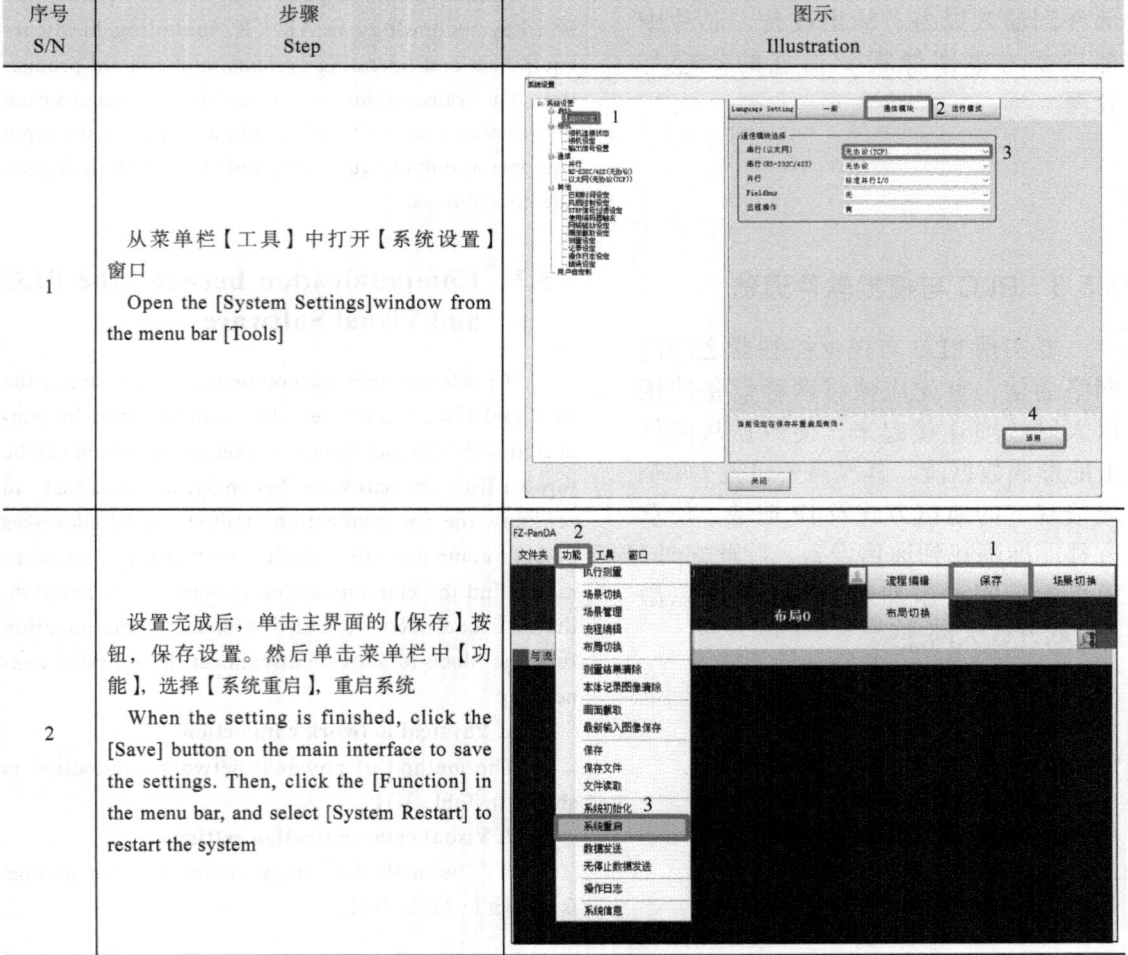

（续）

序号 S/N	步骤 Step	图示 Illustration
3	重新启动后，在系统设置窗口的以太网通信设置中，可以修改或设定视觉系统的 IP 地址为 192.168.0.110，并且修改输入/输出的端口号为 2000 After restarting, one can modify or set the IP address of the visual system as 192.168.0.110, and modify the input/output port number as 2000 in the Ethernet communication settings of the system settings window	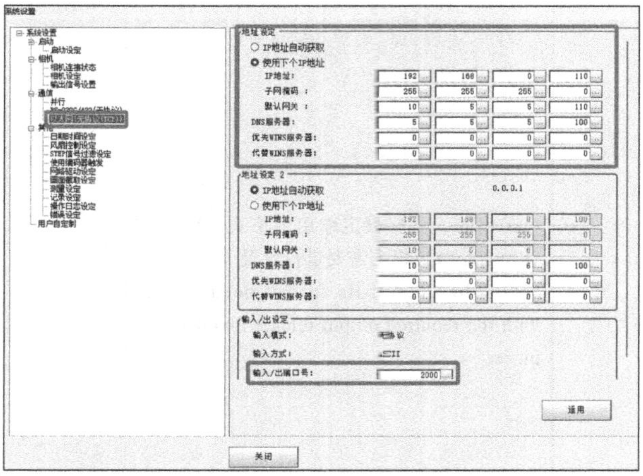
4	视觉系统想要将检测结果上传给上位机时，可以在流程编辑窗口的【结果输出】一栏中选择合适的流程项目 When it needs to upload the detection results from the visual system to the upper machine, one can select appropriate process items from the [Result Output] bar in the process editing window	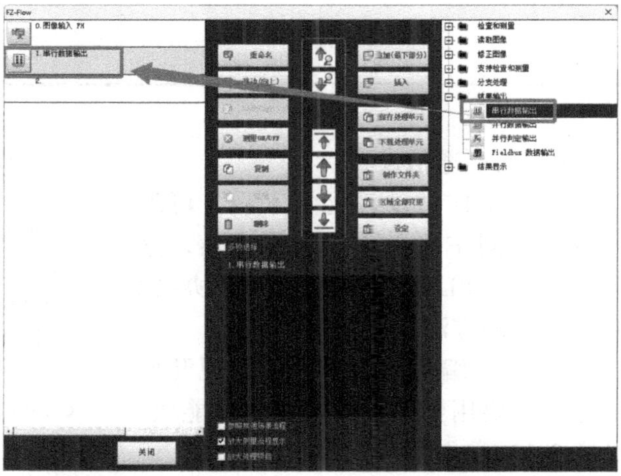
5	在流程项目的属性设置中可以设定具体的输出表达式 Specific output expressions can be set in the attribute settings of the process items	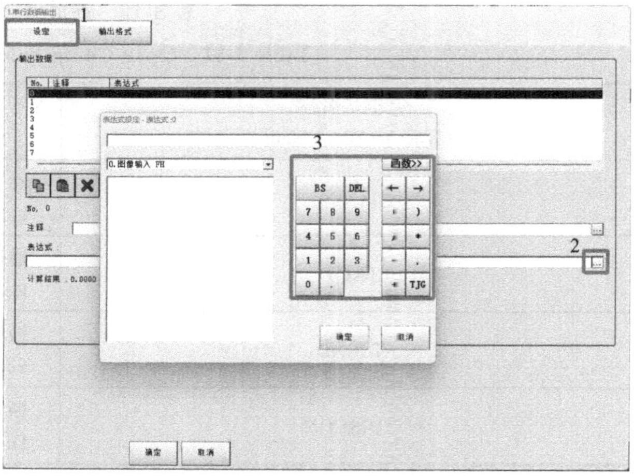

（续）

序号 S/N	步骤 Step	图示 Illustration
6	输出时，需要设定输出的格式，输出整数位与小数位根据需要进行设置 For the output, its format needs to set with the required output integer and decimal places	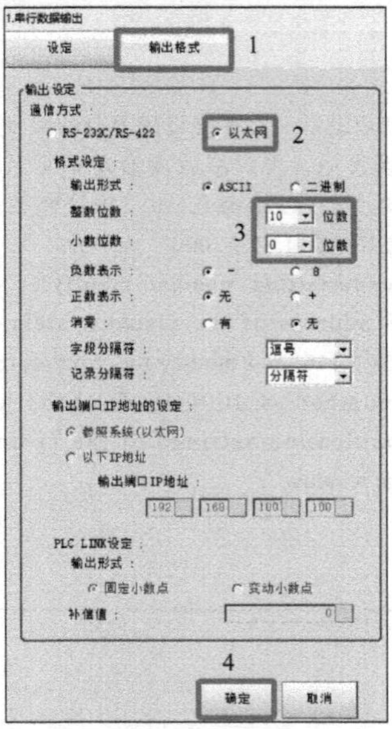

(2) PLC 通信设置

从本工作站硬件配置来看，可编程设备之间主要采用以太网通信，这里主要讲解 PLC 实现开放式（无协议）TCP 通信的方法。

从 PLC 编程角度来说，实现开放式通信需要用到以下两个指令功能块，见表 3-12。

(2) PLC communication settings

With regard to the hardware configuration of this workstation, Ethernet is used mainly for the communication between the programmable equipment. Here, we explain the method of achieving open (protocol free) TCP communication with PLC.

From the perspective of PLC programming, open communication requires the following two instruction function blocks, as shown in Table 3-12.

表 3-12 开放式通信指令
Table 3-12 Open Communication Instructions

功能块名称 Function Block Name	功能描述 Function Description
TCON	建立通信连接 Establish communication connection
TRCV	通过通信连接接收数据 Receive data through communication connection
TSEND	通过通信连接发送数据 Send data through communication connection
TDISCON	断开通信连接 Disconnect communication connection

项目 3　典型工业机器人搬运工作站系统的设计及应用

在通信过程中，视觉控制系统作为机器人或 PLC 的下位机，需要接收上位机发来的控制指令。能够使用到的控制指令有三种：选择场景组、选择场景和执行测量。OMRON FH 系列视觉控制器默认的系统通信代码见表 3-13。

The visual control system, as the lower machine of the robot or PLC, needs to receive control instructions from the upper machine in the communication process. There are three types of control instructions that can be used: select the scene group, select the scene, and execute the measurement. The default system communication code for the visual controller of the OMRON FH series is shown in Table 3-13.

表 3-13　视觉系统通信代码
Table 3-13　Visual System Communication Codes

命令格式 Command format	功能 Function	响应格式 Response format
SG 0	切换所使用的场景组编号 Switch the scene group number used	OK
S 0	切换所使用的场景编号 Switch the scene number used	OK
M	执行一次测量 Execute the measurement for one time	OK+ 测量结果 OK+measurement results

3.5.4　视觉流程与通信程序设计

1. OMRON 视觉流程设计

其设计方法见表 3-14。

2. PLC 通信程序设计

其设计方法见表 3-15。

3.5.4　Visual Process and Communication Program Design

1. OMRON visual flow design

The design method is shown in Table 3-14.

2. PLC communication program design

The design method is shown in Table 3-15.

表 3-14　OMRON 视觉流程设计方法
Table 3-14　Method of OMRON Visual Flow Design

序号 S/N	步骤 Step	图示 Illustration
1	创建一个新的视觉流程编辑界面，单击"与流程显示流动"左边的深绿色方框，车门饰板出现在视图框的情况下，将动态改为静态 Create a new visual flow editing interface, click the dark green box on the left of "Display flow with process", and change the dynamic to static when the car door trim panel appears in the view box	

（续）

序号 S/N	步骤 Step	图示 Illustration
2	检测被测对象的个数，对指定颜色的标签进行计数，计算指定标签号的面积、重心位置。可以通过图像文件选择被测对象的图像，或者通过相机测量获取被测对象的图像 Detect the number of the detected objects, count the labels of the specified color, and calculate the area and position of center of gravity of the specified label number. The image of the detected object can be selected through the image file or obtained by camera measurement	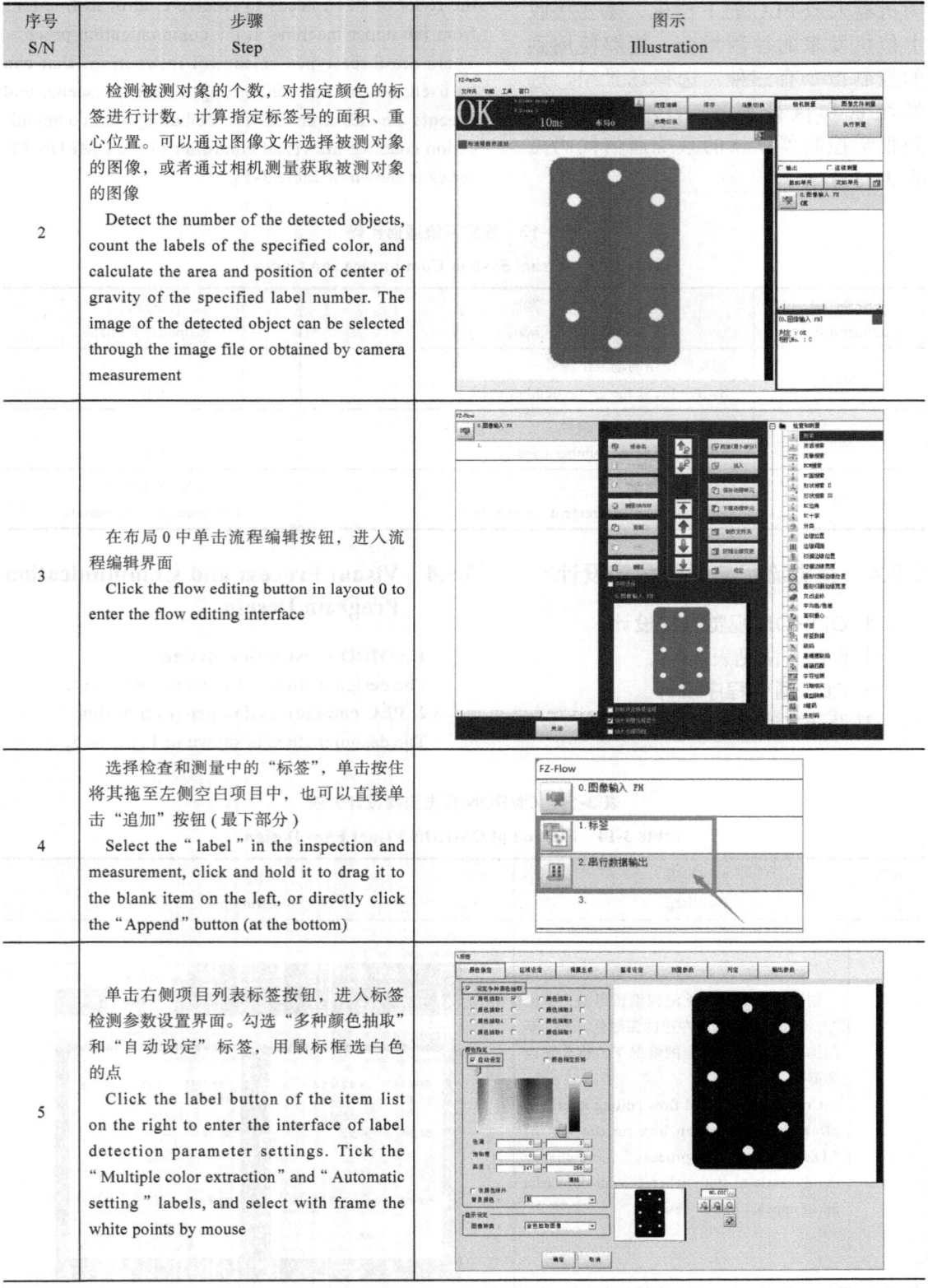
3	在布局 0 中单击流程编辑按钮，进入流程编辑界面 Click the flow editing button in layout 0 to enter the flow editing interface	
4	选择检查和测量中的"标签"，单击按住将其拖至左侧空白项目中，也可以直接单击"追加"按钮（最下部分） Select the "label" in the inspection and measurement, click and hold it to drag it to the blank item on the left, or directly click the "Append" button (at the bottom)	
5	单击右侧项目列表标签按钮，进入标签检测参数设置界面。勾选"多种颜色抽取"和"自动设定"标签，用鼠标框选白色的点 Click the label button of the item list on the right to enter the interface of label detection parameter settings. Tick the "Multiple color extraction" and "Automatic setting" labels, and select with frame the white points by mouse	

（续）

序号 S/N	步骤 Step	图示 Illustration
6	打开区域设定，把要检测的区域框选出来 Open the area settings and select with frame the area to be detected	
7	打开测量参数，添加抽取条件，选择面积，将最大面积调到 260，筛选掉多余的小型区域，之后单击"确定"按钮，如图 a) 所示；打开判定，将标签数的最大值和最小值修改为 8，如图 b) 所示 Open the measurement parameters, add the extraction conditions, select the area, adjust the maximum area to 260, filter out excess small areas, and click the "Confirm" button, as shown in Figure a); open the judgment and modify the maximum and minimum values of the number of label to 8, as shown in Figure b)	

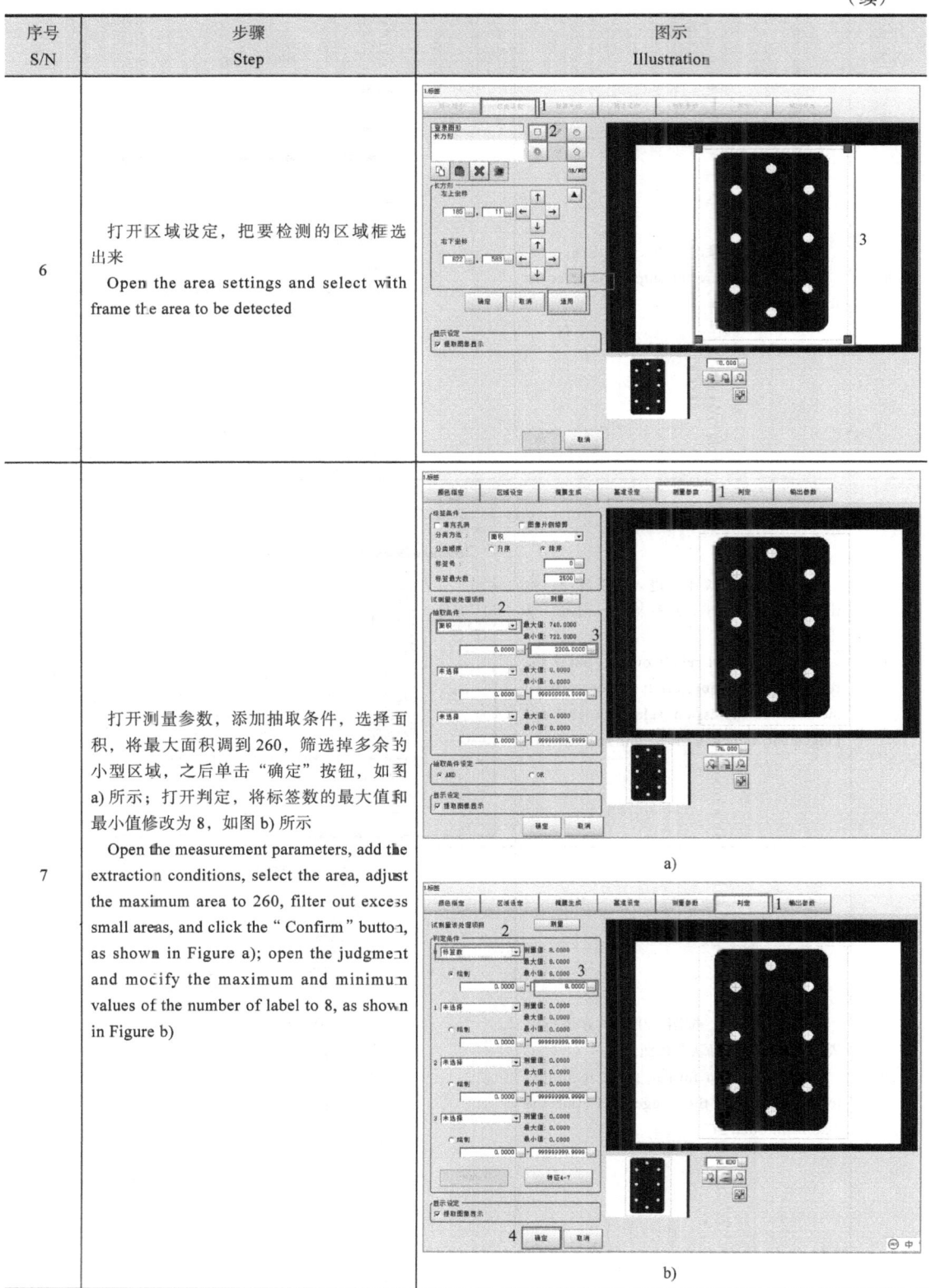

a)

b)

（续）

序号 S/N	步骤 Step	图示 Illustration
8	打开串行结果输出，进入以下界面 Open the serial result output and enter the following interface	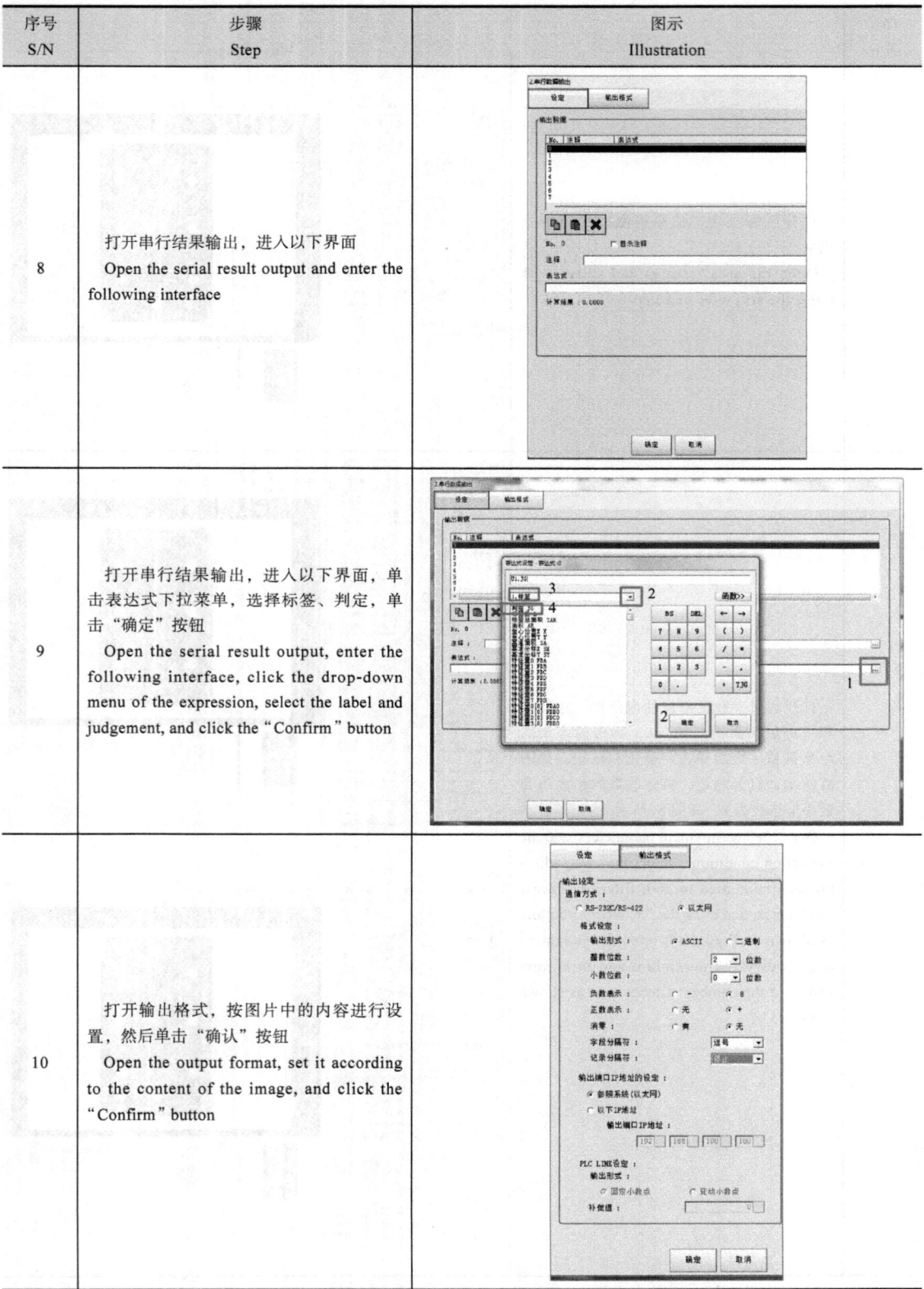
9	打开串行结果输出，进入以下界面，单击表达式下拉菜单，选择标签、判定，单击"确定"按钮 Open the serial result output, enter the following interface, click the drop-down menu of the expression, select the label and judgement, and click the "Confirm" button	
10	打开输出格式，按图片中的内容进行设置，然后单击"确认"按钮 Open the output format, set it according to the content of the image, and click the "Confirm" button	

表 3-15 PLC 通信程序设计方法
Table 3-15 Method of PLC Communication Program Design

序号 S/N	步骤 Step	图示 Illustration
1	任务采用的是 S7-1200 PLC 的开放式用户通信，无论是基于 UDP 协议还是 TCP 协议，西门子 PLC 开放式以太网通信的第一步都是用 TCON 指令建立连接 The open user communication of S7-1200 PLC is adopted for the task. The first step of the open Ethernet communication of Siemens PLC is to call the TCON instruction to establish a connection, whether the communication is based on the UDP or TCP protocol	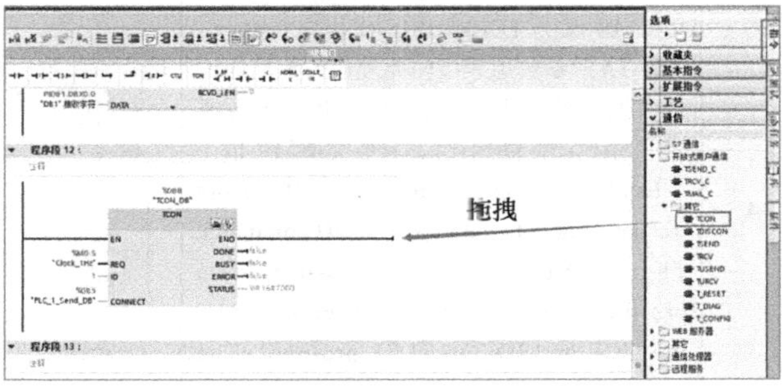 （图示：TCON 指令拖拽）
2	在 TCON 指令的属性窗口中，选择需要的通信伙伴，通信伙伴可以是项目中已有的 CPU，或者不指定。对于 OMRON 视觉伙伴选择不指定，然后单击【连接参数】右侧的列表框 In the attribute window of the TCON instruction, select the desired communication partner, which can be an existing CPU in the program or unspecified. For the partner of OMRON vision, select the unspecified, and then click the list box on the right of [Connection parameters]	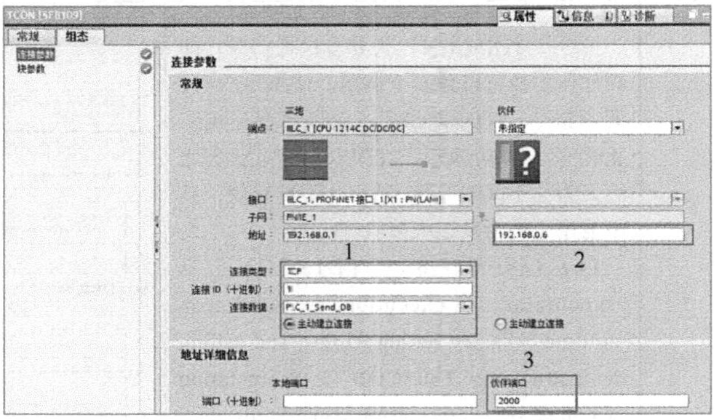 （图示：TCON 属性连接参数）
3	TRCV 指令用来完成对 TCP、ISO-ON-TCP 协议的数据接收（不支持 UDP 协议），当接收到有效数据时，NDR 参数会被设置 1，RCVD_LEN 的值表示实际接收到的数据长度（单位为字节） The TRCV instruction is used to receive data from TCP and ISO-ON-TCP protocols (UDP protocol not supported). When valid data is received, the NDR parameter will be set to 1, and the value of RCVD_LEN represents the actual length of the data received (in bytes)	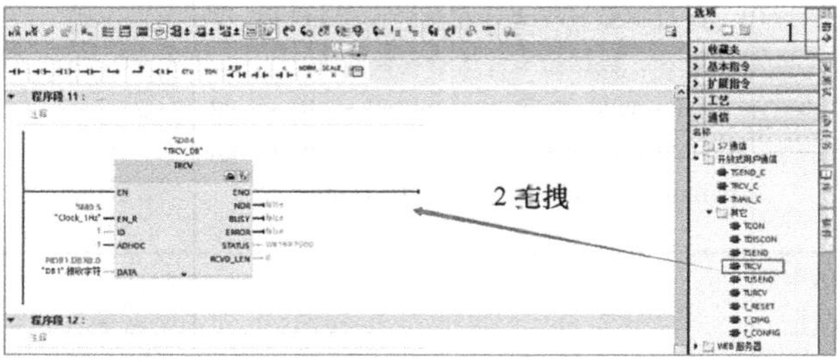 （图示：TRCV 指令拖拽）

（续）

序号 S/N	步骤 Step	图示 Illustration
4	在使用 TSEND 指令发送数据之前，要首先使用 TCON 来建立连接。TSEND 指令基于已经建立好的连接来发送数据，使用的协议为 TCP 或 ISO-ON-TCP 协议。"DATA"参数用来指向要发送的数据的地址，发送数据类型不能是"bit"（位）或者"bit"（位）的数组，其他类型都可以 Establish a connection with TCON before transmitting data with the TSEND command. The TSEND instruction transmits data based on the established connections under the TCP or ISO-ON-TCP protocol. The "DATA" parameter is used to point to the address of the data to be transmitted. The transmit data can be in all types except "bit" or an array of "bit"	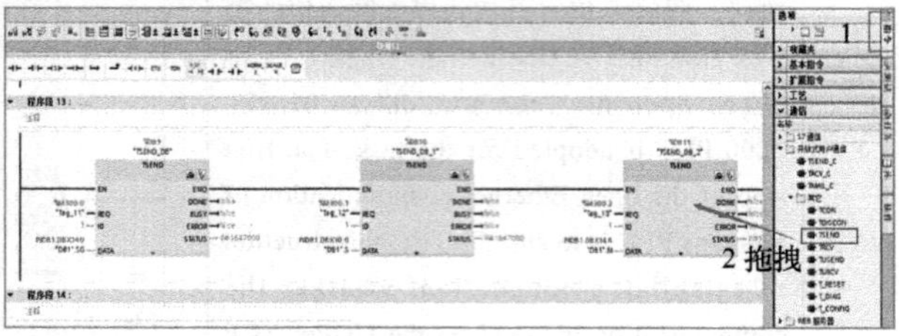
5	指令"TDISCON"为异步执行指令，即该作业的执行可跨多个调用。使 REQ = 1，调用指令"TDISCON"，可以启动连接终止作业。成功执行"TDISCON"指令之后，为"TCON"指定的 ID 不再有效，且不能用于进行发送或接收 The instruction "TDISCON" is asynchronous in execution, as it can span multiple calls. By setting REQ=1 and calling the instruction "TDISCON", the operation of connection termination can be initiated. When the "TDISCON" instruction is successfully executed, the ID specified for "TCON" is no longer valid and cannot be used for transmitting or receiving	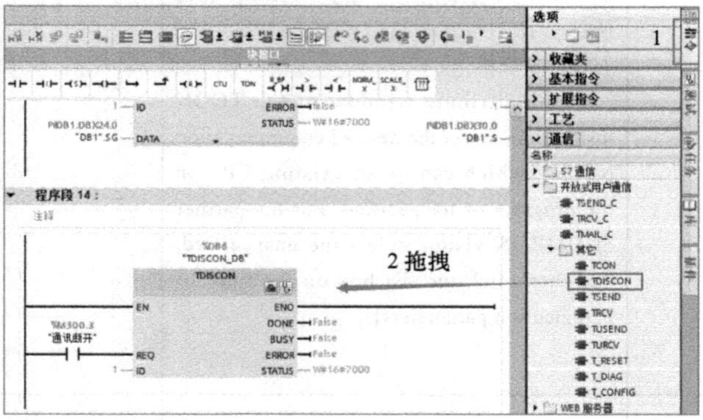
6	通过视觉流程设计参数的设置和 PLC 程序设计，数据需要发送给视觉系统 Data needs to be transmitted to the visual system for the setting of visual flow design parameters and PLC program design	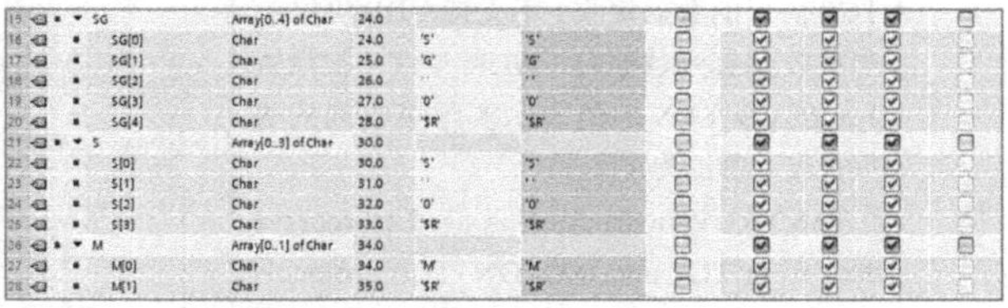
7	接收检测成功数据 Receive the data of successful detection	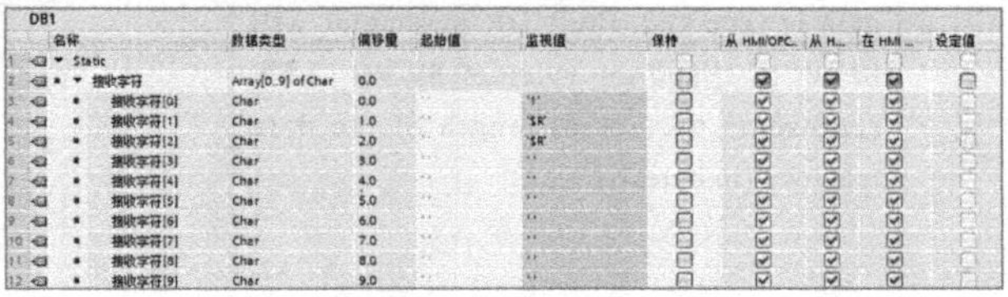

（续）

序号 S/N	步骤 Step	图示 Illustration
8	接收检测失败数据 Receive the data of failed detection	

任务 3.6　工作站安装与调试及程序的编写
Task 3.6　Workstation Installation, Commissioning and Their Programming

【知识目标】

1. 掌握编写 PLC 工艺流程的方法和步骤。
2. 掌握人机界面设计的方法和步骤。
3. 掌握搬运工作站的硬件接线原理及方法。
4. 掌握搬运工作站联机调试的方法。

[Knowledge Objectives]

1. To master the methods and steps for writing PLC process flow.
2. To master the methods and steps of human-machine interface design.
3. To master the principles and methods for hardware wiring of the handling workstations.
4. To master the methods for online commissioning of the handling workstations.

【技能目标】

1. 能够编写符合工艺要求的 PLC 程序。
2. 能够设计出搬运工作站控制系统的人机界面。
3. 能够根据工艺要求联机调试设备。

[Skill Objectives]

1. To be able to write PLC programs that meet process requirements.
2. To be able to design the human-machine interface of the control system of the handling workstations.
3. To be able to conduct online commissioning of equipment according to process requirements.

【素质目标】

1. 具备根据工艺要求完成程序编写的能力。
2. 善于沟通、协调完成任务。

[Competence Objectives]

1. To be able to write programs according to process requirements.
2. To be proficient in communication and coordination to complete tasks.

【任务情景】

某汽车企业为加快生产速度、提高生产质量，在汽车门饰板加工搬运工艺环节采用工业机器人搬运工作站系统实现加工及搬运自动化生产，前期已经完成了方案讨论、施工图样设计、机器人工作站仿真，现在需要对该搬运工作站进行系统安装、编程及调试工作，请根据要求完成工业机器人搬运工作站的安装、编程及调试。

【任务分析】

任务4中完成了工业机器人搬运工作站的仿真，验证了搬运工作站实际工作的可执行性，通过软件仿真对工业机器人搬运工作站提出修改意见并调整仿真工作站的不足。在完成系统仿真后可进入工作站安装与调试环节，工作站的安装调试必须通过前期设计好的相关施工图纸进行规范施工。本项目将前面的5个任务进行综合并实现，首先完成搬运工作站的硬件安装与调试，再按照搬运工作站的工艺要求编写PLC控制程序并设计组态人机界面，最后进行系统的联调。

【知识准备】

3.6.1 工作站安装

1. 工装夹具及气路安装

根据任务3的施工图纸完成搬运工作站的非标准零件及气路的安装，工具及气路的安装如图3-30所示，立体仓库的安装如图3-31所示。

[Task Scenario]

In order to improve productivity and production quality, an automobile manufacturer plans to adopt an industrial robot handling workstation system to achieve automated machining and handling of the car door trim panel. The scheme discussion, detailed drawing design, and robot workstation simulation have been completed at the early stage, and it is necessary to conduct installation, programming, and commissioning of the handling workstation system now. Please complete the said installation, programming and commissioning of the industrial robot handling workstation as required.

[Task Analysis]

In Task 4, such work has been completed as the simulation of the industrial robot handling workstation, the execution verification of its actual operation, the proposal of suggestions on its modification, and the adjustment to avoid shortcomings of the simulation workstation. When the system simulation is completed, the installation and commissioning the workstation can be started and must be carried out in accordance with the detailed drawings previously designed and relevant codes for construction. This program integrates the previous five tasks and realizes their targets, in that, the following work is carried out in order: the hardware installation and commissioning of the handling workstation, the PLC control programming according to the process requirements of the handling workstation, the design and configuration of the human-machine interface, and the on-line commissioning of the system.

[Assumed Knowledge]

3.6.1 Workstation Installation

1. Installation of fixture and gas circuit

Complete the installation of non-standard parts and gas circuit of the handling workstation according to the detailed drawings of Task 3. The installation of tools and gas circuit is shown in Figure 3-30, and the installation of warehouse is shown in Figure 3-31.

项目3 典型工业机器人搬运工作站系统的设计及应用

图 3-30 工具及气路安装
Figure 3-30 Installation of Tools and Gas Circuit

图 3-31 立体仓库的安装
Figure 3-31 Installation of Warehouse

2. 电气安装

按照电气原理图及硬件安装位置，完成电气设备安装及接线，如图 3-32 所示。

2. Electrical installation

Complete the installation and wiring of electrical equipment according to the electrical schematic diagram and hardware installation positions, as shown in Figure 3-32.

图 3-32 电气设备安装及接线
Figure 3-32 Installation and Wiring of Electrical Equipment

3.6.2 PLC 程序编写

1. 建立 I/O 变量分配

首先根据规划创建 PLC 的 I/O 变量，见表 3-16 ～ 表 3-20，然后分别编写各部分程序。

3.6.2 PLC Programming

1. Create I/O variable allocation

Firstly, create I/O variable for the PLC according to the plan, as shown from Table 3-16 to Table 3-20, and then write programs for each part separately.

表 3-16　4A1 模块的 I/O 变量表
Table 3-16　I/O Variable table of the 4A1 Module

序号 S/N	变量名 Variable name	数据类型 Data type	逻辑地址 Logical address	注释 Notes	模块号 Module number
1	1 号气缸运行位 Running position of cylinder 1	Bool	%I0.0	1 号气缸运行位 Running position of cylinder 1	4A1 模块 4A1 module
2	2 号气缸运行位 Running position of cylinder 2	Bool	%I0.1	2 号气缸运行位 Running position of cylinder 2	
3	1 号气缸原位 Home position of cylinder 1	Bool	%I0.2	1 号气缸原位 Home position of cylinder 1	
4	2 号气缸原位 Home position of cylinder 2	Bool	%I0.3	2 号气缸原位 Home position of cylinder 2	
5	上料台急停 Emergency stop of loading table	Bool	%I0.7	上料台急停 Emergency stop of loading table	
6	安全插销 Safety pin	Bool	%I1.3	安全插销 Safety pin	
7	安全光栅 Safety grating	Bool	%I1.4	安全光栅 Safety grating	
8	安全门急停 Emergency stop of safety door	Bool	%I1.5	安全门急停 Emergency stop of safety door	

表 3-17　4A2 模块的 I/O 变量表
Table 3-17　I/O Variable table of the 4A2 Module

序号 S/N	变量名 Variable name	数据类型 Data type	逻辑地址 Logical address	注释 Notes	模块号 Module number
1	传送带_左气缸运行位 Running position of conveyor belt_ left cylinder	Bool	%I2.0	传送带_左气缸运行位 Running position of conveyor belt_ left cylinder	4A2 模块 4A2 module
2	传送带_右气缸运行位 Running position of conveyor belt_ right cylinder	Bool	%I2.1	传送带_右气缸运行位 Running position of conveyor belt_ right cylinder	
3	传送带_左气缸原位 Home position of conveyor belt_ left cylinder	Bool	%I2.2	传送带_左气缸原位 Home position of conveyor belt_ left cylinder	

（续）

序号 S/N	变量名 Variable name	数据类型 Data type	逻辑地址 Logical address	注释 Notes	模块号 Module number
4	传送带_右气缸原位 Home position of conveyor belt_ right cylinder	Bool	%I2.3	传送带_右气缸原位 Home position of conveyor belt_ right cylinder	4A2模块 4A2 module
5	传送带左工件传送位 Left workpiece conveying position of conveyor belt	Bool	%I2.4	传送带左工件传送位 Left workpiece conveying position of conveyor belt	
6	传送带右工件传送位 Right workpiece conveying position of conveyor belt	Bool	%I2.5	传送带右工件传送位 Right workpiece conveying position of conveyor belt	
7	传送带急停 Emergency stop of conveyor belt	Bool	%I2.6	传送带急停 Emergency stop of conveyor belt	
8	气压检测 Air pressure detection	Bool	%I2.7	气压检测 Air pressure detection	
9	启动按钮 Start button	Bool	%I3.0	启动按钮（带灯绿色） Start button (with green light)	
10	停止按钮 Stop button	Bool	%I3.1	停止按钮（带灯红色） Stop button (with red light)	
11	复位按钮 Reset button	Bool	%I3.2	复位扫钮（蓝色） Reset scan button (blue)	
12	暂停按钮 Pause button	Bool	%I3.3	暂停按钮（黄色） Pause button (yellow)	
13	操作面板急停 Emergency stop of operation panel	Bool	%I3.4	操作面板急停 Emergency stop of operation panel	

表 3-18　4A3 模块的 I/O 变量表
Table 3-18　I/O Variable table of the 4A3 Module

序号 S/N	变量名 Variable name	数据类型 Data type	逻辑地址 Logical address	注释 Notes	模块号 Module number
1	CMDENBL/接收输入信号 CMDENBL/Receive input signal	Bool	%I4.0	CMDENBL/接收输入信号 CMDENBL/Receive input signal	4A3模块 4A3 module

（续）

序号 S/N	变量名 Variable name	数据类型 Data type	逻辑地址 Logical address	注释 Notes	模块号 Module number
2	SYSRDY/ 系统准备就绪信号 SYSRDY/System readiness signal	Bool	%I4.1	SYSRDY/ 系统准备就绪信号 SYSRDY/System readiness signal	
3	PROGRUN/ 程序执行中信号 PROGRUN/Program in running signal	Bool	%I4.2	PROGRUN/ 程序执行中信号 PROGRUN/Program in running signal	
4	PAUSED/ 暂停中信号 PAUSED/in pause signal	Bool	%I4.3	PAUSED/ 暂停中信号 PAUSED/in pause signal	
5	HELD/ 保持中信号 HELD/in holding signal	Bool	%I4.4	HELD/ 保持中信号 HELD/in holding signal	
6	FAULT/ 报警信号 FAULT/alarm signal	Bool	%I4.5	FAULT/ 报警信号 FAULT/alarm signal	
7	ATPERCH/ 基准点信号 ATPERCH/reference point signal	Bool	%I4.6	ATPERCH/ 基准点信号 ATPERCH/reference point signal	
8	TPENBL/ 示教操作盒信号 TPENBL/Teach pendant operation box signal	Bool	%I4.7	TPENBL/ 示教操作盒信号 TPENBL/Teach pendant operation box signal	
9	BATALM/ 电池异常信号 BATALM/battery abnormal signal	Bool	%I5.0	BATALM/ 电池异常信号 BATALM/battery abnormal signal	4A3 模块 4A3 module
10	BUSY/ 处理中信号 BUSY/ in process signal	Bool	%I5.1	BUSY/ 处理中信号 BUSY/in process signal	
11	ASK1/SN01 选择程序 1 ASK1/SN01 Select Program 1	Bool	%I5.2	ASK1/SN01 选择程序 1 ASK1/SN01 Select Program 1	
12	ASK2/SN02 选择程序 2 ASK2/SN02 Select Program 2	Bool	%I5.3	ASK2/SN02 选择程序 2 ASK2/SN02 Select Program 2	
13	ASK3/SN03 选择程序 3 ASK3/SN03 Select Program 3	Bool	%I5.4	ASK3/SN03 选择程序 3 ASK3/SN03 Select Program 3	
14	机器人允许 PLC 上料平台伺服工作 Robot allows the PLC loading table servo to work	Bool	%I5.5	ASK4/SN04 选择程序 4 ASK4/SN04 Select Program 4	
15	机器人允许传送带动作 Robot allows the conveyor belt to act	Bool	%I5.6	ASK5/SN05 选择程序 5 ASK5/SN05 Select Program 5	
16	上一工作站允许本工作站机器人工作 The previous workstation allows the robot of this workstation to work	Bool	%I5.7		

表 3-19　4A4 模块的 I/O 变量表
Table 3-19　I/O Variable table of the 4A4 Module

序号 S/N	变量名 Variable name	数据类型 Data type	逻辑地址 Logical address	注释 Notes	模块号 Module number
1	机器人运行灯 Robot running light	Bool	%Q2.0	机器人运行灯（绿色） Robot running light (green)	
2	机器人暂停灯 Robot pause light	Bool	%Q2.1	机器人暂停灯（红色） Robot pause light (red)	
3	三色灯_绿色 Tricolor light_ green	Bool	%Q2.2	塔灯_绿色 Tower light_ green	
4	三色灯_黄色 Tricolor light_ yellow	Bool	%Q2.3	塔灯_黄色 Tower light_ yellow	
5	三色灯_红色 Tricolor light_ red	Bool	%Q2.4	塔灯_红色 Tower light_ red	
6	上料台夹具气缸打开 Loading table fixture cylinder open	Bool	%Q2.5	上料台夹具定位气缸打开 Loading table fixture positioning cylinder open	
7	传送带左气缸 Left cylinder of conveyor belt	Bool	%Q2.6	传送带左气缸 Left cylinder of conveyor belt	
8	传送带右气缸 Right cylinder of conveyor belt	Bool	%Q2.7	传送带右气缸 Right cylinder of conveyor belt	4A4 模块 4A4 module
9	允许下一工作站机器人联动 Allow robots of the next workstation to take joint actions	Bool	%Q3.0	允许下一工作站机器人抓取工件 Allow robots of the next workstation to pick up workpieces	
10	塔灯蜂鸣器 Tower light buzzer	Bool	%Q3.1	塔灯蜂鸣器 Tower light buzzer	
11	变频器 DI 1 Frequency converter DI 1	Bool	%Q3.2	变频器使能 Frequency converter enabling	
12	变频器 DI 2 Frequency converter DI 2	Bool	%Q3.3	变频器多段速 1 Frequency converter multi stage speed 1	
13	变频器 DI 3 Frequency converter DI 3	Bool	%Q3.4	变频器多段速 2 Frequency converter multi stage speed 2	
14	变频器 DI 4 Frequency converter DI 4	Bool	%Q3.5	变频器多段速 3 Frequency converter multi stage speed 3	

表 3-20　4A5 模块的 I/O 变量表
Table 3-20　I/O Variable table of the 4A5 Module

序号 S/N	变量名 Variable name	数据类型 Data type	逻辑地址 Logical address	注释 Notes	模块号 Module number
1	IMSTP/ 急停信号 IMSTP/Emergency stop signal	Bool	%Q4.0	IMSTP/ 急停信号 IMSTP/Emergency stop signal	4A5 模块 4A5 modular
2	HOLD/ 暂停 HOLD/Pause	Bool	%Q4.1	HOLD/ 暂停 HOLD/Pause	
3	SFSPD/ 机器人减速信号 SFSPD/Robot deceleration signal	Bool	%Q4.2	SFSPD/ 机器人减速信号 SFSPD/Robot deceleration signal	
4	CSTOP1/ 循环停止信号 CSTOP1/Cycle stop signal	Bool	%Q4.3	CSTOP1/ 循环停止信号 CSTOP1/Cycle stop signal	
5	FAULT RESET/ 解除报警 FAULT RESET/Alarm reset	Bool	%Q4.4	FAULT RESET/ 解除报警 FAULT RESET/Alarm reset	
6	START/ 外部启动信号 START/External start signal	Bool	%Q4.5	START/ 外部启动信号 START/External start signal	
7	HOME/ 复位 HOME/RESET	Bool	%Q4.6	HOME/ 复位 HOME/RESET	
8	ENBL/ 允许机器人动作 ENBL/Allow robot to act	Bool	%Q4.7	ENBL/ 允许机器人动作 ENBL/Allow robot to act	
9	RSR0001	Bool	%Q5.0	RSR0001/ 程序 1 RSR0001/Program 1	
10	RSR0002	Bool	%Q5.1	RSR0002/ 程序 2 RSR0002/Program 2	
11	RSR0003	Bool	%Q5.2	RSR0003/ 程序 3 RSR0003/Program 3	
12	RSR0004	Bool	%Q5.3	RSR0004/ 程序 4 RSR0004/Program 4	
13	RSR0005	Bool	%Q5.4	RSR0005/ 程序 5 RSR0005/Program 5	
14	允许机器人抓传送带工件 Allow the robot to pick up workpieces from the conveyor belt	Bool	%Q5.5	允许机器人抓传送带工件 Allow the robot to pick up workpieces from the conveyor belt	
15	允许机器人抓上料平台工件 Allow the robot to pick up workpieces from the loading table	Bool	%Q5.6	允许机器人抓上料平台工件 Allow the robot to pick up workpieces from the loading table	
16	允许机器人放料到传送带 Allow the robot to place materials onto the conveyor belt	Bool	%Q5.7	允许机器人放料到传送带 Allow the robot to place materials onto the conveyor belt	

2. PLC 与工业机器人通信

(1) 硬件网络连接

网线直连，普通网线的一头插 S7-1500 PLC 的 PROFINET 通信口，另一头插机器人的 PROFINET 通信板的通信口。机器人与 PLC 对应的通信，在机器人控制柜都需要安装对应的板卡以及软件功能，如机器人柜子：R-30IB Mate；机器人主体：M-10IA12。硬件网络连接方法见表 3-21。

2. Communication between PLC and industrial robots

(1) Hardware network connection

For direct connection of network cables, one end of the ordinary network cable is plugged into the PROFINET communication port of the S7-1500 PLC, and the other end into the communication port of the PROFINET communication board of the robot. The communication between the robot and PLC requires corresponding board cards and software to be installed in the robot control cabinet, such as the robot cabinet: R-30IB Mate, the robot body: M-10IA12. The method of hardware network connection is shown in Table 3-21.

表 3-21 硬件网络连接方法
Table 3-21 Method of Hardware Network Connections

序号 S/N	步骤 Step	图示 Illustration
1	FANUC 机器人以太网通信板 Ethernet communication board of the FANUC robot	
2	4 个以太网口功能 4 Ethernet port functions	
3	机器人作从站时 RJ45 口 RJ45 port when robot serves as slave station	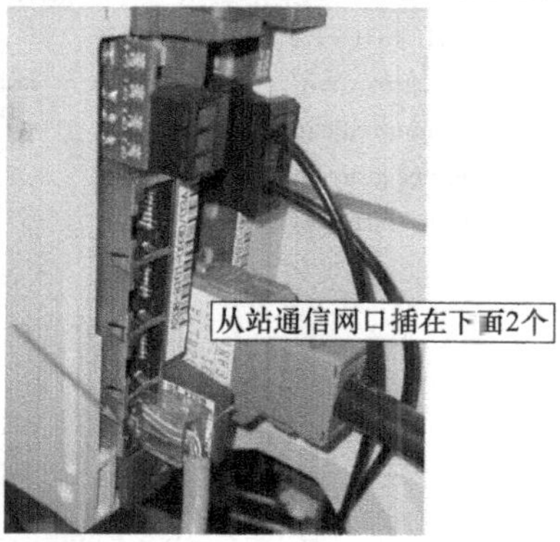

(2) PLC 网络设置
其方法见表 3-22。

(2) PLC network settings
The method is shown in Table 3-22.

表 3-22　PLC 与机器人通信网络设置方法
Table 3-22　Method for PLC and Robot Communication Network Settings

序号 S/N	步骤 Step	图示 Illustration
1	双击 PLC 网络端口，建立以太网地址（PLC 的 IP 地址要与机器人的 IP 地址在同一网段类） Double click the PLC network port to establish an Ethernet address (the IP address of the PLC shall be in the same network segment class as that of the robot)	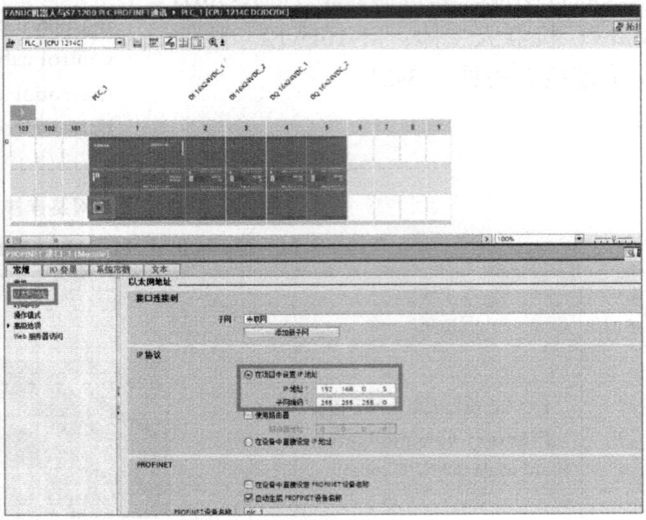
2	当博途软件需要与第三方设备进行 PROFINET 通信时（如与 FANUC 机器人通信），需要安装第三方设备的 GSD 文件 When the PORTAL software needs PROFINET communication with third-party device (such as communication with FANUC robots), it is necessary to install the GSD file of the third-party device	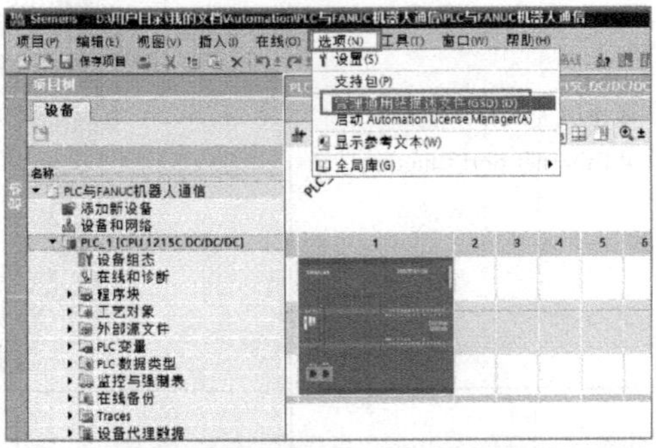 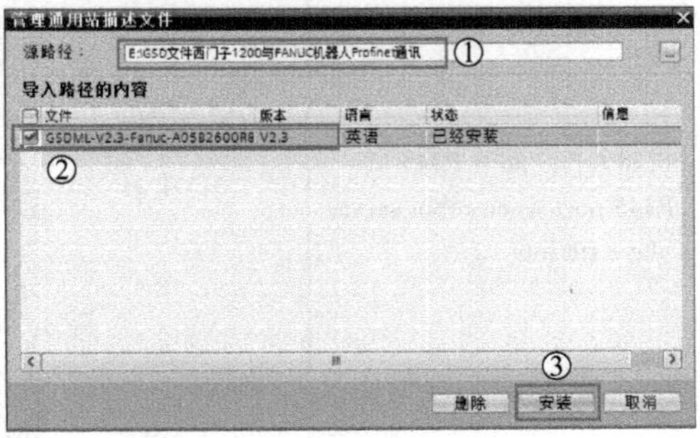

项目3 典型工业机器人搬运工作站系统的设计及应用　167

（续）

序号 S/N	步骤 Step	图示 Illustration
3	在右侧选择硬件目录→其他现场设备→PROFINET I/O→I/O→FANUC→R-30ib EF2→A05B-2600-R834:FANUC Robot Controller (1.0) Select Hardware Directory on the right side → Other on-site equipment → → PROFINET I/O → I/O → FANUC → R-30ib EF2 — A05B-2600-R834:FANUC Robot Controller (1.0)	
4	进行组网。双击机器人硬件端口，添加子网，添加机器人IP地址、PROFINET设备名称（机器人IP地址应与机器人本体设置的IP地址一致、设备名称与机器人本体设置的名称一致） Conduct networking. Double click the robot hardware port, add subnet, add robot IP address, and PROFINET device name (the robot IP address shall be consistent with that of set on the robot body, and the device name shall be consistent with that of set on the robot body)	
5	根据项目需求添加通信I/O字节数。此项目添加3字节的输入/输出模块（与机器人设置的输入/输出一致） Add communication I/O bytes according to program requirements. For this Program, add the module of 8 input/output bytes (consistent with the input/output set on the robot)	

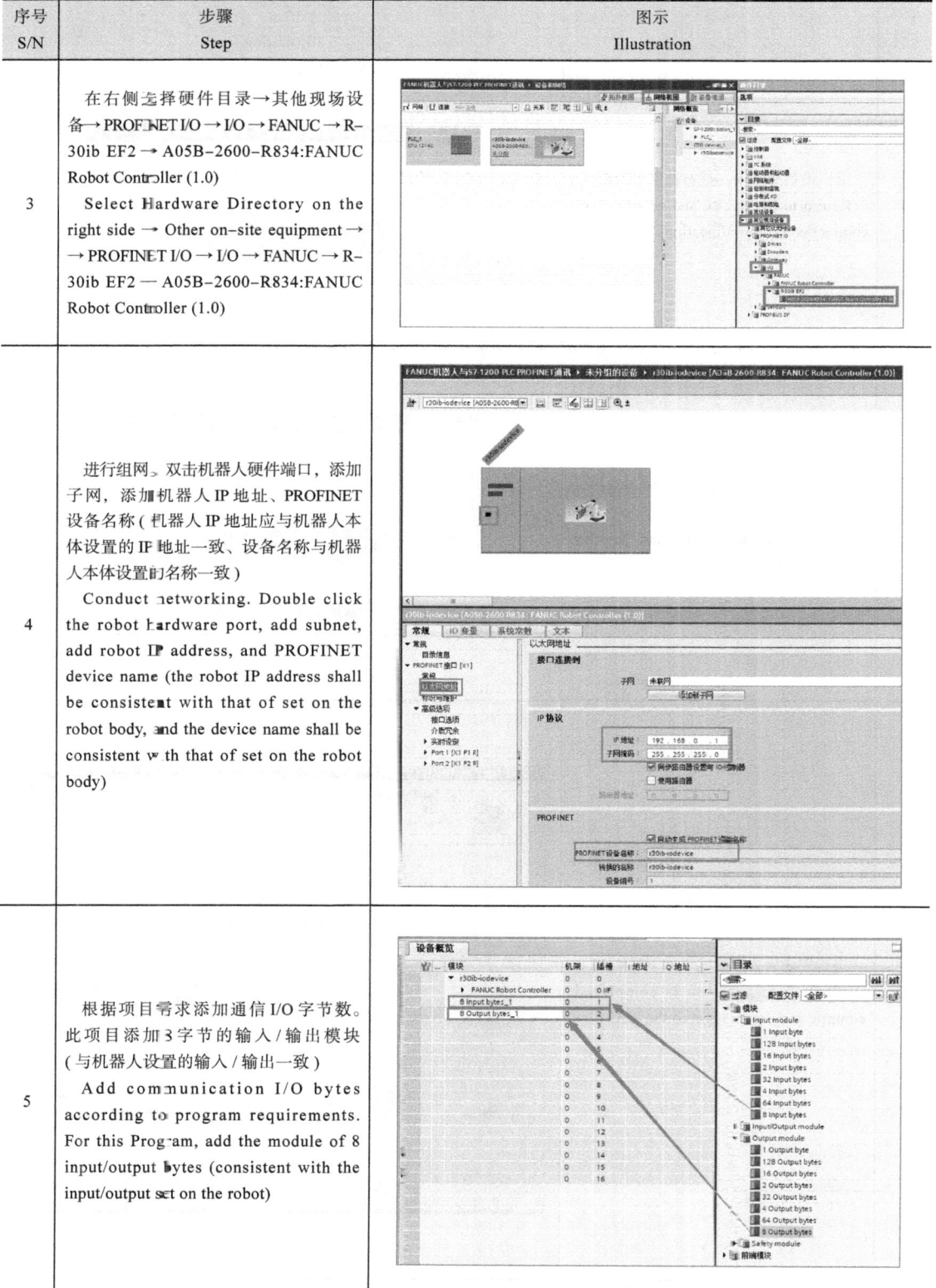

(续)

序号 S/N	步骤 Step	图示 Illustration
6	返回设备和网络，进行控制器分配 Return to the device and network, and conduct controller allocation	
7	分配后，PLC 与机器人之间会链接在一起 After allocation, the PLC and the robot will be linked together	
8	组态完成后，进行编译并下载到 PLC When configuration is completed, compile and download it to the PLC	

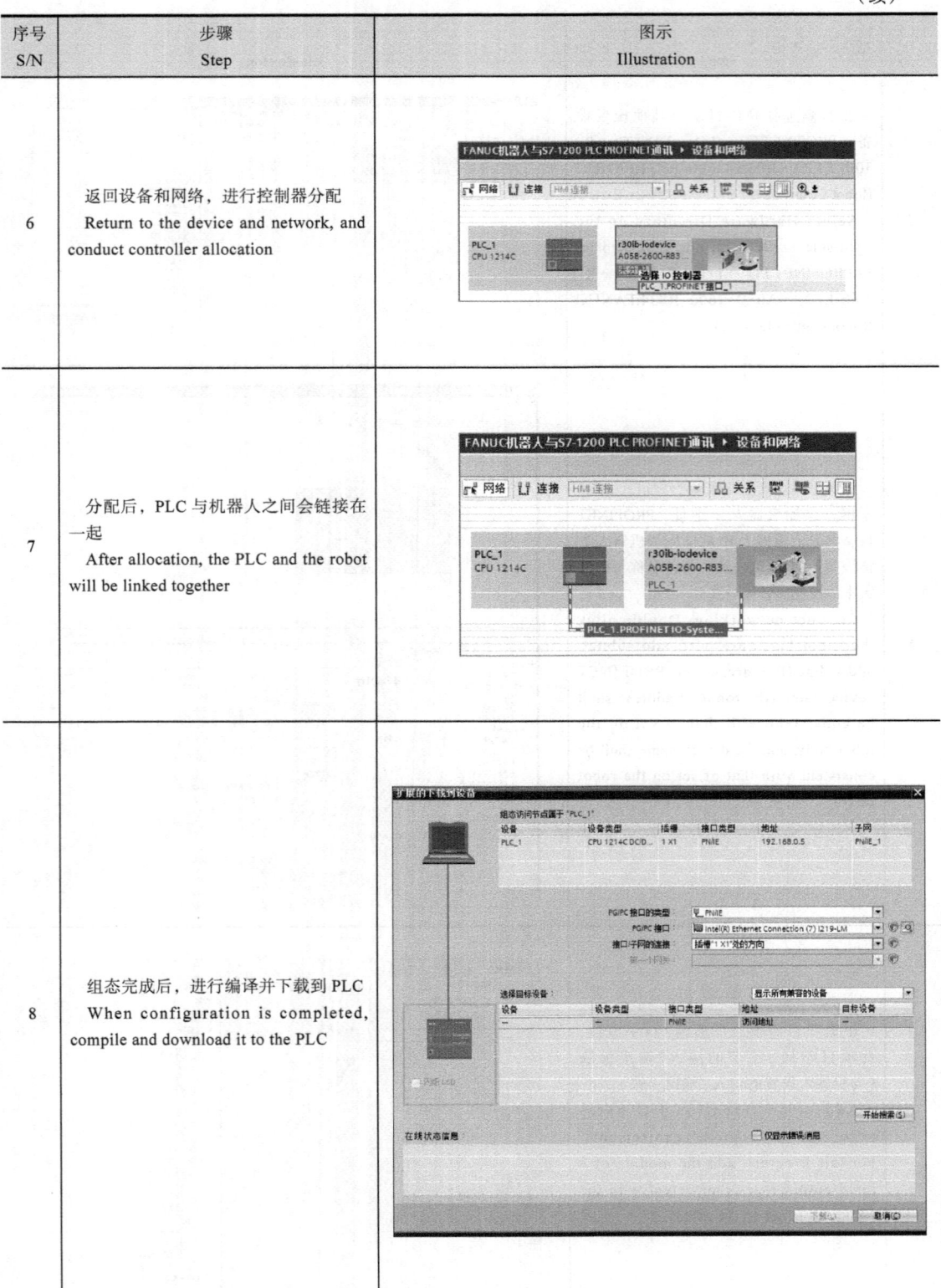

(续)

序号 S/N	步骤 Step	图示 Illustration
9	设置机器人 PROFINET 地址：按下示教器上 MEUN 键→ 5I/O →选中 I/O 页面→ PROFINET (M) → ENTER →选中 2 频道 (2 频道是机器人做从站)→按下 DISP 键→ 定址模式选择 DCP →选中 IP 地址→按下 F4 键 (编辑)→编辑完成后→按下 F1 键→ (适用)→完成 IP 地址编辑 (2 频道需要单击 F5，为有效后 2 频道 方可使用)。注意：此处地址、名称要与 PLC 组态时地址一致 Set the PROFINET address of the robot: Press the MEUN button on the teach pendant → 5I/O → select the I/O page → PROFINET (M) → ENTER → select channel 2 (channel 2 is for the robot serving as slave station) → press the DISP button → select DCP in the addressing mode → select the IP address → press the F4 button (edit) → when edit is completed → press the F1 button → (applicable) → IP address editing completed (F5 needs to be clicked for Channel 2 to become valid for use). Note: The address and name here shall be consistent with that of in the PLC configuration	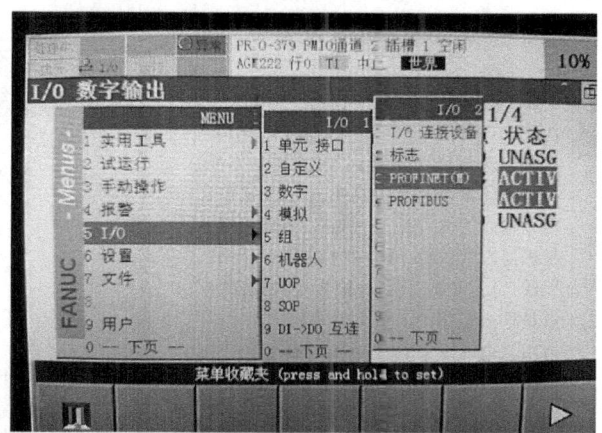 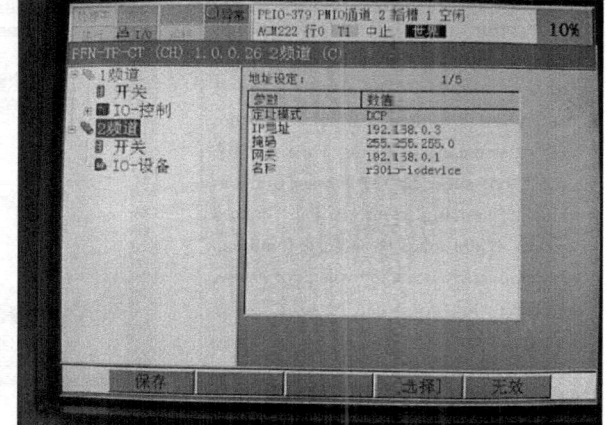
10	编辑插槽类型和字节长度：按下示教器上 MEUN 键→ 5 I/O → 选中 I/O 页面→ PROFINET (M) → ENTER →按下 F4 键→选中输入/输出插槽→编辑完成后→按下 F1 键→光标移到插槽大小→按下 F4 键选中字节大小→按下 F1 键→编辑完成 Edit the slot type and byte length: Press the MEUN button on the teach pendant → 5 I/O → select the I/O page → PROFINET (M) → ENTER → press the F4 button → select the input/output slot → when edit is completed → press the F1 button → move the cursor to the slot size → press the F4 button to select the byte size → press the F1 button → edit is completed	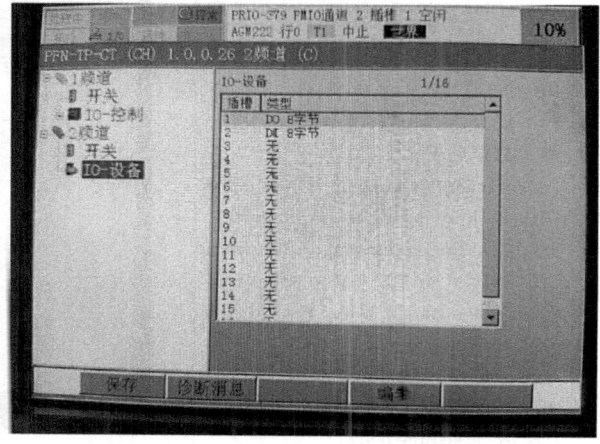

（续）

序号 S/N	步骤 Step	图示 Illustration
11	按下示教器上的 MEUN 键→ 5 I/O →选中 I/O 页面→数字→ ENTER → I/O 数字输入→ F2（分配）。DI 范围：本项目组态了 8 个字节输入/8 个字节输出，所以输入的范围是 1～64；机架：102 是机器人做从站，101 是机器人做主站；插槽：1；开始点：1。机器人的前面 18 点是作为专用的，在不用的情况下暂时占用，也可以从 19 点作为开始点 Press the MEUN button on the teach pendant → 5 I/O → select the I/O page → digit → ENTER → I/O digit input → F2 (allocation), DI range: this program is configured with 8 input bytes/8 output bytes; therefore, the input range is between 1–64; rack: for 102, the robot serves as the slave station, while for 101, the robot serve as the master station; slot: 1; starting point: 1. The first 18 points of the robot is dedicated and can be temporarily occupied when not in use, and the 19th point can be used as the starting point	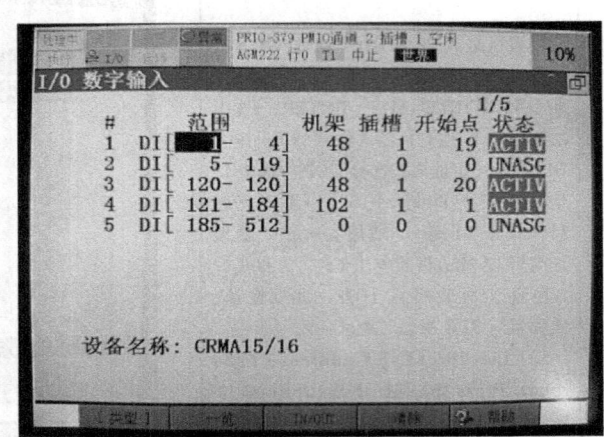 
12	DO 范围：本项目组态了 8 个字节输入/8 个字节输出；所以输出的范围是 1～64；机架：102 是机器人做从站，101 是机器人做主站；插槽：1；开始点：21，机器人的前面 20 点是作为专用的，所以从 21 点开始 DO range: this program is configured with 8 input bytes/8 output bytes; therefore, the output range is between 1–64; rack: for 102, the robot serves as the slave station, while for 101, the robot serve as the master station; slot: 1; starting point: 21, as the first 20 points of the robot are dedicated, the 21st point is used as the starting point.	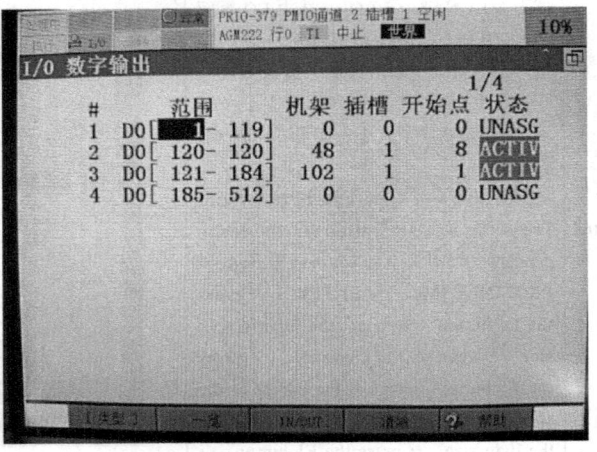

3. 编写 FLC 程序

根据工业机器人搬运工作站工艺流程要求,以 I/O 分配表为基础编写汽车门饰板搬运工作站控制程序。

1) 程序结构如图 3-33 所示。
2) 主程序如图 3-34 所示。
3) 子程序。

其中的报警程序、单机程序、联机程序请根据要求自行设计编写。

3. Compilation of the PLC program

The control program of the car door trim panel handling workstation is compiled according to the process requirements of the industrial robot handling workstation, and based on the I/O distribution list.

1) The program structure is shown in Figure 3-33.
2) The program is shown in Figure 3-34.
3) Subprogram.

Please design and compile the alarm, standalone, and online programs respectively according to the requirements.

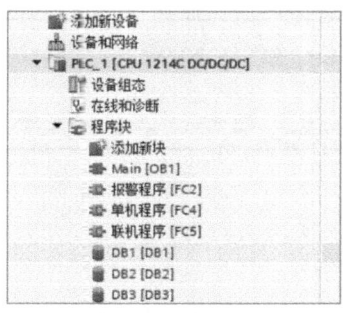

图 3-33 程序结构

Figure 3-33 Program Structure Diagram

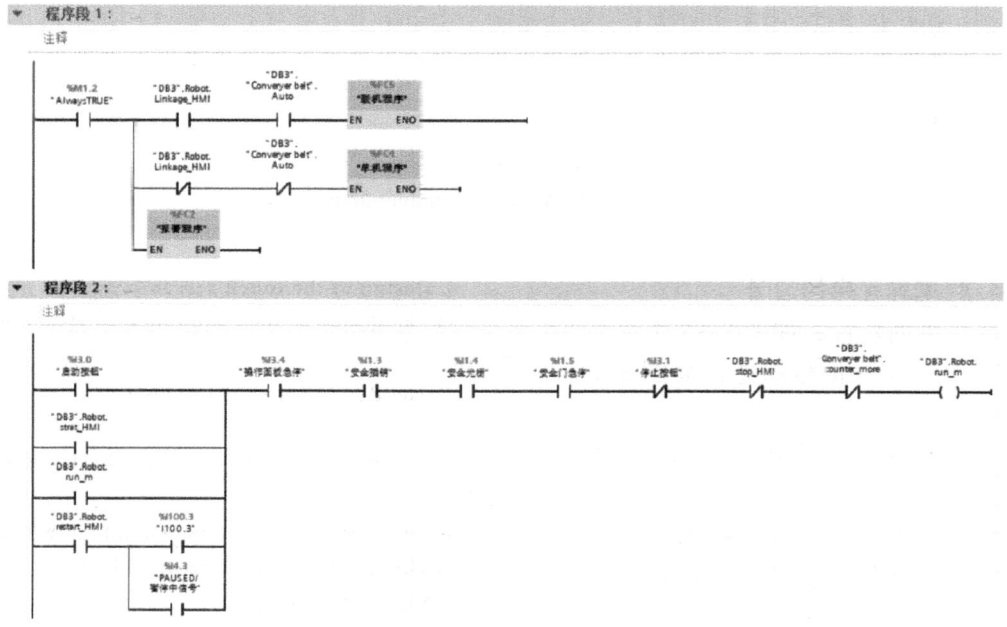

图 3-34 主程序

Figure 3-34 Main Program

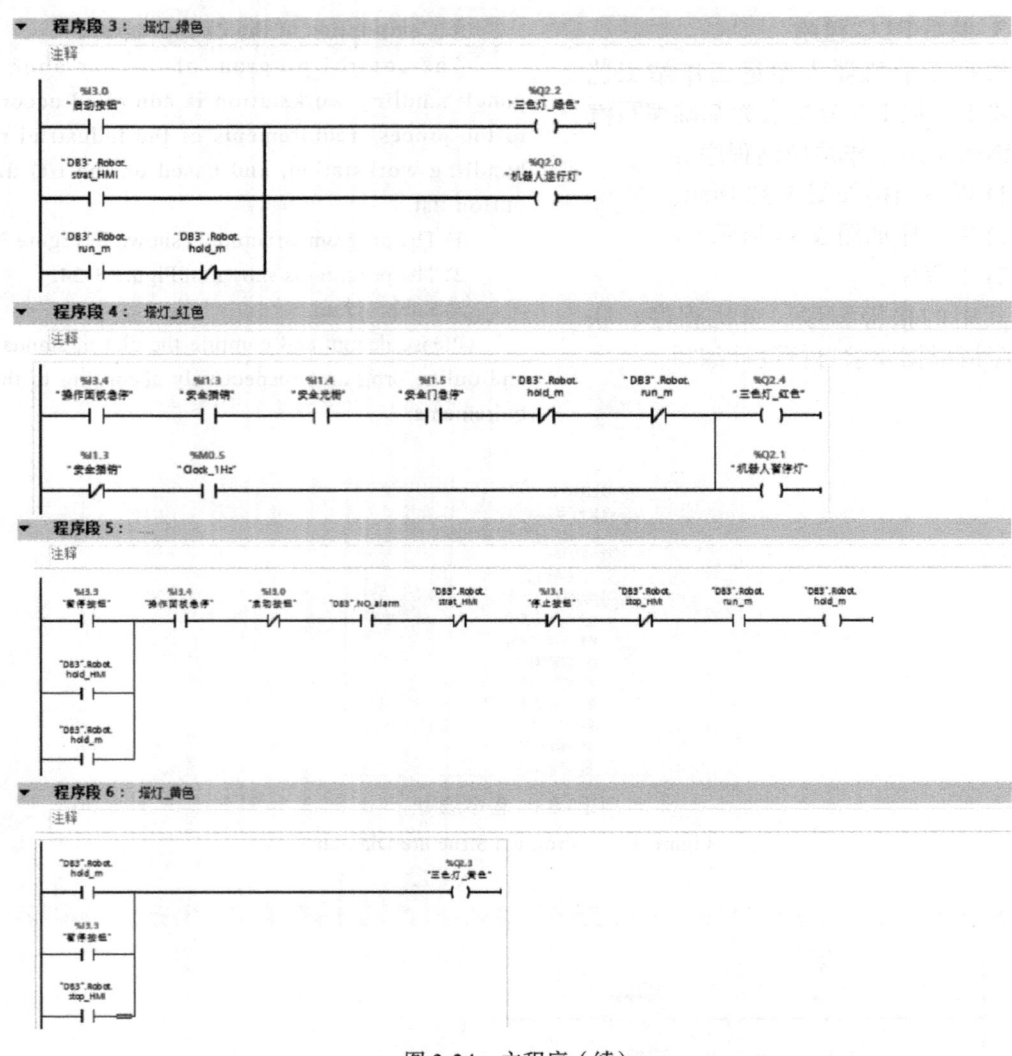

图 3-34 主程序（续）
Figure 3-34　Main Program(continue)

4. 触摸屏程序的设计

1) 主画面设计，如图 3-35 所示。

2) PLC 输入信号状态，如图 3-36 所示。

3) PLC 输出信号状态，如图 3-37 所示。

4) 控制画面，如图 3-38 所示。

5) 报警画面，如图 3-39 所示。

5. 视觉流程设计

该设计方法见表 3-14。

4. Design of the touch screen program

1) The design of the main screen is shown in Figure 3-35.

2) The PLC input signal status is shown in Figure 3-36.

3) The PLC output signal status is shown in Figure 3-37.

4) The control screen is shown in Figure 3-38.

5) The Alarm screen is shown in Figure 3-39.

5. Visual flow design

The design method is shown in Table 3-14.

项目3 典型工业机器人搬运工作站系统的设计及应用 173

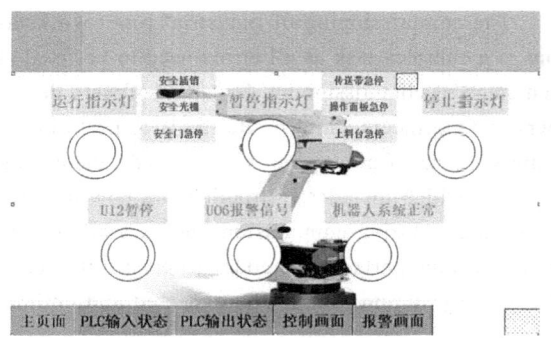

图 3-35 主画面设计
Figure 3-35 Design of Main Screen

图 3-36 PLC 输入信号状态
Figure 3-36 PLC Input Signal Status

图 3-37 PLC 输出信号状态
Figure 3-37 PLC Output Signal Status

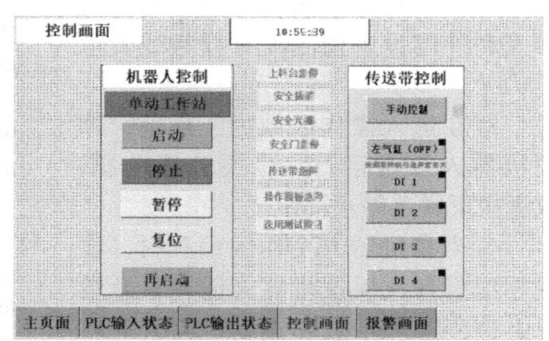

图 3-38 控制画面
Figure 3-38 Control Screen

图 3-39 报警画面
Figure 3-39 Alarm Screen

3.6.3 工作站调试

工业机器人工作站调试是一项复杂的任务，因为需要在调试的每个阶段修复所有错误。软件开发人员和工程师经常使用数字工具来调试程序，以查看和编辑编码语言，其中包含有关程序如何工作的说明。调试允许他们处理代码的各个部分，以确保程序的每个部分都以预期和最佳的方式运行。

1. 调试工业机器人搬运工作站注意事项

1) 所有机器人系统的操作者，都应该参加本系统的培训，学习防护措施和使用机器人的功能。

2) 在开始运行机器人前，确认机器人和外围设备周围没有异常或者危险状况。

3) 在进入操作区域内工作前，即便机器人没有运行，也要关掉电源，或者按下紧急停止按钮。

4) 当在机器人工作区编程时，设置相应看守，保证机器人能在紧急情况迅速停车。示教和点动机器人时不要带手套操作，点动机器人尽量采用低速操作，遇异常情况时可有效控制机器人停止。

5) 必须知道机器人控制器和外围控制设备上的紧急停止按钮的位置，以便在紧急情况下能准确地按下这些按钮。

6) 永远不要认为机器人处于静止状态时其程序就已经完成，因为此时机器人很有可能是在等待让它继续运动的输入信号。

7) 在开机运行前，必须知道机器人根据所编程序将要执行的全部任务。

8) 必须知道所有会左右机器人移动的开关、传感器和控制信号的位置和状态。

3.6.3 Workstation Commissioning

The commissioning of industrial robot workstations is a complex task as all errors need to be fixed at each stage of commissioning. Usually, software developers and engineers adopt digital tools to perform the commissioning of program to view and edit the coding language, which includes instructions on how the program works. During commission, they are allowed to handle various parts of the codes to ensure that each part of the program runs in the expected and optimal manner.

1. Precautions for commissioning of industrial robot handling workstations

1) All operators of the robot system should participate in the training of the system to learn about the protective measures and functions of the robot.

2) Before starting to operate the robot, make sure that there are no abnormal or dangerous conditions around the robot and peripheral devices.

3) Before entering the operating area to work, turn off the power or press the emergency stop button even if the robot is not running.

4) When programming in the robot workspace, set up corresponding guards to ensure that the robot can quickly stop in an emergency. Do not wear gloves when teaching or jogging the robot. Jog the robot at a speed when possible so that the robot can be effectively controlled to stop in case of an abnormal situation.

5) One has to know the positions of the emergency stop button on the robot controller and peripheral control equipment, and accurately press these buttons in emergency situations.

6) Never assume that its program has been completed when the robot is stationary, as it is likely waiting for an input signal to continue its movement at this point.

7) Before starting the running, one has to know all the tasks that the robot will perform according to the program.

8) One has to know the positions and states of all switches, sensors, and control signals that will control the movement of the robot.

2. 准备工作

(1) 上电前短路与电压检查

利用万用表检查主回路是否存在短路故障，检查主回路与各设备供电电压等级是否正常。

(2) 气路与气压检查

启动工位前检查夹具气路情况，检查各连接器是否存在漏气，检查气管是否有磨损漏气，检查压力表观察压力是否足够。

(3) 工位内安全检查

检查机器人工作区域是否有阻挡物体，有则进行清理。

3. 设备上电

首先合上电源总开关2Q0。

然后根据调试要求分别接通需要调试设备的电源。2Q1是控制工业机器人的电源；3Q1是PLC及插排的电源；3Q2是24V开关电源的控制断路器，控制着触摸屏及各种24V用电设备；Q1是松下伺服驱动器的控制断路器；Q2是V20变频器的控制断路器。

4. 单机调试

1) 单站运行工业机器人条件：

① TP开关置于ON。

② 单步或者非单步执行状态（根据需要进行选择）。

③ 模式开关打到T2模式。

④ 自动模式为本地控制。

⑤ ENABLE UI SIGNAL (UI信号有效)：TRUE（有效）。

⑥ UI[1]—UI[3]为ON。

⑦ UI[8] *ENBL为ON。

2) 检查仓库原料工件是否充足，每次重新运行机器人搬运工作要注意补充原料工件。如果是中断运行程序后再次启动时不需要补充原料工件，否则工业机器人手爪会和工件产生碰撞。

2. Preparation work

(1) Short circuit and voltage check before power on

Use a multimeter to check if there is a short circuit fault in the main circuit, and if the voltage level of the main circuit and each equipment is normal.

(2) Gas circuit and pressure check

Before starting the station, check the gas circuit of the fixture to see whether there is any gas leakage in the connector, the gas tube is worn or leaking, and check the pressure gauge to see if the pressure is sufficient.

(3) Safety inspection inside the station

Check if there are any obstacles in the robot's work area, and clean them if there are any.

3. Power on the equipment

First, turn on the main power switch 2Q0.

Then, connect separately the power supply of the equipment, whose commissioned is conducted according to the commissioning requirements. 2Q1 is the power supply for controlling the industrial robot; 3Q1 is the power supply for the PLC and socket; 3Q2 is the control circuit breaker of the 24V switching power supply, which controls the touch screen and various equipment that require 24V voltage; Q1 is the control circuit breaker of Panasonic servo drive; and Q2 is the control circuit breaker of the V20 frequency converter.

4. Single machine commissioning

1) Conditions for single station operation of industrial robots:

① Set the TP switch to ON.

② Single step or non-single step execution status (select as needed).

③ Turn the mode switch to T2 mode.

④ The automated mode is local control.

⑤ ENABLE UI SIGNAL: (UI signal effective) TRUE (effective).

⑥ UI [1]–UI [3] is ON.

⑦ UI [8] * ENBL is ON.

2) Check whether the raw materials of workpieces in the store are sufficient, and replenish them every time the robot is restarted for handling. If the program is interrupted and restarted, do not replenish the raw materials of workpieces, otherwise the gripper of the industrial robot will collide with the workpieces.

3) 注意仿真程序与实际程序的区别。仿真程序每次搬运的原料工件是重复在某个点进行搬运，实际编写工业机器人搬运程序时，应该使用偏置指令 offset 改变下一次搬运时的位置。

5. 联机调试

1) 联机运行工业机器人条件：

① TP 开关置于 OFF。
② 非单步执行状态。
③ 模式开关打到自动模式。
④ 自动模式为远程控制。
⑤ ENABLE UI SIGNAL（UI 信号有效）：TRUE（有效）。
⑥ UI[1]—UI[3] 为 ON。
⑦ UI[8] *ENBL 为 ON。
⑧ 系统变量 $RMT_MASTER 为 0（默认值是 0）。

2) 在单机调试完成后，检查安全插销、安全光栅、安全门急停、传动带急停、操作面板急停、上料台急停是否被按下。

3) 检查设备上的报警是否已经消除。

4) 检查仓库原料工件是否充足，每次重新运行机器人搬运工作要注意补充原料工件。

5) 工业机器人的 I/O 分配机架号使用是否正确。

6) 上一台机器人与下一台机器人之间的联机允许接线是否已经连接，且信号能正常通断。

7) 确定工位不存在危险因素后，若不存在报警即按启动、停止、急停按钮对工位进行操作。

3) Pay attention to the difference between simulation programs and actual ones. For simulation programs, the raw materials of workpieces are repeatedly handled at a certain same point, while writing actual programs of the industrial robot handling, the offset command shall be used to change the position of the next handling.

5. Online commissioning

1) Conditions for online operation of industrial robots:

① Set the TP switch to OFF.
② Non single step execution status.
③ Turn the mode switch to the automated mode.
④ The automated mode is remote control.
⑤ ENABLE UI SIGNAL (UI signal effective): TRUE (effective).
⑥ UI [1]–UI [3] is ON.
⑦ UI [8] * ENBL is ON.
⑧ System Variable $RMT_ MASTER is 0 (default value is 0).

2) When the single machine commissioning is completed, check whether the following things have been pressed: safety pin, safety grating, the emergency stop of the safety door, conveyor belt, and operation panel, as well as the loading table.

3) Check if the alarm on the device has been eliminated.

4) Check whether the raw materials of workpieces in the store are sufficient, and replenish them every time the robot is re-started for handling.

5) Whether the use of the I/O allocation rack number of the industrial robot is correct.

6) Whether the wiring of online permit between the previous and the next robot is connected, and whether the signal can be connected or disconnected normally.

7) When confirming that there are no hazardous factors at the station and there is no alarm, press the start, stop, or emergency stop buttons to operate the station.

任务 3.7 技术交底材料的整理和编写
Task 3.7 Organization and Compilation of Technical Disclosure Materials

【知识目标】

[Knowledge Objectives]

1. 掌握工业机器人搬运工作站技术交底材料的编写步骤与方法。

2. 掌握工业机器人搬运工作站技术交底材料的清单整理。

1. To master the compilation steps and methods of technical disclosure materials for industrial robot handling workstations.

2. To master the organization of list of technical disclosure materials for industrial robot handling workstations.

【技能目标】

[Skill Objectives]

1. 根据工业机器人搬运工作站技术交付材料整理全套说明书。

2. 能够按照交底材料的要求完成方案整理。

1. To be able to organize a complete set of manuals according to the delivered technical materials for the industrial robot handling workstations.

2. To be able to complete the plan organization according to the requirements of the disclosure materials.

【素质目标】

[Competence Objectives]

1. 使用规范的行文格式整理资料。
2. 培养良好的行为习惯。

1. To organize documents with normative formats.
2. To cultivate good behavioral habits.

【任务情景】

[Task Scenario]

某工业机器人系统集成企业为汽车企业建立了一条工业机器人搬运工作站,已全部完成安装调试等施工任务,现需要提供给汽车生产企业关于该工作站的技术资料,你作为工业机器人系统集成商的技术员,请完成该工作站的技术资料整理与编写。

An industrial robot system integrator has built an industrial robot handling workstation for an automobile manufacturer, and completed the tasks of installation and commissioning. Now it is necessary to provide the automobile manufacturer with technical documents about the workstation. As a technician of the industrial robot system integrator, you are required to complete the organization and compilation of the technical documents of the workstation.

【任务分析】

[Task Analysis]

技术交底是企业极为重要的一项技术管理工作,是施工方案的延续和完善,也是项目质量预控的最后一道

Technical disclosure is an extremely important technical management task for enterprises, which is the continuation and improvement of construction schemes, and also the final checkpoint for pre-control of the pro-

关口。其目的是使参与项目施工的技术人员熟悉和了解所承担的项目的特点、设计意图、技术要求、施工工艺及应注意的问题。

通过技术交底，使参与项目施工操作的每一个工人了解自己所要完成的分项工程的具体工作内容、操作方法、施工工艺、质量标准和安全注意事项等，做到施工操作人员任务明确、心中有数，达到有序地施工，以减少各种质量通病，提高施工质量的目的。

ject quality. The purpose of technical disclosure is to make the technicians involved in the project construction get familiar with and understand the characteristics, design intent, technical requirements, and construction techniques of the project and the precautions of its implementation.

Through technical disclosure, every worker involved in the project construction understands the specific work content, operation methods, construction techniques, quality standards, and safety precautions of the the works they need to complete, so that they can carry out the construction in order as they are clear of their tasks and what to be done properly, which reduces various common quality problems and improves construction quality.

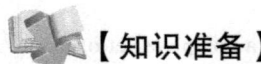【知识准备】

[Assumed Knowledge]

3.7.1 主要技术资料

3.7.1 Main Technical Documents

工业机器人搬运工作站常见的技术资料主要有以下这些：

(1) 工作站操作说明书

工作站操作说明书要包含以下内容：

1) 工作站概况和基本软硬件组成。

2) 基本操作流程、关键性的技术及操作中可能会存在的问题。

3) 特殊设备的操作处理细节及其操作须知。

4) 工作站开关机流程及注意事项。

5) 工作站常见故障的现象描述及处理方法。

6) 如果可以最好能将工作站的程序也列出，别做出注解。

(2) 工作站全套图样

应将方案、设计、施工阶段的所有相关图样都整理好交付给使用方。

1) 工作站布局图及网络拓扑图。

2) 电气原理图及接线图。

The common technical documents of industrial robot handling workstations mainly include the following:

(1) Workstation operating manual

The workstation operating manual should include the following content:

1) Overview of the workstation and composition of basic software and hardware.

2) Basic operating procedures, key technologies, and potential problems in operation.

3) Details and instructions for the operation and handling of special equipment.

4) The startup and shutdown processes of the workstation and their precautions.

5) Description and handling methods of common faults in workstations.

6) If possible, it would be better to also list the programs of the workstation without making annotations.

(2) Complete set of drawings of the workstation

All relevant drawings for stages of scheme, design, and construction shall be organized and delivered to the user.

1) Workstation layout and network topology diagram.

2) Electrical schematic diagram and wiring diagram.

3) 工作站系统安装图。
4) 非标准零件图及装配图。
(3) 工作站内设备程序

应将调试后的设备程序如工业机器人程序、PLC 程序、HMI 工程文件以及变频设置文件等全部整理并标注好后交付使用方。

(4) 工作站设备说明书

应将工作站中使用的成品设备说明书整理好后交付给使用方，以确保使用方在使用过程中方便地查阅资料。

3.7.2 操作说明书的编写

工作站的技术资料编写工作，主要是操作说明书的编写，使用户通过工作站的操作说明书能够操作设备、调试与维护设备、处理设备简单故障。以 FANUC 工业机器人搬运工作站操作说明书为例，了解工作站操作说明书的编写方法。

1. 目录

典型工业机器人搬运工作站操作说明书如图 3-40 所示。

2. 正文

这里以《FANUC 工业机器人搬运工作站设备使用和维护手册》第一章的内容编写为例，为大家演示如何进行操作说明书的编写。操作说明书的其他内容请同学们按要求自己编写。

3) Installation diagram of workstation system.
4) Drawings of non-standard parts and their assembly.
(3) Programs of the workstation equipment

All equipment programs whose commissioning has been done such as industrial robot programs, PLC programs, HMI engineering files, frequency conversion and servo settings files shall be organized and labeled before being delivered to the user.

(4) Workstation equipment manual

The manuals for the finished equipment used in the workstation shall be organized and delivered to the user to ensure easy access to the information during use.

3.7.2 Compilation of Operating Manual

The compilation of technical documents of the workstation mainly involves the compilation of the operating manual, which enables the user to operate equipment, and conduct commissioning, maintenance, and simply troubleshooting of the equipment. The operating manual of the FANUC industrial robot handling workstation is taken as an example to help understand the compilation method of the workstation operating manual.

1. Table of Contents

The table of contents of the operating manual for typical industrial robot handling workstations is shown in Figure 3-40.

2. Text

The compilation of the content of the first chapter in the *Equipment Use and Maintenance Manual of the FANUC Industrial Robot Handling Workstation* is taken as an example here to demonstrate how to compile the operating manual. Other content of the operating manual shall be compiled by the student as required.

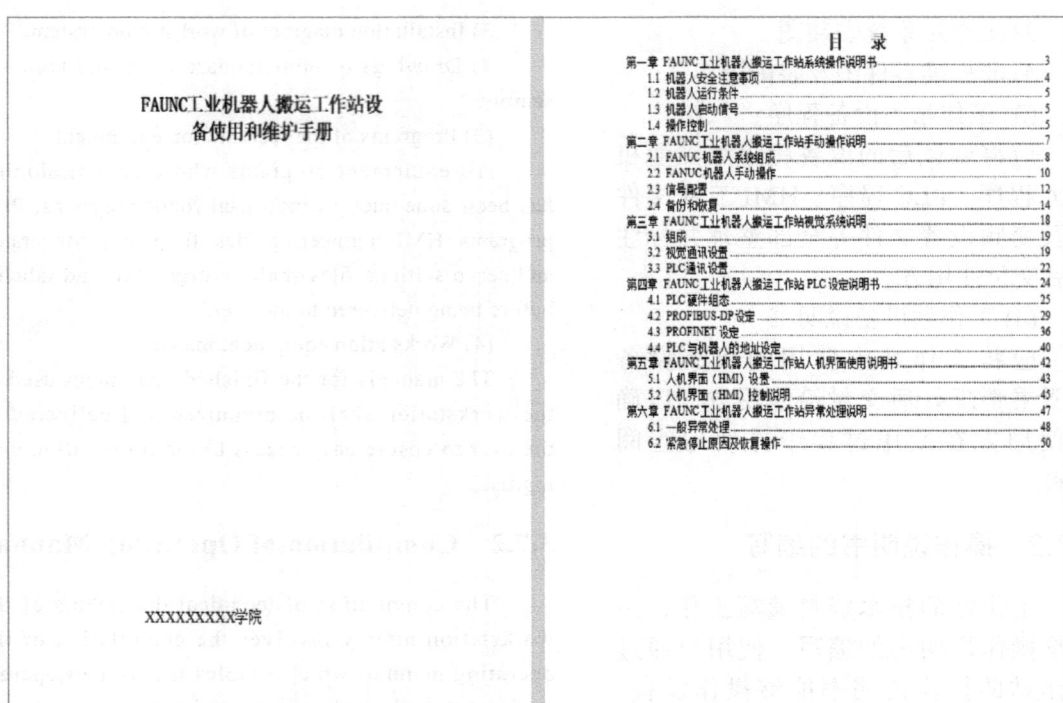

图 3-40 工作站操作说明书
Figure 3-40 Workstation Operating Manual

操作说明书编写示例
Example of compilation of the operating manual

1.1 机器人安全注意事项

本节中将会介绍操作机器人、机器人系统及外围设备应遵守的安全原则和规程。

⚠关闭总电源！在机器人及外围设备的安装、维修、保养时切记要将总电源关闭，带电作业可能会产生致命性后果。如果不慎遭高压电击会导致心搏停止、烧伤或其他严重伤害。

⚠与机器人保持足够安全距离！调试与运行机器人及外围设备时，它可能会执行一些意外的或不规范的运动。并且所有的运动都会产生很大的力量，从而严重伤害个人或损坏机

1.1 Safety Precautions for Robots

This section will introduce the safety principles and procedures that shall be followed when operating robots, robot systems or peripheral devices.

Cut off the main power supply! The main power supply has to be cut off before installation, repair, or upkeep of robots or peripheral equipment can be carried out, as the performance of such operation with power on may lead to fatal consequences. If hit accidentally by high-voltage electric shock, one will suffer from cardiac arrest, burns, or other serious injuries.

Maintain a sufficient safe distance from the robot! When the robot or peripheral device is under commissioning or running, they may make unexpected or non-regulated movements. Moreover, all move-

器人工作范围内的任何设备。所以请时刻警惕与机器人保持足够的安全距离。

⚠️静电放电危险！搬运部件未接地的人员可能会传导大量的静电荷，这一放电过程可能会损坏灵敏的电子设备。

⚠️紧急停止！紧急停止优先于任何其他机器人操纵控制、外围设备控制，因为它会断开机器人及转台的驱动回路，停止所有部件的控制指令。出现以下情况请立即按下任意紧急停止按钮：

1）机器人运动中，工作区域有工作人员。

2）机器人伤害工作人员或损伤了机器或设备。

3）转台旋转时，工作区域有工作人员。

4）机器人与转台没有协调运动，可能会导致相撞。

⚠️灭火！发生火灾时，请确保人员安全撤离后再进行灭火。应首先处理受伤人员。

①工作中的安全。机器人在运动过程中会产生很大的力。停顿或停止都会产生危险，即使可以预测轨迹，但外部信号可能会改变机器人的操作，或者会在没有任何警告的情况下产生意料不到的运动。

因此，进入保护空间时，务必遵守以下所有条例：

1）如果保护空间内有工作人员，请手动操作机器人。

2）当进入保护空间时，请准备好手操器，以便随时控制机器人或拍急停。

3）在靠近机器人时请确保夹具上的设备不会动作或停止动作。

ments generate significant force, which can seriously harm persons or damage any equipment within the robot's working range. Therefore, one shall always be vigilant and maintain a sufficient safe distance from the robot.

Hazardous electrostatic discharge! Persons carrying components that are not grounded may conduct a large amount of static charge, the process of which may damage sensitive electronic devices.

Emergency stop! Emergency stop takes priority over any other robot and peripheral device control, as it will disconnect the drive circuit of the robot and turntable, and stop the control commands of all components. Please press any emergency stop button immediately if:

1) There are workers in the working area when the robot is in movement.

2) The robot has harmed any worker or damaged any machine or equipment.

3) There are workers in the working area when the turntable rotate.

4) There is no coordinated movement between the robot and the turntable, which may cause a collision.

Fire fighting! When a fire occurs, please ensure that persons have safely evacuated before conducting firefighting. The injured person shall be treated firstly.

Safety at work. Robots generate significant force during their movement. Pause or stop may lead to danger, as external signals may change the operation of a robot or cause it make unexpected movements without any warning, even if the trajectory can be predicted.

Therefore, one has to comply with all the following regulations when entering a protected space.

1) If there is any worker in the protected space, please operate the robot manually.

2) When entering the protected space, please bring the hand controller so that it is possible to control the robot or make an emergency stop at any time.

3) Please ensure that the equipment on the fixture does not move or stop when approaching the robot.

4) 注意工件表面，长时间运转会导致机器人表面及电动机表面温度过高。

5) 注意夹具并确保夹好工件。如果夹具松动或掉落，工件脱落会导致人员损伤。

① 示教器的安全！示教器是一种高品质的手持终端，为了避免故障与损伤，请在操作时遵循以下说明：

1) 小心操作，不要摔打、重击、抛掷。

2) 定期清洁触摸屏，切勿使用溶剂，使用软布蘸少量水或中性清洁剂进行擦拭。

3) 切勿用锋利或尖锐物操作触摸屏。

① 手动模式下的安全！在保护空间内工作，请保持手动操作，且电控柜的控制面板使用手动模式。

① 自动模式下的安全！

启动自动模式前，请确定保护空间没有工作人员，所有外围设备处于就位状态。

1.2 机器人运行条件

1) 插上安全门插销。

2) 护栏上急停按钮复位。

3) 护栏上安全光栅复位。

4) 操作面板上的急停按钮复位。

5) 机器人控制箱的急停按钮复位。

6) 示教器上的急停按钮复位。

7) 机器人控制箱的J5A接线端子的急停信号正确接线。

信号条件可以通过"PLC输入"画面来查看，也可以通过"报警信息"来查看。

4) Pay attention to the surface of the workpiece, as prolonged operation can cause excessive temperature on the surface of the robot and motor.

5) Pay attention to the fixture and ensure that it clamps the workpiece properly, due to the fact that if the fixture becomes loose or falls, glass detachment may happen and cause personal injury.

Safety of the teach pendant! Please follow the following instructions when operating a teach pendant which is a high-quality hand-held terminal to avoid its malfunctions and damage:

1) Operate with caution and avoid falling, hitting, or throwing.

2) Regularly clean the touch screen by wiping with a soft cloth dipped in a small amount of water or neutral cleaning agent. Do not use solvents.

3) Do not operate the touch screen with sharp or keen-edged objects.

Safety under the manual mode! Please maintain manual operation when operating in the protected space and the control panel of the electric control cabinet is under the manual mode.

Safety under the automated mode!

Before starting the automated mode, please ensure that there is no any worker in the protected space and that all peripheral devices are in place.

1.2 Operation Conditions of the Robot

1) Insert the safety door latch.

2) Reset the emergency stop button on the guardrail.

3) Reset the safety grating on the guardrail.

4) Reset the emergency stop button on the operation panel.

5) Reset the emergency stop button of the robot control box.

6) Reset the emergency stop button on the TP teach pendant.

7) The emergency stop signal of the J5A terminal of the robot control box is correctly wired.

Signal conditions can be viewed through the "PLC Input" screen or the "Alarm Information" screen.

1.3 机器人启动信号

1. 前提

机器人与PLC通信正常，具备运行条件。

2. 机器人自动运行必备信号

1) UI[1] 急停信号。
2) UI[2] 暂停信号。
3) UI[3] 安全速度信号。
4) UI[8] 使能信号。

3. 启动方式

1) PNS 启动。
2) RSR 启动。

启动方式中的信号均是脉冲型，本次使用了UI[9]RSR1作为启动信号。

1.4 操作控制

1. 单站控制

1) 操作电箱的按钮与HMI触摸屏的按钮功能一致，如"启动""停止""暂停"和"复位"。

2) 按启动按钮，满足机器人的启动条件后，机器人运行，绿色按钮灯亮的同时实训平台上的三色灯的绿色灯亮。

3) 由于机器人没有所谓的停止信号，因此，单击暂停与单击停止的功能是一样的。机器人暂停时，三色灯的黄色灯亮。

4) 机器人启动后，接着暂停，可以通过暂停再启动运行机器人。

5) "单动控制"则是机器人本身作为一个独立的工作站运行。

2. 联动控制

1) 单击机器人控制栏"单动控制"切换到"联动控制"，机器人会作为其中一个搬运工动作。

2) 单击传送带控制栏"手动控制"切换到"自动控制"，传送带作为一个搬运桥梁。

1.3 Robot Start Signal

1. Precondition

The robot and PLC communicate normally, with readiness for running.

2. Necessary signals for the automatic running of robot

1) UI [1] emergency stop signal.
2) UI [2] pause signal.
3) UI [3] safe speed signal.
4) UI [8] enabling signal.

3. Starting mode

1) PNS start.
2) RSR start

The signals in both the starting modes are pulse type, UI [9] RSR1 is used as the start signal this time.

1.4 Operating Control

1. Single workstation control

1) The functions of the buttons of the operating electrical box are the same as that of the buttons of the HMI touch screen, i. e. " Start", " Stop", " Pause", and "Reset".

2) Press the " Start" button and, when the starting conditions of the robot are met, the robot will run, and the green button light and the green light of the tricolor light on the training platform will be on.

3) Since there is not a so-called stop signal for the robot, the function of clicking " Pause " is the same as that of clicking " Stop", and the yellow light of the tricolor light is on when the robot pauses.

4) When the robot starts and pauses, it is possible to restart the robot by pressing the "Pause" button.

5) Single action control refers to the robot itself running as an independent workstation.

2. Online control

1) Click " Single Action Control " on the robot control bar to switch to " Online Control ", and the robot will act as one of the porters.

2) Click " Manual Control " on the conveyor belt control bar to switch to " Automatic Control ", and the conveyor belt will serves as a handling bridge.

3) When starting, the 1# robot takes out the car door trim panel with the suction cup tool from the store and place it on the welding platform, where the panel

3）启动时，1#机器人使用吸盘工具将汽车门饰板从仓库中取出放到焊接平台上由夹具夹紧。然后，1#机器人更换工具对汽车门饰板进行点焊作业。点焊结束、夹具打开后，吸盘工具放回 1#机器人，1#机器人将汽车门饰板放在传送带上。传送带检测到有物料，然后开始传送汽车门饰板至另一端，另一端检测到有物料，传送带停止。2#机器人将汽车门饰板从传送带取走，并搬运到视觉检测位置进行焊点质量及数量的检查。检查完成后，2#机器人将汽车门饰板放至对应的入库位置。

注意：机器人的控制与机器人的编程有关系。

【项目测试】

在两个工业机器人工作站协作完成控制任务：工作站中 PLC 与机器人采用 PROFINET（或者 Profibus）通信，PLC 与触摸屏采用 PROFINET 通信。工业机器人工作站可以实现单站完成搬运、视觉检测（颜色或者形状）、码垛或拆垛工作，也可以两个工业机器人工作站配合共同完成。具体要求如下：

1. 单站控制

1）操作电箱的按钮与 HMI 触摸屏的按钮功能一致，如"启动""停止""暂停"和"复位"。

2）按启动按钮，满足机器人的启动条件后，机器人运行，绿色按钮灯亮的同时实训平台上的三色灯的绿色灯亮。

3）由于机器人没有所谓的停止信号，因此，单击暂停与单击停止的功能是一样的。机器人暂停时，三色灯的黄色灯亮。

will be clamped with the fixture. Then, the 1# robot replaces the tool to perform spot welding on the car door trim panel. When the spot welding is completed and the fixture is opened, the suction cup tool will be put back to the 1# robot which will then place the car door trim panel on the conveyor belt. When detecting the existence of material, the conveyor belt begins to convey the car door trim panel to the other end, and will stop when the existence of material is detected there. The 2 # robot takes the car door trim panel from the conveyor belt and handles it to the visual detection position for inspection of the quality and quantity of the welding points. When detection is completed, the 2 # robot will place the car door trim panel to a corresponding position for store-input.

Attention: the control of the robots is related to its programming.

[Program Testing]

Complete the control task in the collaboration between two industrial robot workstations: for the workstations, the PLC communicates with the robot by PROFINET (or Profibus), and with the touch screen by PROFINET. The work of handling, visual detection (color or shape), and stacking or unstacking can be completed either by a single industrial robot workstation or by collaboration between the two workstations. Specific requirements are as follows:

1. Single workstation control

1) The functions of the buttons of the operating electrical box are the same as that of the buttons of the HMI touch screen, i. e. "Start", "Stop", "Pause", and "Reset".

2) Press the "Start" button and, when the starting conditions of the robot are met, the robot will run, and the green button light and the green light of the tricolor light on the training platform will be on.

3) Since there is not a so-called stop signal for the robot, the function of clicking "Pause" is the same as that of clicking "Stop", and the yellow light of the tricolor light is on when the robot pauses.

4) When the robot starts and pauses, it is possible to restart the robot by pressing the "Pause" button.

5) Single action control refers to the robot itself

4) 机器人启动后,接着暂停,可以通过暂停再启动运行机器人。

5) "单动控制"则是机器人本身作为一个独立的工作站运行。

2. 联动控制

1) 单击机器人控制栏"单动控制"切换到"联动控制",机器人会作为其中一个搬运工动作。

2) 单击传送带控制栏"手动控制"切换到"自动控制",传送带作为一个搬运桥梁。

3) 启动时,1#机器人使用吸盘工具将汽车门饰板从仓库中取出放到焊接平台上由夹具夹紧。然后,1#机器人更换工具对汽车门饰板进行点焊作业。点焊结束、夹具打开后,吸盘工具放回1#机器人,1#机器人将汽车门饰板放在传送带。传送带检测到有物料,然后开始传送汽车门饰板至另一端,另一端检测到有物料,传送带停止。2#机器人将汽车门饰板从传送带取走,并搬运到视觉检测位置进行焊点质量及数量的检查。检查完成后,2#机器人将汽车门饰板放至对应的入库位置。

running as an independent workstation

2. Online control

1) Click "Single Action Control" on the robot control bar to switch to "Online Control", and the robot will act as one of the porters.

2) Click "Manual Control" on the conveyor belt control bar to switch to "Automatic Control", and the conveyor belt will serves as a handling bridge.

3) When starting, the 1# robot takes out the car door trim panel with the suction cup tool from the store and place it on the welding platform, where the panel will be clamped with the fixture. Then, the 1# robot replaces the tool to perform spot welding on the car door trim panel. When the spot welding is completed and the fixture is opened, the suction cup tool will be put back to the 1# robot which will then place the car door trim panel on the conveyor belt. When detecting the existence of material, the conveyor belt begins to convey the car door trim panel to the other end, and will stop when the existence of material is detected there. The 2# robot takes the car door trim panel from the conveyor belt and handles it to the visual detection position for inspection of the quality and quantity of the welding points. When detection is completed, the 2# robot will place the car door trim panel to a corresponding position for store-input.

项目 4　典型工业机器人弧焊工作站系统的设计及应用

Program 4　Design and Application of a Typical Industrial Robot Arc Welding Workstation System

【项目场景】

智能制造焊接车间为了完成焊接任务需要设计一个有 3 台 6 轴 FANUC M-10iA 弧焊焊接机器人，配有焊接电源、送丝机及专用焊枪，可分别实现点焊或者 TIG 焊接。其可以独立完成焊接工作，也可以使用在自动化生产线上，作为焊接工序的一个工艺部分，成为生产线上一个具有焊接功能的一个工作"站"。典型的机器人弧焊工作站主要包括：机器人系统（机器人本体、机器人控制柜、示教盒）、焊接电源系统（焊机、送丝机、焊枪、焊丝盘支架）、焊枪防碰撞传感器、变位机、焊接工装系统（机械、电控、气路/液压）、清枪器、控制系统(PLC 控制柜、HMI 触摸屏、操作台）、安全系统（围栏、安全光栅、安全锁）和排烟除尘系统（自净化除尘设备、排烟罩、管路）等。常见的机器人弧焊工作站如图 4-1 所示。

[Program Scenario]

In order to complete welding tasks, it is required to design a robot arc welding workstation for an intelligent manufacturing welding workshop. The workstation consists of three 6-axis FANUC M-10iA arc welding robots, and is equipped with a welding power supply, wire feeder, and dedicated welding gun, for the realization of spot welding or TIG welding. The workstation can complete welding independently, or be used on an automated production line as a part of the welding process, thus becoming a work " station" with welding functions on the production line. A typical robot arc welding workstation mainly includes: robot system (robot body, robot control cabinet, teach box), welding power supply system (welding machine, wire feeder, welding gun, welding wire reel support), anti-collision sensor for the welding gun, positioner, tooling system (mechanical, electrical control, pneumatic/hydraulic) for welding, gun cleaner, control system (PLC control cabinet, HMI touch screen, operating console), safety systems (fence, safety grating, safety lock), and smoke exhaust system, as well as dedusting system (selfcleaning dedusting equipment, smoke exhaust hood, pipelines).A common robot arc welding workstation is shown in Figure 4-1.

项目 4　典型工业机器人弧焊工作站系统的设计及应用

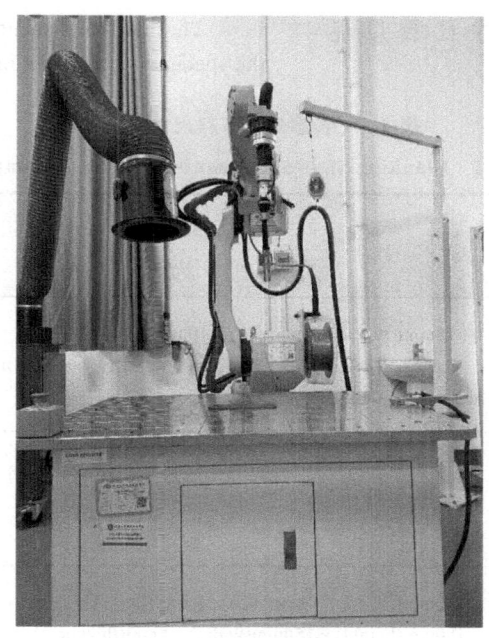

图 4-1　机器人弧焊工作站
Figure 4-1　A Robot Arc Welding Workstation

 【项目描述】

智能制造焊接车间接到一个任务单，需要给学校实训室焊接一批全由低碳钢板、管组件组装焊接成的全封闭压力容器。钢板厚度有 8mm、10mm 两种，钢管规格为 $\phi 51\text{mm} \times 4\text{mm}$。容器结构示意图如图 4-2 所示。

[Program Description]

An intelligent manufacturing welding workshop has received a task order from a training room for the welding of a batch of fully enclosed pressure vessels. The vessels shall be made entirely of low-carbon steel plates and pipes which are to be assembled and welded together. The thickness of the steel plate are 8mm and 10mm respectively, the specifications of the steel pipe are $\phi 51\text{mm} \times 4\text{mm}$, and the schematic diagram of the vessel structure is shown in Figure 4-2.

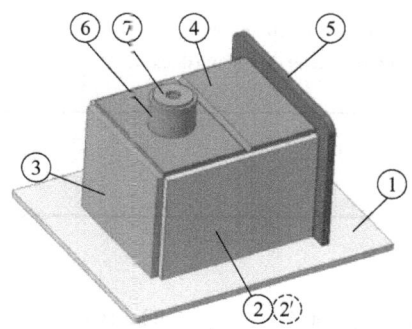

图 4-2　碳钢容器试件
Figure 4-2　A Carbon Steel Vessel Specimen

试件材料下料尺寸及数量见表 4-1。

The blanking quantity and size requirements for the specimen are shown in Table 4-1.

表 4-1 各零部件下料尺寸及数量
Table 4-1 Blanking Sizes and Quantities of Each Component

编号 S/N	名称 Name	尺寸描述 (mm) Size Description (mm)	数量 Qty	材质 Material
1	底板 Base plate	318mm(长)×266mm(宽)×10mm(厚) 318mm (length) × 266mm (width) × 10mm (thickness)	1	Q235/Q345B
2	左右立板 Left and right vertical plates	200mm(长)×180mm(高)×8mm(厚) 200mm(length) × 180mm(height) × 8mm(thickness)	2	Q235B/Q345B
3	前立板 Front vertical plate	180mm(高)×150mm(宽)×8mm(厚) 180mm(height) × 150mm(width) × 8mm(thickness)	1	Q235B/Q345B
4	盖板 Cover plate	150mm(长)×100mm(宽)×8mm(厚), 150mm (length) × 100mm (width) × 8mm (thickness), 在长边单侧开单 V 型坡口 (30°±2.5°) Open a single V-shaped groove of 30° ± 2.5 °on one of the long sides	2	Q235B/Q345B
5	后立封板 Rear vertical sealing plate	226mm(长)×218(宽)×10mm(厚) 226mm(length) × 218 (width) × 10mm (thickness)	1	Q235B/Q345B
6	管 Pipe	51mm(管径)×4mm(壁厚) 51mm (pipe diameter) × 4mm (wall thickness)	1	20g
7	管端盖板 Pipe end cover plate	43mm(直径)×10mm(厚) 43mm (diameter) × 10mm (thickness)	1	Q235B/Q345B
合计 Total			9	

试件二维图如图 4-3 ~ 图 4-5 所示。

The two-dimensional drawing of the specimen is shown from Figure 4-3 to Figure 4-5.

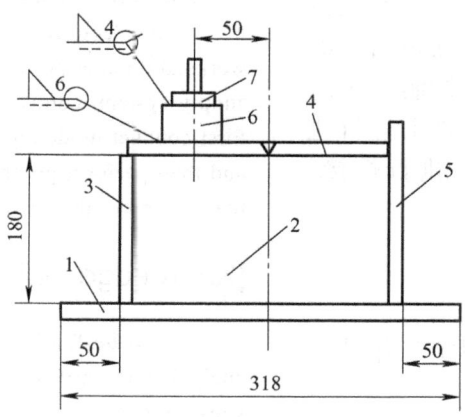

图 4-3 零件序号与焊缝代号 1
Figure 4-3 Part Number and Welding Seam Code 1

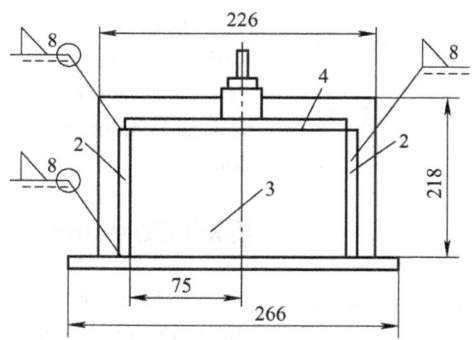

图 4-4 零件序号与焊缝代号 2
Figure 4-4 Part Number and Welding Seam Code 2

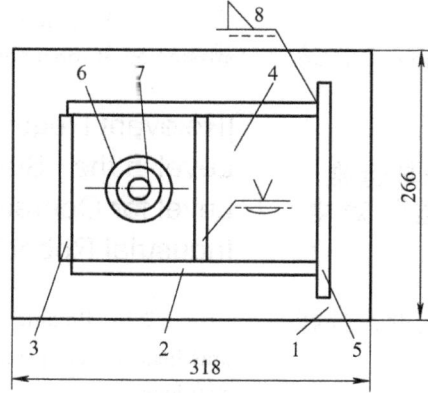

图 4-5 零件序号与焊缝代号 3
Figure 4-5 Part Number and Welding Seam Code 3

现在需要根据焊接任务进行工艺分析及硬件选型，设计一个典型机器人弧焊焊接工作站。工作站包括机器人系统、焊接电源系统、控制系统以及安全系统等。首先在软件上进行建模仿真，然后编程调试并进行焊接，最终完成任务。

Now, it is required to design a typical robot arc welding workstation based on the process analysis and hardware selection according to the welding task. The workstation consist of the robot system, welding power supply system, control system, and safety system, etc. First conduct modeling and simulation on the software, and then perform programming and commissioning for the welding of the device to complete the task.

【知识目标】

1. 熟悉机器人弧焊焊接工作站工艺要求分析及硬件选型。

2. 熟悉机器人弧焊焊接工作站设计方案的编写。

3. 了解机器人弧焊焊接工作站施工图的设计及建模。

4. 掌握机器人弧焊焊接工作站系统的仿真。

5. 熟悉焊接机器人弧焊工作站程序的编写及安装与调试。

[Knowledge Objectives]

1. To be familiar with the process requirement analysis and hardware selection for robot arc welding workstations.

2. To be familiar with the compilation of design schemes for robot arc welding workstations.

3. To understand the design and modeling of detailed drawings for robot arc welding workstations.

4. To master the simulation of robot arc welding workstation systems.

5. To be familiar with the programming, installation, and commissioning of robot arc welding workstations.

【技能目标】

1. 能说出弧焊机器人的工艺要求及硬件选择依据。

2. 能编写机器人弧焊焊接工作站的设计方案。

3. 能利用仿真软件对机器人弧焊焊接工作站进行仿真。

4. 能编写机器人弧焊焊接工作站程序并进行调试。

[Skill Objectives]

1. To be able to describe the process requirements and hardware selection criteria for arc welding robots.

2. To be able to compile design schemes for robot arc welding workstations.

3. To be able to conduct the simulation of arc welding robot workstations by simulation software.

4. To be able to conduct programming and commissioning of robot arc welding workstations.

【"工业机器人操作与运维职业技能等级标准"对中级的相关要求】

2.2.1 能识读机械装配图，选择机械零部件并规划位置。

2.2.2 能识读电气线路图，选择电气元件并规划位置。

[Relevant Requirements for Intermediate Level in the "Standard of Vocational Skill Level for Operation and Maintenance of Industrial Robot"]

2.2.1 Be able to recognize mechanical assembly drawings, select mechanical components, and plan their positions.

2.2.2 Be able to recognize electrical circuit drawings, select electrical components, and plan their positions.

2.2.4 能根据工业机器人典型工作站工艺指导文件完成装配。

2.1.5 能通过编程完成对装配物品的定位、夹紧和固定。

2.1.3 能通过手动或自动模式控制机器人末端执行器对工件进行焊接、打磨抛光等操作。

2.1.4 能通过编程控制焊接、打磨抛光等复杂工艺周边外围设备进行协同运动。

2.2.4 Be able to complete assembly according to the process guidance documents of typical industrial robot workstations.

2.1.5 Be able to conduct positioning, clamping, and fixing of the assembled items through programming.

2.1.3 Be able to control the robot end effector to perform such operations as welding, and polishing on the workpiece under the manual or automatic mode.

2.1.4 Be able to control collaborative movement by peripheral devices through programming for complex processes such as welding, and polishing.

任务 4.1 工艺要求分析及硬件选型
Task 4.1 Process Requirement Analysis and Hardware Selection

【知识目标】

1. 了解弧焊机器人的特点。
2. 熟悉弧焊机器人的工艺要求分析。
3. 熟悉弧焊机器人的硬件选型依据。

[Knowledge Objectives]

1. To understand the characteristics of arc welding robots.
2. To be familiar with the process requirement analysis of arc welding robots.
3. To be familiar with the hardware selection basis of arc welding robots.

【技能目标】

1. 能说出弧焊机器人的工艺要求。
2. 能选择合适的弧焊机器人硬件。

[Skill Objectives]

1. To be able to describe the process requirements for arc welding robots.
2. To be able to choose suitable hardware for arc welding robots.

【素质目标】

1. 培养学生具有一定的全局观念，以及信息收集和处理能力，分析、解决问题能力和交流、合作能力。
2. 引导学生树立职业理想，增强学生的家国情怀。

[Competence Objectives]

1. To cultivate students to have a proper holistic perspective, and be able to collect and process information, analyze and solve problems, as well as communicate and cooperate with others.
2. To guide students to establish their career aspirations when enhancing their patriotism.

【任务情景】

焊接车间接到一批来料焊接任务，需要把两块尺寸为 50mm × 200mm × 3mm 的 Q235 钢板，通过对接 I 形坡

[Task Scenario]

A welding workshop has received a task of welding of a batch of materials, which are two steel plates of Q235 with the size of 50mm × 200mm × 3mm, with

口焊接方式进行焊接，焊接的位置是水平位置焊接。

技术要求：

1) 采用 CO_2 作为保护气体，使用 $\phi 1.0mm$ 的 H08Mn2SiA 焊丝，通过在线示教编程操作机器人完成焊接作业。

2) 焊缝质量要求。焊缝外观质量要求见表 4-2。

the horizontal part of I-shaped grooves to be butt welded.

Technical requirements:

1) Operate the robot to complete welding through programming and online teaching. For the welding, CO_2 is used as the protective gas, and H08Mn2SiA welding wires of $\phi 1.0mm$ adopted.

2) Quality requirements for the welding seam. The quality requirements for the welding seam appearance are shown in Table 4-2.

表 4-2 焊缝外观质量要求

Table 4-2 Quality Requirements for the Welding Seam Appearance

检查项目 Inspection item	标准值 /mm Standard value/mm	检查项目 Inspection item	标准值 /mm Standard value/mm
焊缝余高 Reinforcement of welding seam	0～2	焊缝高低差 Height difference of welding seam	0～1
焊缝宽度 Width of welding seam	4～6	错边量 Misalignment amount	0～1
焊缝宽窄差 Width difference of welding seam	0～1	角变形 Angular deformation	0°～3°
咬边 Undercut	深度≤0.5，长度≤15 Depth ≤ 0.5, length ≤ 15	焊缝外观成形 Forming of welding seam appearance	波纹均匀整齐，焊缝成形良好 The corrugation is uniform and neat, and the welding seam is well formed

【任务分析】

在焊接车间，每一个弧焊焊接机器人都由机器人本体、控制柜、焊接电源、示教器、送丝机构、供气系统、焊枪、电缆等组成。所选用的焊接机器人配套设备一般均具有与机器人本体通信的相应接口，以便与机器人本体交换信号，顺利被机器人焊接控制系统调用。已知焊件结构的技术要求、结构尺寸、母材牌号及规格（板厚、管径与壁厚）、接头形式、焊接位置、焊接方法、焊材、气体等。请根据焊接

[Task Analysis]

In the welding workshop, each arc welding robot is composed of the robot body, control cabinet, welding power supply, teach pendant, wire feeding mechanism, gas supply system, welding gun, cable, etc. The selected welding robot supporting equipment generally has corresponding interfaces for communication with the robot body, in order to exchange signals with the robot body and be successfully called by the robot welding control system. What have been known are as follows: the technical requirements, and structural dimensions, as well as the base metal grade and specifications (plate thickness, pipe diameter, and wall thickness) of the weldment structure, joint form, welding position, welding method,

任务进行弧焊机器人焊接工艺分析及弧焊机器人硬件的选择。

【知识准备】

4.1.1 焊接机器人基础知识

工业机器人弧焊工作站由机器人系统、焊枪、焊接电源、送丝装置、焊接变位机等组成，如图 4-6 所示。

welding material, gas, etc. Please analyze the welding process of and select the hardware for the arc welding robot based on the welding task.

[Assumed Knowledge]

4.1.1 Basic Knowledge of Welding Robots

The industrial robot arc welding workstation consists of a robot system, welding gun, welding power supply, wire feeding device, welding positioner, etc., as shown in Figure 4-6.

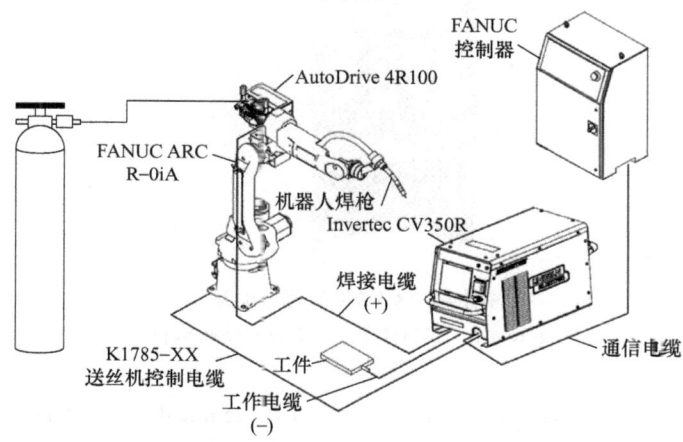

图 4-6 工业机器人弧焊工作站的组成
Figure 4-6 Composition of an Industrial Robot Arc Welding Workstation

(1) 弧焊机器人

弧焊机器人包括 FANUC R-0iA 机器人本体、R-30iA Mate 控制柜以及示教器。FANUC R-0iA 机器人如图 4-7 所示。

(1) Arc welding robot

The arc welding robot includes the FANUC R-0iA robot body, R-30iA Mate control cabinet, and a teach pendant. The FANUC R-0iA robot is shown in Figure 4-7.

图 4-7 FANUC R-0iA 机器人
Figure 4-7 FANUC R-0iA Robot

FANUC R-0iA 机器人为 6 轴弧焊专用机器人，由驱动器、传动机构、机械手臂、关节以及内部传感器组成，可以精确地保证机械手末端执行器所要求的位置、姿态和运动轨迹。焊枪与机器人手臂可直接通过法兰连接。

(2) 弧焊焊接电源

弧焊焊接电源是为电弧焊提供电源的设备。Lincoln Invertec® CV350-R 焊接电源如图 4-8 所示。

The FANUC R-0iA robot is a specialized 6-axis arc welding robot, composed of a driver, transmission mechanism, robotic arm, joints, and internal sensors, which can ensure the accurate position, posture, and motion path required by the end effector of the robotic arm. The welding gun can be directly connected to the robotic arm through a flange.

(2) Arc welding power supply

Arc welding power supply is the equipment that supplies power for arc welding. The Lincoln Invertec® CV350-R welding power supply is shown in Figure 4-8.

图 4-8　CV350-R 焊接电源
Figure 4-8　CV350-R Welding Power Supply

Lincoln Invertec® CV350-R 焊接电源的技术参数见表 4-3。

The technical parameters of Lincoln Invertec® CV350-R welding power supply are shown in Table 4-3.

表 4-3　Lincoln Invertec® CV350-R 焊接电源的技术参数
Table 4-3　Technical Parameters of Lincoln Invertec ® CV350-R Welding Power Supply

项目 Item	规格 Specifications
额定输出电压、相数 Rated output voltage, number of phase	AC 380V、三相 AC380 V three-phase
额定频率 Rated frequency	50 Hz
额定输出功率 Rated output power	18 kV·A
输出电流范围 Output current range	60～350A, 350A/60%
通信方式 Communication mode	ArcLink
焊接波形 Welding waveform	CV
熔接法（焊接方法） Fusion welding method (welding method)	CO_2 短路焊接、MAG/MIG 短路焊接、脉冲焊接 CO_2 short-circuit welding, MAG/MIG short-circuit welding, pulse welding
适用母材 Applicable base material	碳钢、不锈钢、铝 Carbon steel, stainless steel, aluminum

机器人控制柜 R-30iA Mate 通过焊接指令电缆向焊接电源发出控制指令，如焊接参数(焊接电压、焊接电流)、起弧、息弧等。

(3) 焊枪

焊枪将焊接电源的大电流产生的热量聚集在焊枪的终端来熔化焊丝，熔化的焊丝渗透到需焊接的部位，冷却后，被焊接的物体牢固地连接成一体。

FANUC R-0iA 机器人安装的焊枪型号为 SRCT-308R，内置防撞传感器，如图 4-9 所示。

The robot control cabinet R-30iA Mate sends control commands such as welding parameters (welding voltage, welding current), arc starting, and arc extinguishing to the welding power supply through the welding command cable.

(3) Welding gun

The welding gun gathers the heat generated by the high current of the welding power supply at the terminal of the welding gun to melt the welding wire. The melted welding wire penetrates into the area to be welded and, when cooled, the welded objects are firmly joint together.

The FANUC R-0iA robot is equipped with the SRCT-308R welding gun which has a built-in anti-collision sensor. The appearance of the welding gun is shown in Figure 4-9.

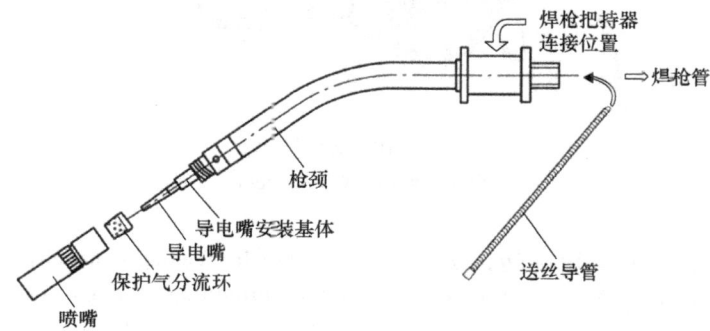

图 4-9 焊枪

Figure 4-9 Welding Gun

SRCT-308R 型焊枪的技术参数见表 4-4。

The technical parameters of the SRCT-308R welding gun are shown in Table 4-4

表 4-4 SRCT-308R 型焊枪的技术参数

Table 4-4 Technical Parameters of the SRCT-308R Welding Gun

项目 Item	参数 Parameter
额定电流 Rated current	350A(CO_2) 300A(MAG)
使用率 Usage rate	60%
适用焊丝直径 Applicable wire diameter	0.8～1.2mm
冷却方式 Cooling method	空冷 Air cooling
电缆长度 Cable length	0.8～5m

(4) 送丝机

送丝机是在微机控制下可以根据设定参数连续稳定地送出焊丝的自动化送丝装置。送丝机如图4-10所示，主要由送丝电动机、压紧机构、送丝滚轮（主动轮、从动轮）等组成。

(4) Wire feeder

The wire feeder is an automated wire feeding device controlled by a microcomputer that can continuously and stably send welding wires according to the set parameters. The wire feeder is shown in Figure 4-10, mainly consists of a wire feeding motor, a compression mechanism, wire feeding rollers (driving wheel, driven wheel), etc.

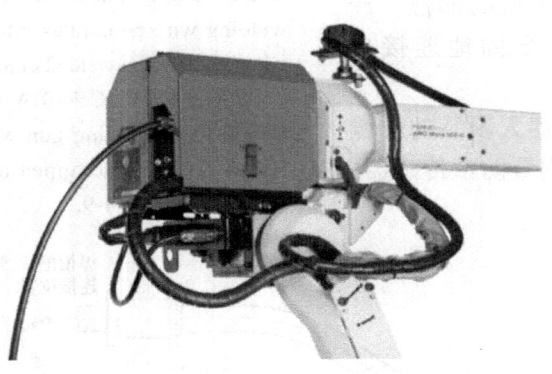

图 4-10 送丝机
Figure 4-10 Wire Feeder

送丝电动机驱动主动轮旋转，为送丝提供动力。从动轮将焊丝压入送丝轮上的送丝槽，增大焊丝与送丝轮的摩擦，将焊丝修整平直，平稳送出，使进入焊枪的焊丝在焊接过程中不会出现卡丝现象。

(5) 焊接变位机

焊接变位机承载工件及焊接所需工装，主要作用是实现焊接过程中将工件进行翻转变位，以便获得最佳的焊接位置，可缩短辅助时间，提高劳动生产率，改善焊接质量，是机器人焊接作业不可缺少的周边设备。焊接变位机如图4-11所示。

如果采用伺服电动机驱动变位机翻转，焊接变位机可作为机器人的外部轴，与机器人实现联动，达到同步运行的目的。

The wire feeding motor drives the driving wheel to rotate, providing an impetus for wire feeding. The driven wheel presses the welding wire into the wire feeding groove on the wire feeding roller, which increases the friction between the welding wire and the wire feeding roller and trims straight the welding wire before sending it out smoothly, so that the welding wire entering the welding gun will not get stuck during the welding process.

(5) Welding positioner

The welding positioner carries the workpieces and required tools for welding, with the main function of flipping and repositioning the workpieces during the welding process for optimal welding positions. It is an indispensable peripheral equipment for robot welding as it can shorten the auxiliary time, and improve labor productivity and welding quality, with the appearance as shown in Figure 4-11.

If a servo motor is used to drive the positioner to flip, the welding positioner can serve as the external axis of the robot and achieve synchronous operation with the robot.

图 4-11　焊接变位机
Figure 4-11　Welding Positioner

(6) 焊丝盘架

盘状焊丝可装在机器人 S 轴上，也可装在地面上的焊丝盘架上。焊丝盘架用于焊丝盘的固定，如图 4-12 所示。焊丝从送丝套管中穿入，通过送丝机构送入焊枪。

(6) Welding wire reel holder

The reel shaped welding wire can be installed on the S-axis of the robot or on the welding wire reel holder which is on the ground and used for fixing the welding wire reel, as shown in Figure 4-12. The welding wire passes through the wire feeding sleeve and is fed into the welding gun through the wire feeding mechanism.

图 4-12　焊丝盘
Figure 4-12　Welding Wire Reel

(7) 保护气气瓶总成

气瓶总成由气瓶、减压器、PVC 气管等组成，如图 4-13 所示。气瓶出口处安装了减压器，减压器由减压机构、加热器、压力表和流量计等部分组成。气瓶中装有 $80\%CO_2+20\%Ar$ 的保护焊气体。

(7) Protective gas cylinder assembly

The gas cylinder assembly is composed of a gas cylinder, a pressure reducer, and PVC gas pipes, etc. as shown in Figure 4-13. The pressure reducer is installed at the outlet of the gas cylinder, and consists of a pressure reducing mechanism, a heater, a pressure gauge, and a flow meter. The gas cylinder contains the gas of $80\% CO_2+20\% Ar$ for shielded welding.

图 4-13 焊接气瓶
Figure 4-13 Welded Gas Cylinder

4.1.2 机器人焊接工艺的制订及硬件选型

1. 弧焊机器人的选型依据

选择弧焊机器人时，应根据焊接工件的形状和大小来选择机器人的工作范围，一般保证一次将工件上的所有焊点都焊到为准；其次考虑效率和成本，选择机器人的轴数和速度以及负载能力。

在其他情况同等的情况下，应优先选择具备内置弧焊程序的工业机器人，便于程序的编制和调试；应优先选择能够在上臂内置焊枪电缆，底部还可以内置焊接地线电缆、保护气气管的工业机器人，这样在减少电缆活动空间的同时，也延长了电缆的寿命。

对于焊接机器人，还要考虑焊接用的专用技术指标。

1）可以适用的焊接方法。这对弧焊机器人尤为重要。这实质上反映了机器人控制和驱动系统抗干扰的能力。一般弧焊机器人只采用熔化极气体保护焊方法，因为这些焊接方法不需采用高频引弧起焊，机器人控制和驱动系统没有特殊的抗干扰措施。能

4.1.2 Formulation of Robot Welding Process and Hardware Selection

1. Selection basis of arc welding robots

When selecting an arc welding robot, the working range of the robot shall be considered based on the shape and size of the welding workpiece, and that all welding points on the workpiece can be welded at once shall be ensured in general; then choose the quantity of the axis, speed and load capacity of the robot by considering efficiency and cost.

When other situations are equal, priority shall be given to the industrial robots with built-in arc welding programs for easier programming and commissioning, and to those that can have welding gun cables built into the upper arm and welding ground cables and protective gas pipes built into the bottom, which not only reduces the space for cable movement, but also extends the lifespan of the cable.

For welding robots, special technical indicators for welding should also be considered.

1) Applicable welding method, which is particularly important for arc welding robots, as it essentially reflects the anti-interference ability of the control and driving system of the robot. In general, ordinary arc welding robots adopt the gas metal arc welding methods only, as such methods do not require high-frequency arc striking, and the robot control and driving systems are not equipped with special anti-interference measures. The arc welding robot that can use tungsten chloride arc

采用钨极氩弧焊的弧焊机器人是近几年的新产品，它有一套特殊的抗干扰措施。

2) 摆动功能。关系到弧焊机器人的工艺性能。目前弧焊机器人的摆动功能差别很大，有的机器人只有固定的几种摆动方式，有的机器人只能在 x-y 平面内任意设定摆动方式和参数。最佳的选择是能在空间 (x-y,z) 范围内任意设定摆动方式和参数。

3) 焊接工艺故障自检和自处理功能。对于常见的焊接工艺故障，如弧焊的粘丝、断丝等，如不及时采取措施，则会发生损坏机器人或报废工件等大事故。因此，机器人必须具有检出这类故障并实时自动停车报警的功能。

4) 引弧和收弧功能。焊接时起弧、收弧处特别容易产生二生气孔、裂纹等缺陷。为确保焊接质量，在机器人焊接中，通过示教应能设定和修改引弧和收弧参数，这是弧焊机器人必不可少的功能。

5) 焊接尖端点示教功能。一种在焊接示教时十分有用的功能，即在焊接示教时，先示教焊缝上某一点的位置，然后调整其焊枪或焊钳姿态，在调整姿态时，原示教点的位置完全不变。

2. FANUC R-0iA 焊接机器人

FANUC R-0iA 机器人本体结构：FANUC R-0iA 是多功能智能六轴机器人，机器人各部和动作轴的运动范围如图 4-14 所示。

FANUC R-0iA 工业机器人本体的技术参数见表 4-5。

welding is a new product emerging in recent years, and is equipped with a set of special anti-interference measures.

2) Swing function, which is related to the process performance of arc welding robots. At present, arc welding robots vary greatly in terms of the swing function. Some robots have fixed swing modes only, and some others can be set with swing modes and parameters freely only in the x-y plane, which the best choice is the possibility that the swing modes and parameters can be set freely in the space (x-y, z).

3) Self-check and self-processing function for welding process faults. Common welding process faults such as sticking or breaking wires in arc welding, if corresponding measures are not taken in a timely manner, may lead to major accidents such as damaging the robot or workpiece. Therefore, robots have to be able to detect such faults, and stop to alarm automatically in real time.

4) Arc striking and extinguishing functions. During welding, defects such as secondary pores and cracks are particularly easy to occur at the art starting and extinguishing points. To ensure welding quality, it shall be possible to set and modify arc starting and extinguishing parameters through teaching for robot welding, which is an essential function of arc welding robots.

5) Welding tip point teaching function, a very useful function in welding teaching, that is, during welding teaching, first teach the position of a certain point on the welding seam, and then adjust the posture of the welding gun or welding pliers while keeping the position of the original teaching point completely unchanged.

2. The FANUC R-0iA welding robot

Structure of the FANUC R-0iA robot body: The FANUC R-0iA is a multifunctional intelligent 6-axis robot, and the motion range of each part and action axis of the robot is shown in Figure 4-14.

The technical parameters of the FANUC R-0iA industrial robot body are shown in Table 4-5.

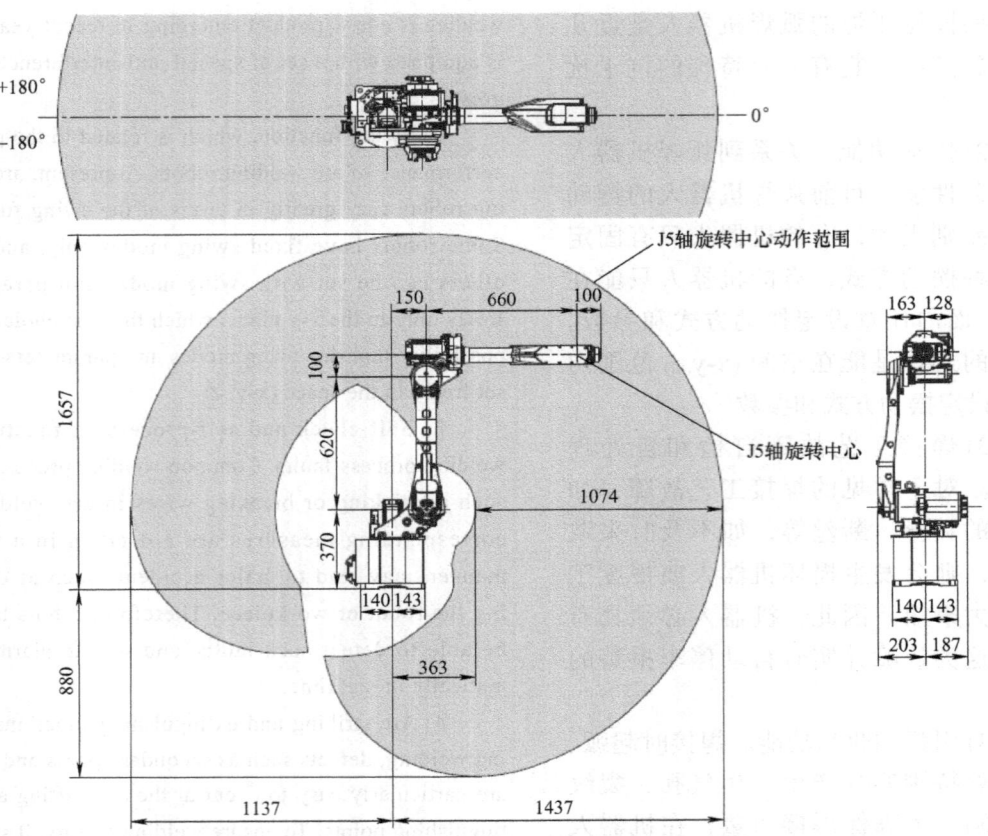

图 4-14 FANUC R-0iA 机器人动作轴的运动范围
Figure 4-14 Motion Range of the Action axis of FANUC R-0iA Robot

表 4-5 FANUC R-0iA 机器人技术参数
Table 4-5 Technical Parameters of the FANUC R-0iA Robot

安装方式 Installation method		地面、壁挂、倒挂 Ground, wall mounted, hung upside down
自由度 Degree of freedom		6 轴 (J1, J2, J3, J4, J5, J6) 6-axis (J1, J2, J3, J4, J5, J6)
可达半径 Reachable radius		1437mm
重复定位精度 Repetitive positioning precision		±0.08mm
运动范围 （最大旋转速度） Range of motion (Maximum rotational speed)	J1	360°(225°/s)
	J2	250°(215°/s)
	J3	455°(225°/s)
	J4	380°(425°/s)
	J5	280°(425°/s)
	J6	720°(625°/s)
手腕部可搬运质量 Transportable mass of wrist		3kg

(续)

安装方式 Installation method	地面、壁挂、倒挂 Ground, wall mounted, hung upside down	
手腕允许负载转矩 Torque of allowable load of wrist	J4	3.9 N·m
	J5	3.9 N·m
	J6	3.0 N·m
手腕允许负载转动惯量 Moment of inertia of allowable load of wrist	J4	0.230 kg·m²
	J5	0.230 kg·m²
	J6	0.035 kg·m²
机器人质量 Robot mass	99kg	
安装环境要求 Environment requirements for installation	环境温度 Ambient temperature	0～45℃
	环境湿度 Ambient humidity	通常在75%RH以下（无结霜现象） Below 75%RH usually (without frost formation) 短期在95%RH以下（1个月内） Below 95% RH for short term (within 1 month)
	振动加速度 Vibration acceleration	≤0.5g

3. FANUC R-0iA 机器人的特点

FANUC R-0iA 机器人机身设计紧凑、纤巧，整体结构超轻量。该款机器人最大亮点是具有卓越的高性价比，性能更优越。FANUC R-0iA 特点诸多：

1) 设计极致：其手臂既具有负载能力，又轻量、紧凑。

2) 重量轻：与同系列机器人相比较，FANUC R-0iA 机器人的本体重量进一步降低，仅 110kg。

3) 动作性能更优越：在同系列中具有最高性能的动作能力，平均提速 6%，其中 J2 轴最高提速高达 13%。

4) 高性能：采用最新的伺服技术，重复定位精度高达 ±0.08mm，实现高精度和高可靠的性能。

5) 安装多样：可实现地装式、天吊式、倾斜式安装，而且吊装时能实现有效地反转运动。

6) 拥有广阔的动作范围：针对

3. Characteristics of the FANUC R-0iA robot

The FANUC R-0iA robot is designed to have a compact and delicate body with an ultra lightweight structure. The highlight of the robot is its excellent cost-effectiveness and superior performance, with many characteristics as follows:

1) Ultimate design: represented by its arms that have load-bearing capacity while being lightweight and compact.

2) Lightweight: compared with robots of the same series, the weight of the FANUC R-0iA robot body is further reduced to 110kg only.

3) Better action performance: it has the action ability of the highest performance among the robots of the same series, with an average speed increase of 6%, and a maximum speed increase of 13% on the J2 axis.

4) High performance: latest servo technology is adopted to have a repetitive positioning precision of ±0.08mm, achieving high-precision and highly reliable performance.

5) Diversified installation: ground mounted, ceiling mounted, or inclined installation are applicable, and effective backward movement can be made for hung installation.

6) A wide range of actions: for arc welding appli-

弧焊应用，具有同类级别弧焊机器人最大的手臂长度和行程距离，达1437mm，与原来的 M-10iA 相比，手臂长度和行程都得到有效扩展。

7) 配套协调：针对弧焊应用，可与变位机协调动作，也可在控制器 R-30iA Mate 上追加附加轴机柜，用于驱动附加轴变位机，完美实现各种高质量、高效的弧焊应用。

4.R-30iA Mate 控制柜的构成

R-30iA Mate 控制柜主要由主板、I/O 印制电路板、急停板、MCC 单元、电源单元等构成，R-30iA 内部部件构成如图 4-15 和图 4-16 所示。

5.R-30iA Mate 控制柜构成单元的功能

(1) 主板

主板上安装有微处理器及其外围电路、存储器，以及操作箱控制电路。此外，主板还进行伺服系统的位置控制。

cations, it has the largest arm length and walking distance, i.e. 1437mm, among the arc welding robots of the same class and level, an effective extension of the arm length and walking distance compared with the original M-10iA robot.

7) Supporting coordination: for arc welding applications, it is possible to coordinate with the positioner, or add an additional axis cabinet on the controller R-30iA Mate for the driving of the additional axis positioner, perfectly achieving the purposes of various high-quality and efficient arc welding applications.

4. Composition of the R-30iA Mate control cabinet

The R-30iA Mate control cabinet mainly consists of the main board, I/O printed circuit board, emergency stop board, MCC unit, power supply unit, etc. The composition of internal components of the R-30iA are shown in Figure 4-15 and Figure 4-16.

5. Functions of the composition units of R-30iA Mate control cabinet

(1) Main board

The main board is equipped with a microprocessor and its peripheral circuits, memory, and control circuits of the operation box. In addition, the main board also controls the position of the servo system.

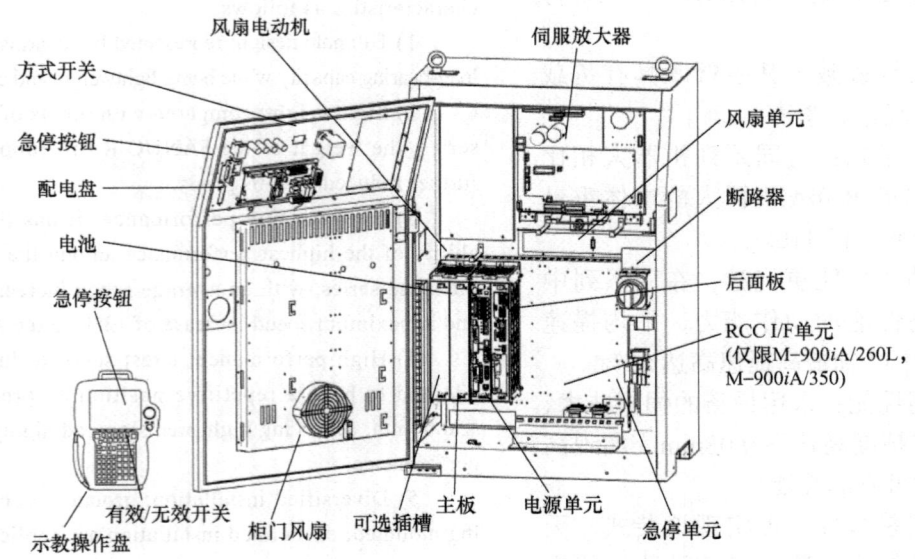

图 4-15　R-30iA Mate 控制柜内部部件安装图（前面）

Figure 4-15　Installation Diagram of the Internal Components of R-30iA Mate Control Cabinet (Front)

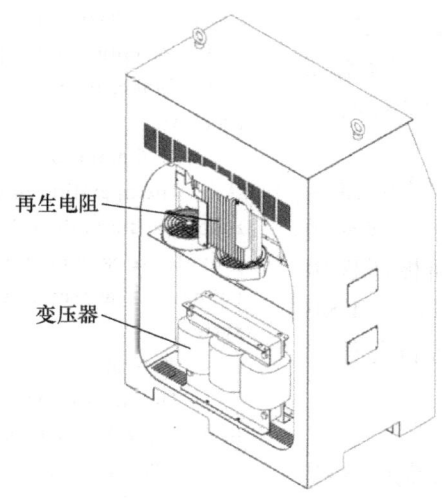

图 4-16 R-30iA Mate 控制柜内部部件安装图（背面）
Figure 4-16 Installation Diagram of the Internal Components of R-30iA Mate Control Cabinet (Back)

(2) I/O 印制电路板、FANUC I/O Unit-MODEL A

根据 I/O 处理等应用备有各类印制电路板。此外，还可以安装 FANUC I/O Unit-MODEL A，在这种情况下，可以选择各类输入/输出类型。全部通过 FANUC I/O Link 来连接。

(3) 急停板、MCC 单元

急停板、MCC 单元用来对急停系统、伺服放大器的电磁接触器以及预备充电进行控制。

(4) 电源单元

电源单元用来将 AC 电源转换为各类 DC 电源。

(5) 后面板

后面板上安装有各类控制板。

(6) 示教操作盘

包括机器人的编程作业在内的所有作业，都通过此示教操作盘进行操作。另外，示教操作盘还通过 LCD(液晶显示屏)进行控制装置的状态、数据的显示。

(2) I/O printed circuit board, FANUC I/O Unit-MODEL A

Various printed circuit boards are available for applications such as I/O processing. In addition, the FANUC I/O Unit-MODEL A can also be installed, in which case various input/output types can be selected. All connections are made through FANUC I/O Link.

(3) Emergency stop board, MCC unit

The emergency stop board and MCC unit are used for the control of the emergency stop system, the electromagnetic contactor of the servo amplifier, and pre-charge.

(4) Power supply unit

The power supply unit is used to convert AC power supply into various DC power supplies.

(5) Rear panel

Various control boards are installed on the rear panel.

(6) Teach pendant operation panel

All operations including robot programming are operated through this teach pendant operation panel. In addition, the panel also displays the status and data of the control device through an LCD (liquid crystal display).

(7) 伺服放大器

伺服放大器进行伺服电动机的控制、脉冲编码器信号的接收、制动器控制、超程、机械手断裂等方面的控制。

(8) 操作箱／操作面板

操作箱／操作面板通过按钮和 LED 进行机器人的状态显示、起动操作。此外，操作箱／操作面板还提供有用来连接外部设备的串行接口、USB 接口。操作箱／操作面板进行急停系统的控制。

(9) 变压器

变压器由输入电源向控制装置提供所需的 AC 电压。

(10) 风扇单元、热交换器

风扇单元和热交换器用来冷却控制装置内部。

(11) 断路器

在控制装置内部的电气系统异常或输入电源异常而流过强电流时，为了保护设备，输入电源连接于断路器。

(12) 再生电阻

再生电阻用来释放伺服电动机的反电动势，连接在伺服放大器上。

6. 焊接机器人标准弧焊功能

(1) 再引弧功能

在工件引弧点处有铁锈、油污、氧化皮等杂物时，可能会导致引弧失败。通常，如果引弧失败，机器人会发出"引弧失败"的信息，并报警停机。当机器人应用于生产线时，如果引弧失败，便有可能导致整个生产线的停机。为此，可利用再引弧功能来有效地阻止这种情况的发生。

再引弧实现的步骤如图 4-17 所示。与再引弧功能相关的最大引弧次数、退丝时间、平移量以及焊接速度、电流、电压等参数均可在焊接辅助条件文件中设定。

(7) Servo amplifier

The servo amplifier is for the control of the servo motor, reception of the pulse encoder signal, and control of the brake, over walking distance, fracture of robotic arm, etc.

(8) Operating box/panel

The operating box/panel displays the status of the robot and starts the robot through buttons and LED, and provides serial and USB interfaces for connecting external devices, as well as the control of the emergency stop system.

(9) Transformer

The transformer provides the required AC voltage to the control device from the input power supply.

(10) Fan unit, heat exchanger

The fan unit and heat exchanger are used to cool the interior of the control device.

(11) Circuit breaker

The input power supply is connected to the circuit breaker in order to protect the equipment when the electrical system inside the control device or the input power supply is abnormal and a strong current flows through.

(12) Regenerative resistor

The regenerative resistor is connected to the servo amplifier to release the back electromotive force of the servo motor.

6. Standard arc welding functions of welding robots

(1) Arc re-striking function

Arc striking may fail when there are impurities such as rust, oil stains, and oxide scales at the arc striking point of the workpiece. Usually, the robot will send a " arc striking failed" message, alarm and shut down if the arc striking fails. When a robot is applied to a production line, the failure of arc striking may cause shutdown of the entire production line. To this end, the arc re-striking function can effectively prevent this situation from occurring.

The steps for realizing arc re-striking are shown in Figure 4-17. The parameters of maximum number of arc re-striking, wire withdrawal time, translation amount, welding speed, current, and voltage related to the re-striking function can be set in the file of auxiliary conditions for welding.

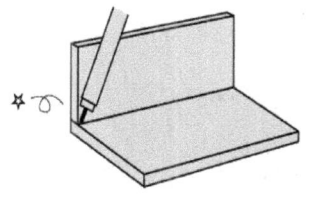 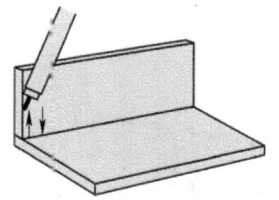 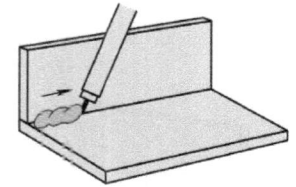

a) 引弧点引弧失败　　　　　b) 从引弧失败点处移开一点，进行再引弧　　　　　c) 引弧成功，返回引弧点，之后继续以正常焊接条件进行焊接作业

a) Arc striking fails at the striking point　　　b) Move a little away from the failed arc striking point, perform arc re-striking　　　c) Against successful arc striking, return to the striking point, and then continue to weld under normal welding conditions

图 4-17　再引弧实现的步骤
Figure 4-17　Steps for Realizing Arc Re-striking

(2) 再启动功能

因为工件缺陷或其他偶然因素，有可能出现焊接中途断弧的现象，并导致机器人报警停机。若在机器人停止位置继续焊接，焊缝容易出现裂纹。

利用再启动功能可有效地预防产生焊缝裂纹。利用再启动功能后，将按照在"焊接辅助条件文件"中指定的方式继续动作。断弧后的再启动方法有三种：

1) 不再引弧，但输出异常信号。输出"断弧、再启动中"的信息，机器人继续动作。走完焊接区间后，输出"断弧、再启动处理完成"的信息，之后继续正常的焊接动作，如图 4-18 所示。

2) 引弧后，以指定搭接量返回一段，之后以正常焊接条件继续动作，如图 4-19 所示。

3) 如果断弧是由机器人不可克服的因素导致的，则停机后必须由操作者手工介入。手工介入解决问题后，使机器人回到停机位置，然后按"启动"按钮，使其以预先设定的搭接量返回，之后再进行引弧、焊接等作业，如图 4-20 所示。

(2) Restart function

Workpiece defects or other accidental factors may lead to arc breakage during welding, and cause the robot to alarm and shut down. If welding is continued at the position where the robot stops, the welding seam is prone to cracking.

The use of the restart function can effectively prevent the occurrence of welding seam cracks. Upon the use of the restart function, the action will continue in the manners specified in the "file of auxiliary conditions for welding". There are three methods for restarting after arc interruption.

1) Do not perform arc re-striking, and output an abnormal signal. Output the message of "arc interruption, restart in progress", and the robot continues to operate. When the welding section is completed, output the message of "arc interruption, restart completed", and then continue with normal welding operations, as shown in Figure 4-18.

2) After arc striking, return to the previous section by a specific overlapping amount, and then continue to operate under normal welding conditions, as shown in Figure 4-19.

3) The operator has to intervene manually after the shutdown if the arc interruption is caused by factors that cannot be overcome by the robot. When the problem is solved by the manual intervention, make the robot go back to the position where it shuts down, then press the "Start" button to make the robot return by the preset overlapping amount, and proceed with operations of arc striking, welding, etc., as shown in Figure 4-20.

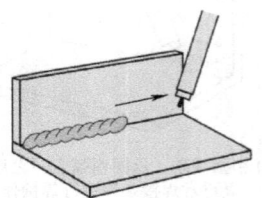 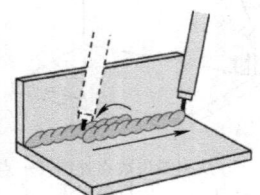 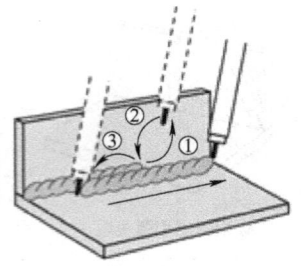

图 4-18 断弧后的再启动方法 1	图 4-19 断弧后的再启动方法 2	图 4-20 断弧后的再启动方法 3
Figure 4-18 Restarting Method 1 after Arc Interruption	Figure 4-19 Restarting Method 2 after Arc Interruption	Figure 4-20 Restarting Method 3 after Arc Interruption

（3）自动解除粘丝功能

对于大多数的自动焊机来说，都具有防粘丝功能。即：在熄弧时，焊机会输出一个瞬间相对高电压以进行粘丝解除。尽管如此，在焊接生产中仍会出现粘丝的现象，这就需要利用机器人的自动解除粘丝功能进行解除。若使用该功能，即使检测到粘丝，也不会马上输出"粘丝中"信号，而是自动施加一定的电压，进行解除粘丝的处理。

自动解除粘丝功能也是利用一个瞬间相对高电压以使焊丝粘连部位爆断。至于自动解除粘丝的次数、电流、电压和时间等参数均可在"焊接辅助条件文件中"设定。

在未使用粘丝自动解除功能时，若发生粘丝或者自动解除粘丝处理失败的情况下，机器人就会进入暂停状态，停机。暂停状态时，示教编程器"HOLD"显示灯亮并且外部输出信号输出"粘丝中"的信息。

自动解除粘丝功能的实现步骤如图 4-21 所示，先是焊丝与工件粘在一起发生粘丝，然后是瞬间的相对高电压进行粘丝解除。经过焊机自身的粘丝解除处理后，粘丝仍未能解除，则利用机器人的。

(3) The function of automatic release of sticking wire

Most of the automated welding machines have the function of anti-sticking wire, that is, the welding machine outputs an instantaneous relatively high voltage to release the sticking wire when the arc is extinguished. Nevertheless, wire sticking still happens in welding, and the sticking wire can be released by the function of automatic release of sticking wire of the robot. When this function is used, the signal of " wire sticking in progress " won't be outputted immediately even if wire sticking is detected. Instead, a certain voltage will be applied to release the sticking wire.

The function of automatic release of sticking wire is to use an instantaneous relatively high voltage to burst the sticking part of the welding wire. Parameters such as the number of times, current, voltage, and time for release of sticking wire can be set in the " file of auxiliary conditions for welding " .

The robot will enter the state of pause, and shut down if wire sticking happens when the function of automatic release of sticking wire is not in use, or the automatic release fails. When in the state of pause, the " HOLD " indicator of the teaching programmer lights up and the external output signal outputs the message of "wire sticking in progress " .

The steps of realizing the function of automatic release of sticking wire are shown in Figure 4-21. The welding wire and the workpiece stick together first, then the wire is released by an instantaneous relatively high voltage. If the sticking wire cannot be released by the action of the welding machine for wire releasing, resort to that of the robot.

项目 4 典型工业机器人弧焊工作站系统的设计及应用

(4) 渐变功能

所谓渐变功能是指在焊接的执行中逐渐改变焊接条件的功能。即在某一区段内将电流/电压由某一数值渐变至另一数值。示意说明如图 4-22 所示。

(4) Gradual change function

The so-called gradual change function refers to the function of changing the welding conditions gradually during the execution of welding. That is, changing the current/voltage gradually from one value to another within a certain section, with the schematic explanation as shown in Figure 4-22.

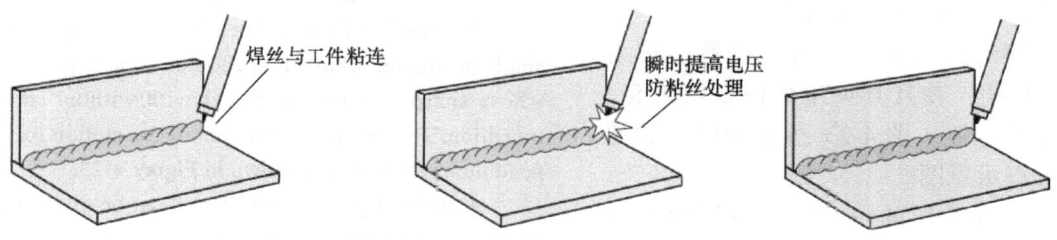

图 4-21 自动解除粘丝功能的实现步骤

Figure 4-21 Steps for Realizing the Function of Automatic Release of Sticking Wire

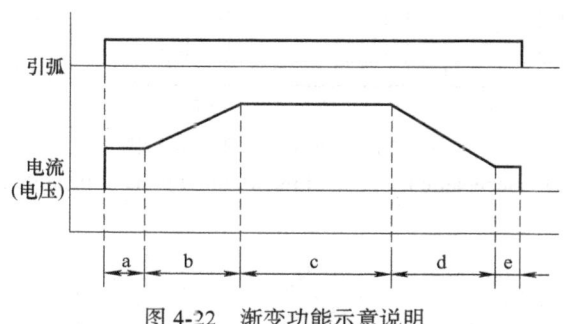

图 4-22 渐变功能示意说明

Figure 4-22 Schematic Explanation of Gradual Change Function

a 段：以引弧条件文件中设定的规范参数引弧。

b 段：焊接电流（电压）由小渐变大。

c 段：以恒定的规范参数焊接。

d 段：焊接电流（电压）由大渐变小。

e 段：以熄弧条件文件中设定的规范参数熄弧。

对于铝材、薄板以及其他特殊材料的焊接，由于其容易导热，特别是焊接到结束点附近时，工件容易发生

Section a: perform arc striking with the standard parameters set in the file of conditions for arc striking.

Section b: the welding current (voltage) increases from a small value to bigger ones gradually.

Section c: welding with standard constant parameters.

Section d: the welding current (voltage) gradually decreases from a big value to smaller ones gradually.

Section e: extinguish arc with the standard parameters set in the file of conditions for arc extinguishing.

When welding the aluminum, thin plates, and other special materials, the workpiece is prone to fracture and burn-through, especially when welding is going on near the end point, as the said materials have a good

断裂、烧穿。若在结束焊接前,逐渐降低焊接条件,则可防止工件断裂、烧穿。

(5) 摆焊功能

摆焊功能的利用提高了焊接生产效率,改善了焊缝表面质量。摆焊条件可在"摆焊条件文件"中设定,例如形态、频率、摆幅以及角度等。摆焊条件文件最多可输入 16 个。

摆焊的动作形态有单振摆、三角摆、L 摆,并且其尖角可被设定为有/无平滑过渡。图 4-23 所示为摆焊的动作形态示意图。

摆焊动作的一个周期可以分为四个或三个区间,如图 4-24 所示。

thermal conductivity. Gradual reduction in the welding conditions before welding end can prevent fracture and burn-through of the workpiece.

(5) Pendulum welding function

The utilization of pendulum welding function has improved the welding efficiency and surface quality of welding seams. The conditions for pendulum welding, maximum 16 conditions such as shape, frequency, pendulum amplitude, and angle can be set in the " file of conditions for pendulum welding " .

The motion forms of pendulum welding include single pendulum, triangular pendulum, and L-pendulum, whose sharp corners can be set with/without smooth transition. The schematic diagram of the motion form of pendulum welding is as shown in Figure 4-23.

A cycle of pendulum welding can be divided into four or three sections, as shown in Figure 4-24.

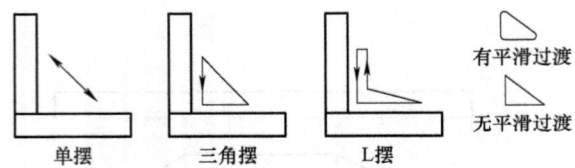

图 4-23 摆焊的动作形态示意图

Figure 4-23 Schematic Diagram of the Motion Forms of Pendulum Welding

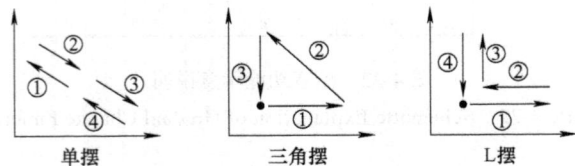

图 4-24 摆焊动作的一个周期

Figure 4-24 A Cycle of the Motion of Pendulum Welding

在区间之间的节点上可以设定延时,延时的方法有两种,即:机器人停止和摆焊停止。可以根据要焊接的母材的可熔性,灵活地选择适当的延时方法,以取得比较理想的熔深。

7. 机器人焊接工艺的选择

机器人焊接工艺主要包括焊接方法、焊接电源、母材、板厚(管径及壁厚)、接头、坡口形式、焊前准备加

Delay can be set with two methods at nodes between sections: either the robot or pendulum welding stops. Suitable delay methods can be selected based on the fusibility of the base material to be welded, in order to achieve desirable penetration depth.

7. Selection of robot welding process

The robot welding process mainly includes welding method, welding power supply, base material, plate thickness (pipe diameter and wall thickness), joint, groove form, pre-welding preparation, assembly, welding position, welding sequence, welding material, gas,

工、装配、焊接位置、焊接顺序、焊材、气体、机器人焊接轨迹点的设置、焊枪角度、焊接参数等。机器人焊接是用焊接机器人代替手工完成焊接作业，因此，同样需要制订切实可行的焊接工艺方案。

(1) 已知条件

焊件结构的技术要求、结构尺寸、母材牌号及规格（板厚、管径与壁厚）、接头形式、焊接位置、焊接方法、焊材、气体等。

(2) 焊件的机器人焊接工艺性分析

对焊件材料的焊接性、下料、成形加工工艺、装配方法的选用以及机器人的焊接轨迹、姿态、焊枪角度、焊接参数等进行分析，确定焊接重点及难点，制订解决措施，以便控制焊接质量，提高效率，降低成本等。

(3) 硬件选型

用于焊接机器人的焊接电源须具备以下特点：

1) 电源功率必须满足机器人自动化焊接所要求的高输出、高稳定性要求。焊接电源的负载持续率是衡量其功率输出性能的重要参数。在选择焊接电流时，一定要结合连续工作的具体情况考虑焊接电源的负载能力。

2) 电源具有机器人控制接口，以满足机器人柔性自动化焊接的需要。

3) 电源具备应对各种焊接辅助功能的能力，如始端检出功能、焊接方法选择功能等，以满足焊接工件对焊接自动化的要求。

适合焊接机器人的焊枪应具备以下特点：

1) 机器人焊枪必须满足机器人自

setting of trajectory points of robot welding, angle of welding gun, welding parameters. It is necessary to formulate practical and feasible welding schemes for robot welding, during which welding robots are adopted to replace manual welding.

(1) Conditions known

The technical requirements, and structural dimensions, as well as the base metal grade and specifications (plate thickness, pipe diameter, and wall thickness) of the weldment structure, joint form, welding position, welding method, welding material, gas, etc.

(2) Analysis of robot welding process for weldments

Conduct the analysis on weldability, blanking, forming process, and selection of assembly method of welding materials, and the welding trajectory, posture, angle of welding gun, welding parameters of the robot, determine the key points and difficulties of welding and formulate corresponding solutions in order to control welding quality, improve efficiency, and reduce costs.

(3) Hardware selection

The welding power supply welding robots must have the following characteristics:

1) The power supply must meet the high output and stability requirements for the automated welding of robots. The load duration of a welding power supply is an important parameter to evaluate its power output performance. When selecting the welding current, the load capacity of the welding power supply has to be considered with reference to the specific situation of continuous operation.

2) The power supply shall be equipped with interfaces for robot control to meet the needs of flexible and automated welding of the robot.

3) The power supply shall be able to satisfy various auxiliary functions of welding, such as functions of detection at the starting end, and selection of welding methods, in order to meet the requirements of welding automation for welding workpieces.

A welding gun suitable for welding robots shall have the following characteristics:

1) The robot welding gun must meet the high load-bearing capacity requirements of automated welding of the robot. The working ability of welding gun is evaluated by load duration, which is similar to welding power supply. When selecting the welding current, the

动化焊接的高承载能力的要求。对于焊枪而言，与焊接电源类似，也通过负载持续率衡量其工作能力。在选择焊接电流时，一定要结合连续工作的具体情况考虑焊枪的负载能力。

2）由于机器人焊接的速度通常比较快，焊枪质量的优劣决定着焊接时电弧的稳定性，从而对焊接质量产生相应的影响。

3）机器人焊接时要求焊枪的TCP点（焊丝的尖端点）具有比较好的稳定性，以保证焊接时电弧位置的精确度。

4）必须保证同一型号焊枪的TCP点的精度一致性，这样在更换旧的焊枪时，才可以保证新旧焊枪的TCP点相一致，才可以尽可能地缩短系统的待机时间，提高工作效率。

根据现场生产条件及焊接技术要求，考虑是否需要翻转变位、机器人的臂伸长（动作范围）能否覆盖整个作业面，以及机器人最大承载重量等，选择机器人及焊接电源类型、系统形式。

(4) 机器人焊接工艺试验与优化

机器人焊接工艺试验是根据焊件的技术要求，通过工艺分析，拟订机器人焊接工艺方案，并将机器人焊接工艺知识应用于示教编程，充分考虑焊接顺序、关键点的处理、焊枪角度及机器人的姿态等。

编程完成后对焊接参数（焊接电流、焊接电压、焊接速度、干伸长、振幅、摆动停留时间、气体流量等）进行设置和调整，完成焊接工艺试验。最终从质量、效率、成本三方面进行机器人焊接工艺方案比较，选定最佳方案。

load capacity of the welding gun has to be considered with reference to the specific situation of continuous operation.

2) Since the welding speed of robots is relatively fast usually, the quality of the welding gun determines the stability of the arc during welding, which further affects the welding quality.

3) For robot welding, it is required that the TCP point (Tool Center Point, the tip of the welding wire) of the welding gun has good stability to ensure the precision of the arc position during welding.

4) The precision consistency of the TCP points of welding guns of the same model shall be ensured, so that the TCP points of the new and old welding guns are consistent when replacing the old welding gun, and the standby time of the system can be minimized to improve work efficiency.

Select the type and system form of the robot and welding power supply, by taking into account such issues as whether flipping and repositioning are required, whether the robot's arm extension (action range) can cover the entire working surface, and the maximum bearing weight of the robot, based on the on-site conditions for production and technical requirements for welding.

(4) Test and optimization of robot welding process

For the test of robot welding process, schemes for robot welding process are formulated based on the technical requirements of the weldment and the analysis of process, and apply the knowledge of robot welding process to the teaching programming, taking into account the welding sequence, treatment of key points, angel of the welding gun, and posture of the robot, etc.

Upon completion of programming, set and adjust the welding parameters (welding current, welding voltage, welding speed, stick out, amplitude, pendulum dwell time, gas flow rate, etc.) to complete the test of welding process. Finally, select the best scheme based on a comparison of the schemes for robot welding process in terms of quality, efficiency, and cost.

4.1.3 弧焊机器人的示教编程

1. 材料焊接性

产品材料为 Q235 钢,属于常用低碳钢,焊接性较好。

2. 焊件装配

焊件为平对接,因焊接过程中焊缝逐渐收缩,易引起焊接缺陷,应考虑后焊间隙比先焊间隙约大 0.5mm;焊件两端定位焊长度约为 20mm。

3. 焊件的焊接工艺与编程要点

1)该焊件属薄板焊接,其接头形式为 I 形坡口对接,焊接位置为水平焊,采用机器人 CO_2 气体保护焊易施焊,操作简单。

2)焊前将焊件坡口两端清理干净。

3)单面焊双面成形焊缝编程时,要根据焊件板厚、坡口间隙,考虑焊缝的熔合性、焊透性、双面焊缝的均匀性及坡口间隙收缩变形,设定合适的焊枪角度和焊接参数。

4)起焊处编程时,考虑焊缝的熔合性、焊透性、焊缝宽窄和高低的均匀性,设定焊接参数时应适当增加焊接电流、电压及控制引弧停留时间。

5)收弧处易产生弧坑及焊穿缺陷,编程中设定焊接参数时应适当减小焊接电流、焊接电压及控制收弧停留时间。

6)机器人焊接方式采用直线行走焊接即可完成。

4.1.3 Teaching Programming of Arc Welding Robots

1. Weldability of material

The material of product is Q235 steel, which is a kind of commonly used low-carbon steel and has a good weldability.

2. The assembly of weldment

Square butt joints are adopted for the assembly of weldments, and the later welding gaps shall be about 0.5mm larger than the previous ones, as the welding seams gradually shrinks during welding, which is easy to cause welding defect; the length of tack welding at both ends of the weldment shall be about 20mm.

3. Key points for the welding process of weldment and programming

1) This weldment belongs to a kind of thin plate, and its welding adopts the form of I-shaped groove butt joint. As horizontal welding is applied to the welding position, it is easy to perform the welding operation with CO_2 gas welding by robot, which is easy to operate.

2) Clean both ends of the welding groove before welding.

3) For the programming of welding seams of one-side welding with back formation, it is necessary to set appropriate angels of the welding gun and parameters of welding, taking into account the fusion and penetration of the welding seam, the uniformity of welding seams at both sides, and the shrinkage of the groove gap based on the plate thickness and groove gap of the weldment.

4) For the programming of welding at the starting point, consideration shall be given to the fusion, penetration, and the uniformity of width and height of the welding seam. When setting welding parameters, the welding current and voltage shall be appropriately increased, while the dwell time of arc striking shall be controlled.

5) Arc pits and welding defects such as burn-through are prone to occur at the position of arc extinguishing. When setting welding parameters in programming, the welding current and voltage shall be appropriately decreased and the dwell time of arc extinguishing reasonably controlled.

6) The method of linear walking welding can be adopted to complete the robot welding.

4. 设备选择

1) 机器人品牌：机器人本体型号选择 FANUC M-10iA。

2) 焊接电源：焊接电源选择 R-30iA Mate 控制柜。

5. 示教编程

1) 示教运动轨迹：示教运动轨迹一般包括原点、前进点或退避点、焊接开始点和结束点、焊枪姿态等。薄板平对接产品的示教运动轨迹如图 4-25 所示，主要由编号为①～⑥的 6 个示教点组成。

①点、⑥点为原点（或待机位置点），其应处于与工件、夹具不干涉的位置，焊枪姿态一般为 45°（相对于 X 轴）。

③点、④点为焊接起始点和结束点，焊枪姿态为平行于焊缝法线且与待焊方向成一夹角（95°～100°）。

②点（进枪点）、⑤点（退枪点）为过渡点，也要处于与工件、夹具不干涉的位置，焊枪角度任意。

2) 焊接参数设置：焊接参数设置包括焊接层数、焊接电流、焊接电压、焊接速度、干伸长度、气体流量等的设置。薄板平对接焊接参数见表 4-6。

4. Equipment selection

1) Robot brand: the FANUC M-10iA as the model of the robot body.

2) Welding power supply: the R-30iA Mate control cabinet as the welding power supply.

5. Teaching programming

1) Teaching motion trajectory: the teaching motion trajectory generally consists of the home point, forward or backward point, starting and ending points of welding, and posture of welding gun, etc. The teaching motion trajectory of products with thin plate square butt joints is shown in Figure 4-25, mainly composed of six teaching points numbered ①-⑥.

Point ① and point ⑥ are the home point (or point of standby position), which shall be in a position that does not interfere with the workpiece or fixture. Generally, the welding gun is in a posture of having an angle of 45° (relative to the X-axis).

Point ③ and point ④ are the starting and ending points of welding respectively, and the welding gun is in a posture of being parallel to the normal of the welding seam and at an angle (95°-100°) with the intended welding direction.

Point ② (entry point of gun) and point ⑤ (exit point of gun) are transition points, which shall be in positions that do not interfere with the workpiece or fixture, while the welding gun can be at angles of any degree.

2) Welding parameter settings: include settings of welding layers, welding current, welding voltage, welding speed, dry extension length, gas flow rate. The parameters for square butt welding of thin plates are shown in Table 4-6.

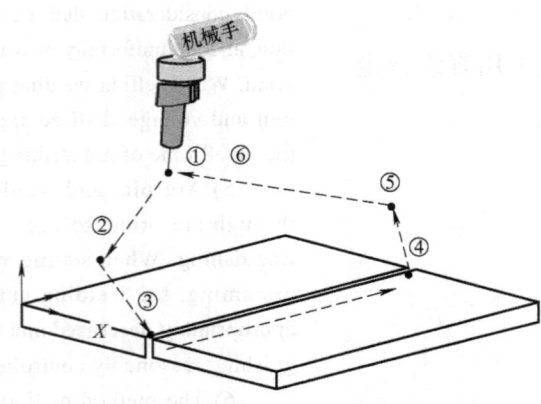

图 4-25 薄板平对接产品的示教运动轨迹

Figure 4-25 Teaching motion trajectory of products with thin plate square butt joints

表 4-6 薄板平对接焊接参数
Table 4-6 Parameters for Square Butt Welding of Thin Plates

焊接层数 Welding layers	焊接电流/A Welding current/A	焊接电压/V Welding voltage/V	焊接速度/(mm/min) Welding speed/(mm/min)	运枪方式 Gun carrying method	振幅 Amplitude	干伸长度/mm Stick out/mm
一层 1 layer	120	18.4	300	直线 Straight line	0	15

6. 焊接效果

焊接效果图如图 4-26 所示。

6. Welding rendering

The welding rendering is shown in Figure 4-26.

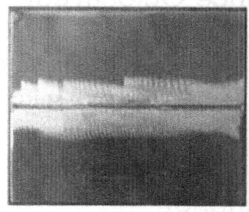
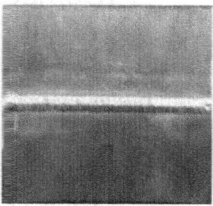
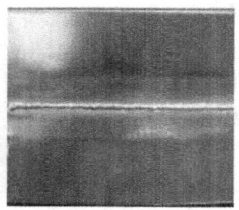

a) 试件装配 b) 试件正面 c) 试件背面

图 4-26 焊接效果图
Figure 4-26 Welding Rendering

【课后巩固】

[Consolidation after Class]

1. 简述弧焊机器人的特点。

2. 简述弧焊机器人工艺分析的要点。

3. 简述弧焊机器人硬件选型的要求。

1. Describe briefly the characteristics of arc welding robots.

2. Describe briefly the key points of process analysis of arc welding robots.

3. Describe briefly the requirements for hardware selection of arc welding robots

任务 4.2　设计方案的编写
Task 4.2　Compilation of Design Schemes

【知识目标】

[Knowledge Objectives]

1. 掌握工业机器人弧焊工作站简介及布局。

2. 掌握工业机器人弧焊工作站工作流程及控制要求。

3. 了解工业机器人弧焊工作站主要设备清单。

4. 掌握整体设计方案的编写方法。

1. To master the introduction and layout of industrial robot arc welding workstations.

2. To master the workflow and control requirements of industrial robot arc welding workstations.

3. To understand the list of main equipment of industrial robot arc welding workstations.

4. To master the compilation method of overall design schemes.

【技能目标】

1. 根据工业机器人弧焊工作站要求选择设备清单。
2. 根据工艺要求说明工作流程及控制要求。
3. 能够设计工业机器人弧焊工作站的整体方案。

【素质目标】

1. 养成良好的自主学习习惯。
2. 培养团队协作精神。

【任务情景】

工业机器人弧焊工作站系统能够有效实现机器人完成全封闭压力容器的焊接,一个好的方案能加快实现弧焊工作站的生产。某生产封闭压力容器的企业需要一份工业机器人弧焊工作站的设计方案实现自动化弧焊工作,该方案包括工作站简介、弧焊工件说明、弧焊工作站设备清单、工作流程。

【任务分析】

在项目实施的过程中,需要编写设计方案交付客户,在设计方案中必须详细叙述出项目实施的优势、项目能够给企业带来的利益、生产效率的提升、项目实施过程中的设备选型和布局、项目的工程预算等。一个好的设计方案对于项目的推进和实施有着重要的意义。

[Skill Objectives]

1. To select the list of equipment according to the requirements of industrial robot arc welding workstations.
2. To explain the workflow and control requirements according to the process requirements.
3. To be able to compile overall design schemes for industrial robot arc welding workstations.

[Competence Objectives]

1. To develop good self-directed learning habits.
2. To cultivate teamwork spirit.

[Task Scenario]

The industrial robot arc welding workstation system can effectively achieve the welding of fully enclosed pressure vessels by robot. A good scheme can improve the productivity of arc welding workstations. An enterprise that produces closed pressure vessels asks for a design scheme for an industrial robot arc welding workstation for automated arc welding, and the scheme consists of an introduction to the workstation, an description of the arc welding workpiece, a list of equipment of the arc welding workstation, and a workflow.

[Task Analysis]

During the program implementation process, it is necessary to prepare a design scheme and submit it to the client. The design scheme must provide a detailed description of the advantages of the program, the benefits that the program can bring to the enterprise, the improvement of production efficiency, the selection and layout of equipment during the program implementation process, the engineering budget of the program, etc. A good design scheme is of great significance for the promotion and implementation of the program.

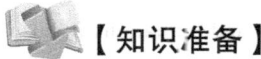【知识准备】

4.2.1 设计方案的结构和要素

1. 设计背景及工作站布局

设计方案编写的目的是说明一个焊接工作站各部分中的每个设备和工作情况的设计考虑。方案重点是详细描述工作站的执行流程和工作情况。

(1) 设计背景

提供的设计背景应包含以下几个方面的内容：待设计工作站名称、该工作站基本概念(如该工作站的类型、从属地位等)和开发项目组名称。

(2) 参考资料

列出详细设计报告引用的文献或资料，包括资料的作者、标题、出版单位和出版日期等信息，必要时说明如何得到这些资料。

(3) 术语定义及说明

列出本文档中用到的可能会引起混淆的专门术语、定义和缩写词的原文。

(4) 工作站任务和目标

说明详细设计的任务及详细设计所要达到的目标。

(5) 需求概述

对所设计工作站的概要描述，包括主要的业务需求、输入、输出、主要功能、性能等，尤其需要描述工作站性能需求。

(6) 运行环境概述

对本工作站所依赖于运行的硬件的描述，包括操作系统、数据库系统、中间件、接口软件、可能的性能监控与分析软件环境及配置要求。

(7) 条件与限制

详细描述系统所受的内部和外部条件的约束和限制说明，包括业务和技术方面的条件与限制，以及进度、

[Assumed Knowledge]

4.2.1 Structure and Elements of Design Schemes

1. Design background and workstation layout

The purpose of compiling a design scheme is to explain what has been considered in the design of each equipment and working situation in each part of a welding workstation. The focus of the scheme is a detailed description of the execution process and working situation of the workstation.

(1) Design background

Provide a design background that should include the following aspects: the name of the workstation to be designed, the basic concept of the workstation (such as the type, subordinate status of the workstation), and the name of the program development team.

(2) References

List the literature or documents cited in the detailed design report, including the information of author, title, publisher, and publication date of the documents and, if necessary, the explanation on the way of obtaining these documents.

(3) Definition and explanation of terms

List the original text of special terms, definitions, and abbreviations used in the document that may cause confusion.

(4) Tasks and objectives of the workstation

Describe the tasks of detailed design and the objectives to be achieved by the detailed design.

(5) Overview of requirements

Provide a brief description of the requirements for the designed workstation, including requirements of the main business, inputs, outputs, main functions, and the performance in particular.

(6) Overview of operating environment

Provide a description of the hardware that this workstation relies on to run, including the operating system, database system, middleware, interface software, and the environment of software for possible performance monitoring and analysis, as well as configuration requirements.

(7) Conditions and limitations

Provide a detailed description of the internal and external constraints and limitations that the system is

管理等其他方面的限制。

(8) 详细设计方法和工具

简要说明详细设计所采用的方法和使用的工具。如 HIPO 图方法、IDEF(I2DEF) 方法、E-R 图、数据流程图、业务流程图、选用的 CASE 工具等，尽量采用标准和规范的辅助工具。

(9) 工作站详细需求分析

主要对工作站的需求进行分析。首先，应对需求分析提出的企业需求进行确认；然后对由于情况变化而带来的需求变化进行较为详细的分析。

(10) 详细需求分析

分析包括：详细功能需求分析、详细性能需求分析、详细资源需求分析、详细系统运行环境及限制条件分析。

(11) 详细的系统运行环境、限制条件及接口需求分析

分析包括：工作站接口需求分析，现有硬、软件资源接口需求分析，引进硬、软件资源接口需求分析。

2. 加工工件说明

说明加工工件毛坯尺寸、加工图样以及零件加工工艺。

3. 工艺动作流程

说明工作站工作流程及控制要求。

4. 工业机器人弧焊工作站主要设备清单

说明工作站主要设备，以及这些的技术配置及参数。

4.2.2 设计方案编写示例

1. 目录

编写工业机器人弧焊工作站系统设计方案的目录样式。

subject to, including conditions and limitations in business and technology, and limitations in progress, management, and other aspects.

(8) Methods and tools for detailed design

Provide a brief description of the methods and tools used in the detailed design. For example, HIPO diagram method, IDEF (I2DEF) method, E-R diagram, data flow diagram, business flow diagram, selected CASE tool. Standard, normative and auxiliary tools shall be adopted wherever possible.

(9) Analysis of detailed requirements for the workstation

Provide an analysis mainly of the requirements for the workstation. First, confirm the requirements for the enterprise proposed in analysis of the requirements; then conduct a detailed analysis of the changes in requirements caused by the changes in circumstances.

(10) Analysis of detailed requirements

Provide an analysis that includes: analysis of detailed function requirement, detailed performance requirement, and detailed resource requirement, as well as detailed system operating environment and constraints.

(11) Analysis of detailed system operating environment, constraints, and interface requirements

Provide an analysis that includes: analysis of interface requirements for the workstation, interface requirements for the existing hardware and software resource, and interface requirements for introducing hardware and software resource.

2. Description of the machining workpieces

Provide a description of the dimensions of the workpiece blanks, machining drawings, and machining process for the parts.

3. Process action flow

Provide a description of the workflow and control requirements of the workstation.

4. List of main equipment for the industrial robot arc welding workstation

Provide a description of the main equipment for the workstation, and their technical configuration and parameters.

4.2.2 An Example of Design Scheme Compilation

1. Table of contents

Compile the style of table of contents of the design scheme for industrial robot arc welding workstation system.

2. 正文

这里以"工业机器人弧焊工作站系统设计方案"的内容编写为例,为大家演示如何进行方案文档编写,方案的其他内容按要求自己编写,具体任务要求及文档编写规则请按照国际规定的流程来写。

示例:

一、弧焊工作站焊接系统的设计

弧焊机器人一般较多采用熔化极气体保护焊(MIG焊、MAG焊、CO_2)或非熔化极气体保护焊(TIG焊、等离子弧焊)方法。机器人弧焊系统主要包括弧焊电源、送丝机和焊枪。

弧焊电源是用来对焊接电弧提供电能的一种专用设备。弧焊电源的负载是电弧,它必须具有弧焊工艺要求的电气性能,如合适的空载电压、一定形状的外特性、良好的动态特性和灵活的调节特性等。

1. 弧焊电源的类型

弧焊电源有各种分类方法。按输出的电流分,有直流、交流和脉冲三类;按输出外特性分,有恒流特性、恒压特性和介于这两者之间的缓降特性三类。

2. 弧焊电源的特点和适用范围

(1) 弧焊变压器式交流弧焊电源

1) 特点:将网路电压的交流电变成适于弧焊的低压交流电,结构简单,易造易修,耐用,成本低,磁偏吹小,空载损耗小,噪声小,但其电流波形为正弦波,电弧稳定性较差,功率因数低。

2) 适用范围:酸性焊条电弧焊、埋弧焊和TIG焊。

2. Text

Here, the compilation of content of "Design Scheme of Industrial Robot Arc Welding Workstation System" is taken as an example to demonstrate how to compile the scheme documents. Other content of the scheme shall be compiled by yourself according to the requirements. For specific task requirements and rules for document compilation, please follow the international procedures.

Example:

I. Design of the welding system of arc welding workstation

Arc welding robots mainly adopt gas metal arc welding (MIG welding, MAG welding, CO_2) or Tungsten Inert Gas Welding (TIG welding, plasma arc welding) methods. The robot arc welding system mainly includes arc welding power supply, wire feeder, and welding gun.

Arc welding power supply is a specialized equipment used to provide electricity for welding arcs. The load of arc welding power supply is arc, which must have the electrical performance required by the arc welding process, such as appropriate no-load voltage, external characteristics of a certain shape, good dynamic and flexible control characteristics.

1. Types of arc welding power supply

There are various classification methods for arc welding power supply, which can be categorized as DC, AC, and pulse power supply according to the output current, and power supply with constant current characteristics, constant voltage characteristics, and slow drop characteristics between the aforementioned two according to the external characteristics of output.

2. Characteristics and application range of arc welding power supply

(1) AC arc welding power supply with arc welding transformer

1) Characteristics: it transforms the AC of the network into low-voltage AC suitable for arc welding, and is durable with a simple structure, easy for construction and repair, while being low in cost, arc blow, and no-load loss, as well as noise. However, its current is in the form of sine wave, with poor arc stability and low power factor.

(2) 矩形波交流弧焊电源

1) 特点：网路电压经降压后运用半导体控制技术获得矩形波的交流电，电流过零点极快，其电弧稳定性好，可调节参数多，功率因数高，但设备复杂、成本高。

2) 适用范围：碱性焊条电弧焊、埋弧焊和TIG焊。

(3) 直流弧焊发电机式直流弧焊电源

1) 特点：由柴(汽)油发动机驱动发电而获得直流电，输出电流脉动小，过载能力强，但空载损耗大，噪声大，效率低。

2) 适用范围：适用于各种弧焊。

(4) 整流器式直流弧焊电源

1) 特点：将网路交流电经降压和整流后获得直流电，与直流弧焊发电机相比，制造方便，省材料，空载损耗小，节能，噪声小，由电子控制的近代弧焊整流器的控制与调节灵活方便，适应性强，技术和经济指标高。

2) 适用范围：适用于各种弧焊。

(5) 脉冲型弧焊电源

1) 特点：由于脉冲型弧焊电源输出幅值大小周期变化的电流，因此，效率高，可调参数多，调节范围宽而均匀，热输入可精确控制，但设备较复杂，成本高。

2) 适用范围：TIG、MIG、MAC焊和等离子焊。

二、数字式逆变焊接电源 CV350-R

机器人：焊接工作站选用FANUC R-0iA焊接机器人，焊接电源为Lincoln Invertec® CV350-R焊接电源。

2) Application range: acid electrode arc welding, submerged arc welding and TIG welding.

(2) AC arc welding power supply with rectangular wave

1) Characteristics: rectangular wave AC is obtained by semiconductor control technology from the network current after voltage drop, and crosses the zero point extremely fast, with good arc stability, many adjustable parameters, and high power factor. However, the equipment is complex and expensive.

2) Application range: alkaline electrode arc welding, submerged arc welding and TIG welding.

(3) DC arc welding power supply with DC arc welding power generator

1) Characteristics: a diesel (gasoline) engine drives to generate DC, which is with small output current ripple and strong overload capacity, but big no-load loss and noise, as well as low efficiency.

2) Application range: arc welding of various kinds.

(4) DC arc welding power supply with rectifier

1) Characteristics: DC is obtained by voltage drop and rectification of the network AC, and is easy for manufacturing with less materials, low in no-load loss, energy consumption, and noise, compared with DC arc welding power generators. The control and adjustment of modern arc welding rectifiers controlled by electronics are flexible and convenient, with strong adaptability, and high technical and economic indicators.

2) Application range: arc welding of various kinds.

(5) Pulse arc welding power supply

1) Characteristics: the pulse arc welding power supply is efficient as it outputs the current with periodic changes in amplitude, while having multiple adjustable parameters with a wide and uniform range for adjustment, and precise control of thermal input. However, the equipment is complex and expensive.

2) Application range: TIG, MIG, MAC welding, and plasma welding.

II. Digital inverter welding power supply CV350-R

Robot: The welding workstation adopts the FANUC R-0iA welding robot and Lincoln Invertec® CV350-R welding power supply.

1. CV350-R 焊接电源技术规格

CV350-R 焊接电源技术规格见表 4-7。

1. Technical specifications of CV350-R welding power supply

The technical specifications for the CV350-R welding power supply are shown in Table 4-7.

表 4-7 CV350-R 焊接电源技术规格

Table 4-7 Technical Specifications of CV350-R Welding Power Supply

项目 Item	参数 Parameter	规格 Specifications
输入 （仅适用于三相） Input (Three-phase only)	标准电压/相/频率 Standard voltage/phase/frequency	380～415V(±10%)、三相、50/60Hz
	额定的输入功率 Rated input power	14kV·A
输出 （仅适用于直流） Output (DC only)	暂载率 Temporary load rate	60%～100%
	焊机电流 Welding machine current	350A 300A
	额定电流下的电压 Voltage at rated current	31V 29V
输出 Output	焊接电流范围 Welding current range	50～390A
	开路电压 Open circuit voltage	70V
	焊接电压范围 Welding voltage range	16～33.5V
输入导线和熔丝规格 Specifications of input wire and fuse	输入电压/频率 Input voltage/frequency	342～456V、50/60Hz 342～456V，50/60Hz
	最大输入电流 Maximum input current	21A
	最大有效供应电流 Maximum effective supply current	17A
	在60℃下套管中铜丝规格 Specification of copper wire in casing at 60℃	12mm²
	熔丝或断路器尺寸（延时型） Fuse or circuit breaker size (delay type)	30A
	接地导线规格 Grounding wire specifications	100mm²
外形尺寸 Overall dimensions	高度×宽度×深度 Height × width × depth	464mm×325mm×823mm
	重量 Weight	56kg
温度范围 Temperature range	工作温度 Working temperature	−10～40℃
	存放温度 Storage temperature	−25～55℃

2.CV350-R 焊接电源的电气连接

电源背后的输入电源保护盒有三根导线穿过输入接线架中的三孔,并分别夹紧和固定。如图 4-27 所示,按照设备背面的"输入接线图"连接 L1、L2 和 L3。在输入回路中安装所推荐的延迟熔丝或延迟型断路器。

将机器人控制器连接到 CV350-R 电源上需要一根 K60036-5 通信电缆。CV350-R 电源需配合机器人送丝机 Auto Drive 4R100 使用。连接送丝机和 CV350-R 需要一根 14 针对 14 针的 K1785-[] (12、16 或 25 in) 控制电缆。

2. Electrical connection of CV350-R welding power supply

The input power protection box behind the power supply has three wires that pass through the three holes in the input wiring frame and are clamped and fixed respectively. Connect L1, L2 and L3 according to the "Input Wiring Diagram" on the back of the device, as shown in Figure 4-27. Install the recommended delay fuse or circuit breaker of delay type in the input circuit.

One K60036-5 communication cable is required for connecting the robot controller to the CV350-R power supply which needs to be used together with the robot wire feeder Auto Drive 4R100, while one 14-pin to 14-pin K1785-[](12, 16, or 25 inch) control cable is required for connecting the wire feeder to CV350-R.

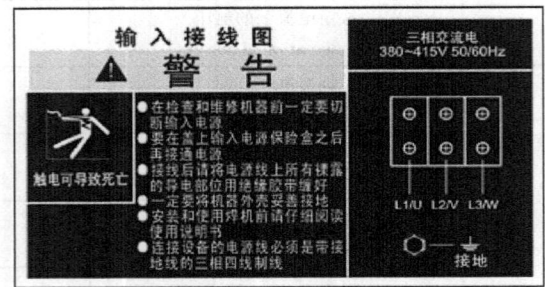

图 4-27 输入接线图
Figure 4-27 Input Wiring Diagram

3.CV350-R 焊接电源的通信

焊接电源的通信连接如图 4-28 所示。

3. Communication of CV350-R welding power supply

The communication connection of the welding power supply is shown in Figure 4-28.

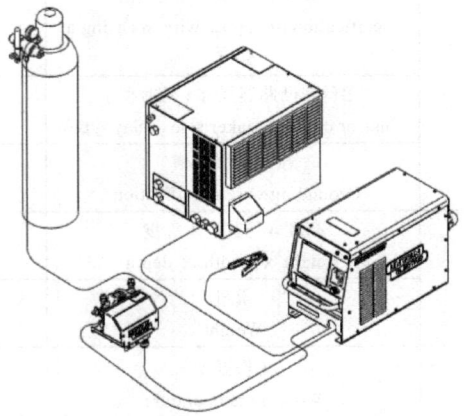

图 4-28 焊接电源的通信连接
Figure 4-28 Communication Connection of Welding Power Supply

(1) 与机器人通信电缆连接

将 CV350-R 电源开关转到"OFF"的位置。将 K60036-5 通信电缆连接到机器人控制器和 4 针连接端口，见表 4-8。

(1) Connection to the communication cable of the robot

Turn the CV350-R power switch to the "OFF" position, and connect the K60036-5 communication cable to the robot controller and the 4-pin connection port, as shown in Table 4-8.

表 4-8 电缆连接端口
Table 4-8 Cable Connection Ports

针式连接端口 Pin connection port	导线颜色 Wire color	描述 Description
A	白色 White	CAN 低 CAN low
B	蓝色或橙色 Blue or Orange	CAN 高 CAN high
C	黑色 Black	电源— Power supply –
D	红色 Red	电源+ Power supply+

(2) 与送丝机通信电缆连接

将焊接电源开关转到"OFF"的位置。将 K1785 控制电缆从送丝机上连接到焊接电源的 14 针端口。将焊接电缆连接焊接电源的"正"输出端和送丝机的接线端子。

(2) Connection to the communication cable of the wire feeder

Turn the welding power switch to the "OFF" position, connect the K1785 control cable from the wire feeder to the 14-pin port of the welding power supply, and connect the welding cable to the "positive" output terminal of the welding power supply and the wiring terminal of the wire feeder.

4. CV350-R 焊接的焊接工艺

CV350-R 焊接是一台支持直流恒压焊接的非一元化机器人电弧焊机。推荐与 GMAW 焊接工艺配合使用，包括一系列材料。表 4-9 为焊接工艺。

4. Welding process of the CV350-R welder

The CV350-R welding adopts a non-unitary robot arc welding machine that supports DC constant voltage welding, and is recommended to be used together with GMAW welding process, including a series of materials. The non-unitary CV welding mode is shown in Table 4-9.

表 4-9 CV350-R 焊接的焊接工艺
Table 4-9 Welding Process of CV350-R Welding

材料 Material	焊丝直径 Diameter of welding wire	气体 Gas
钢 Steel	1.2mm	80%Ar, 20%CO_2
	1.2mm	100%CO_2
	1.0mm	80%Ar, 20%CO_2
	1.0mm	100%CO_2

（续）

材料 Material	焊丝直径 Diameter of welding wire	气体 Gas
钢 Steel	0.9mm	80%Ar，20%CO_2
	0.9mm	100%CO_2
	0.8mm	80%Ar，20%CO_2
	0.8mm	100%CO_2
不锈钢 Stainless steel	1.2mm	98% Ar，2%CO_2
	1.0mm	
	0.9mm	

5.CV350-R 焊接电源面板

CV350-R 焊接电源面板如图 4-29 所示，其操作功能如下：

5. CV350-R welding power panel

The CV350-R welding power supply panel is shown in Figure 4-29, with its operating functions shown as follows.

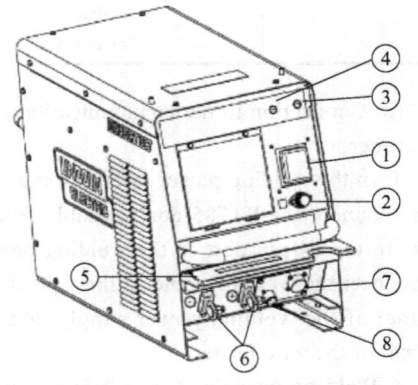 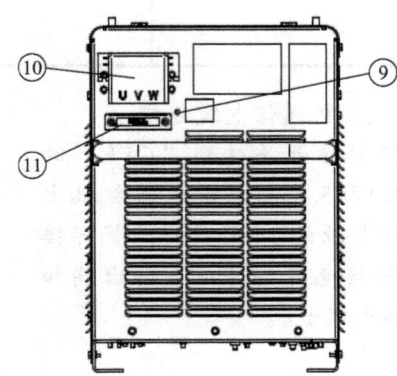

图 4-29　CV350-R 焊接电源面板
Figure 4-29　CV350-R Welding Power Supply Panel

① ON/OFF 电源开关。

② 4A 慢熔丝和熔丝盒。高压熔丝用于保护控制回路。

③ 热保护指示灯。当焊接电源处于温度过热保护状态时，该状态指示灯将亮起。在开机时指示灯也将亮起，然后熄灭。

④ 状态指示灯。机器初始化时此灯将闪烁，稳定的绿色表明初始化完成，电源已准备好。

⑤ 输出端子盖板。保护输出接线端和送丝机连接端口。在关闭电源开

① ON/OFF power switch.

② 4A slow fuse and fuse box. High voltage fuses are used to protect the control circuit.

③ Thermal protection indicator. The indicator will be on when the welding power supply is in a status of thermal protection, and will be on and then off when the power supply is stated.

④ Status indicator. When the machine is initialized, this indicator will flash, and presents a stable green color when initialization is completed and the power supply is ready.

⑤ Output terminal cover, which protects the output terminal and the connection port of the wire feeder. After turning off the power switch, the user can open

关之后，用户可打开盖板连接焊接电缆和 Auto Drive 4R100 控制电缆。

⑥ 正极和负极输出连接端口。

⑦ 送丝机连接端口。

⑧ 机器人通信连接端口。此连接端口的导线为 INVERTEC CV350R 机器人控制器供电并与其通信。

⑨ 接地螺栓。用于连接电源地线。

⑩ 输入电源保护盒。该保护盒罩住输入端子，防止操作人员触电。

⑪ 输入电缆固定夹。用于固定三相电源线。

6. CV350-R 焊接电源故障排除

在焊接过程中出现异常状况时，按照表 4-10 的要点进行检查。

the cover to connect the welding cable and Auto Drive 4R100 control cable.

⑥ Connection ports of positive and negative output.

⑦ Connection port of wire feeder.

⑧ Connection port of robot communication. The wire of this connection port supplies power to and communicates with the INVERTEC CV350R robot controller.

⑨ Grounding bolts, which are used to connect the ground wire of power supply.

⑩ Input power protection box, which covers the input terminals to prevent operators from getting electric shock.

⑪ Input cable fixing clip, used to fix three-phase power supply wires.

6. Troubleshooting of CV350-R welding power supply

When abnormal conditions occur during the welding process, check according to the key points in Table 4-10.

表 4-10　CV350-R 焊接电源故障排除

Table 4-10　Troubleshooting of CV350-R Welding Power Supply

故障 Fault	可能原因 Possible cause	解决方案 Solution
电源输出故障 Power output fault		
打开机器外壳有明显损坏 Obvious damage if found when opening the machine casing	无 None	联系当地电气授权维修部获得技术援助 Contact the local Electric Company authorized maintenance department for technical assistance
熔丝熔断或电源开关跳闸 Fuse blown or power switch tripped	1. 熔丝和电源开关容量不足 1. Insufficient capacity of fuse and power switch 2. 焊接过程输出电流过大，暂载率过高 2. Too big output current and too high temporary load rate during the welding process 3. 电源内部损坏 3. Internal damage to the power supply	1. 参见手册的安装部分，选择合适的熔丝和电源开关容量 1. Please select the appropriate fuse and power switch capacity with reference to the installation section of the manual 2. 减少输出电流和降低暂载率 2. Reduce output current and temporary load rate 3. 如有内部损坏，联系客服 3. If there is internal damage, contact the customer service personnel
机器无供电（状态指示灯不亮，风扇不运转） No power supply to the machine (the status indicator is not on, and the fan is not running)	1. 机器没有正常供电 1. No normal power supply to the machine 2. 前面板 4A 熔丝断开 2. Disconnection of the 4 ampere fuse at the front panel	1. 确保 CV350-R 正常供电 1. Ensure the normal power supply of CV350-R 2. 联系当地林肯电气授权维修部获得技术援助 2. Contact the local Electric Company authorized maintenance department for technical assistance

（续）

故障 Fault	可能原因 Possible cause	解决方案 Solution
电源输出故障 Power output fault		
热保护指示灯亮 The thermal protection indicator is on	1. 风机不运转（焊机启动时，风机应正常运转） 1. The fan is not running (when the welding machine is started, the fan should run normally) 2. 百叶窗异物堵塞 2. Blinds are blocked by foreign objects 3. 机器超负荷运转 3. Overload operation of the machine	1. 清理堵塞扇叶异物或联系服务人员检修风机 1. Clean up foreign objects blocking the fan blades or contact service personnel to repair the fan 2. 清理异物 2. Clean up foreign objects 3. 待机器冷却后，降低负载或暂载率 3. When the machine cools down, reduce the load or temporary load rate
机器无输出不能焊接 Machine cannot work without output	1. 输入电压过低或过高 1. Input voltage too low or too high 2. 如错误代码也显示（错误代码显示为状态指示灯以红色和绿色同时闪烁） 2. If the error code is also displayed (the error code is displayed as the status indicator flashing red and green simultaneously)	1. 确保输入电压符合铭牌规定 1. Ensure that the input voltage complies with the regulations on the nameplate 2. 联系当地电气授权维修部获得技术援助 2. Contact the local Electric Company authorized maintenance department for technical assistance
机器不产生满幅输出 The machine does not have full output	1. 输入电压过低 1. Input voltage too low 2. 输出电流或电压没有正确校准 2. Output current or voltage not calibrated correctly	1. 确保输入电压符合铭牌上的规定 1. Ensure that the input voltage complies with the specifications on the nameplate 2. 联系当地电气授权维修部获得技术援助 2. Contact the local Electric Company authorized maintenance department for technical assistance
焊接质量问题 Welding quality problems		
焊缝熔宽不够（输出限制在100A左右） Insufficient fusion width of the welding seam (output limited to around 100A)	焊机内部次级电流过载，机器进入自保护状态 The machine enters a self-protection state, as its internal secondary current is overloaded	1. 改变焊接工艺 1. Change welding process 2. 降低电流输出 2. Reduce current output
焊接性能总体下降 Overall decline in welding performance	1. 送丝不稳 1. Unstable wire feeding 2. 焊接模式选择不当 2. Improper welding mode 3. 输出电源或电压没有正确校正 3. Incorrect calibration of output power or voltage	1. 检查送丝装置，确保连接畅通 1. Check the wire feeding mechanism to ensure smooth connection 2. 选择与焊接工艺匹配的焊接模式 2. Choose a proper welding mode that matches the welding process 3. 联系当地电气授权维修部获得技术援助 3. Contact the local Electric Company authorized maintenance department for technical assistance

（续）

故障 Fault	可能原因 Possible cause	解决方案 Solution
焊接质量问题 Welding quality problems		
电弧过长且不稳定 The arc is too long and unstable	1. 导电嘴和焊丝不匹配 1. Improper matching between the tip and welding wire 2. 干伸长度不当 2. Improper stick out	1. 选择合适的导电嘴 1. Choose a suitable tip 2. 调整焊丝干伸长度 2. Adjust the stick-out of the welding wire
机器人上出现电弧缺失故障 Arc missing fault on the robot	1. 送丝不畅 1. Poor wire feeding 2. 导丝管扭曲导致送丝速度降低 2. Twisted wire guiding conduit leads to a decrease in wire feeding speed	1. 将通向送丝机的导管拉直 1. Straighten the conduit leading to the wire feeder 2. 使用一根较短的导管 2. Use a shorter conduit

三、机器人送丝机构的选型

弧焊机器人配备的送丝机构包括送丝机、送丝软管和焊枪三部分。弧焊机器人的送丝稳定性是关系到焊接能否连续稳定进行的重要问题。

1. 送丝机的选择

(1) 送丝机的类型

1) 送丝机按安装方式分为一体式和分离式两种。将送丝机安装在机器人上臂的后部上面与机器人组成一体为一体式；将送丝机与机器人分开安装为分离式。

由于一体式送丝机到焊枪的距离比分离式的短，连接送丝机和焊枪的软管也短，所以一体式送丝阻力比分离式的小。从提高送丝稳定性的角度看，一体式比分离式要好一些。

一体式送丝机，虽然送丝软管比较短，但为了方便更换焊丝盘，而把焊丝盘或焊丝桶有时放在远离机器人的安全围栏之外，这就要求送丝机有足够的拉力从较长的导丝管中把焊丝从焊丝盘（桶）拉过来，再经过软管推向焊枪。对于这种情况，和送丝软管比较长的分离式送丝机一样，应选用送丝力较大的送丝机。忽视这一点，往往会出现送丝不稳定甚至中断送丝

III. Selection of the Wire Feeding Mechanism of Robot

The wire feeding mechanism equipped with the arc welding robot includes three parts: a wire feeder, a wire feeding hose, and a welding gun. The stability of wire feeding of arc welding robots is important as it determines the continuous and stable welding.

1. Selection of wire feeder

(1) Type of wire feeder

1) The wire feeder is divided into two types according to installation method: integrated and separated types. For integrated wire feeders, the wire feeder is stalled on the back of the robot's upper arm, integrating itself with the robot as a whole, while separated wire feeders are not installed on the the robot.

The resistance of wire feeding of the integrated wire feeder is smaller than that of the separated ones, due to a shorter distance, and thus a shorter connecting hose, between the integrated wire feeder and the welding gun. From the perspective of improving the stability of wire feeding, the integrated wire feeders are better than the separated ones.

For easier replacement of welding wire reel, the welding wire reel or barrel is sometimes placed outside the safety fence far away from the robot. Therefore, an integrated wire feeder, although with a relatively shorter wire feeding hose, shall have sufficient drawing force to draw the welding wire through the relatively long wire guiding conduit from the reel (barrel), and then push it towards the welding gun through the hose. In this case, a wire feeder with a relatively large wire feeding force shall be selected, just the same as it is selected for a

的现象。

目前，弧焊机器人送丝机采用一体式的安装方式已越来越多了，但对要在焊接过程中进行自动更换焊枪（变换焊丝直径或种类）的机器人，必须选用分离式送丝机。

2）送丝机按滚轮数分为一对滚轮和两对滚轮两种。送丝机的结构有一对送丝滚轮的，也有两对滚轮的；有只用一个电动机驱动一对或两对滚轮的，也有用两个电动机分别驱动两对滚轮的。

3）送丝机按控制方式分为开环和闭环两种。目前，大部分送丝机仍采用开环的控制方法，也有一些采用装有光敏传感器（或编码器）的伺服电动机，使送丝速度实现闭环控制，不受网路电压或送丝阻力波动的影响，保证送丝速度的稳定性。

4）送丝机按送丝动力方向分为推丝式、拉丝式和推拉丝式三种。

① 推丝式。此种类型的送丝机主要用于直径为0.8～2.0mm的焊丝；它是应用最广的一种送丝方式。其特点是焊枪结构简单轻便，易于操作，但焊丝需要经过较长的送丝软管（一般软管长度为3～5m）才能进入焊枪，焊枪在软管中受到较大阻力，影响送丝稳定性。

② 拉丝式。此种类型的送丝机主要用于细焊丝（焊丝直径小于或等于0.8mm），因为细丝刚性小，推丝过程易变形，难以推丝。拉丝时送丝电动机与焊丝盘均安装在焊枪上，由于送丝力较小，所以拉丝电动机功率较小，但尽管如此，拉丝式焊枪仍然较重。可见拉丝式虽保证了送丝的稳定性，但由于焊枪较重，增加了机器人的载荷，而且焊枪操作范围受到限制。

separated wire feeder with a longer wire feeding hose. Otherwise, unstable or even interrupted wire feeding is likely to occur.

At present, there are more and more integrated wire feeders of arc welding robots, but separate wire feeders must be selected for robots that need to automatically replace the welding gun (changing the diameter or type of welding wire) during the welding process.

2) The wire feeder can have either one pair or two pairs of rollers. The structure of the wire feeder can have either a pair or two pairs of rollers for wire feeding; a motor can be adopted to drive either one pair or both of the two pairs of rollers, while two motors can be adopted to drive the two pairs of rollers separately.

3) There are open-loop controlled wire feeders and closed-loop controlled ones according to the control methods. At present, most wire feeders still use the open-loop control method, while others adopt servo motors equipped with photosensitive sensors (or encoders) to achieve closed-loop control of wire feeding speed, which is not affected by network voltage or fluctuations in the wire feeding resistance, ensuring the stability of wire feeding speed.

4) There are three types of wire feeder according to the direction of wire feeding force: wire pushing, wire drawing, and wire pushing-drawing types.

① Wire pushing type. Wire feeders of this type are the most widely used and mainly for welding wires with a diameter of 0.8–2.0mm, the characteristics of which is that the equipped welding guns have a simple and lightweight structure and are easy to operate. However, the welding wires need to pass through a relatively long, 3-5m in general, wire feeding hose to enter the welding gun and are subject to a relatively large resistance in the hose, which affects the stability of wire feeding.

② Wire drawing type. Wire feeders of this type are mainly used for fine welding wires with a diameter of less than or equal to 0.8mm, as such fine wires have low rigidity and are easy to be deformed during the pushing process, and therefore, difficult to be performed with pushing. For wire drawing, both the wire feeding motor and welding wire reel are installed on the welding gun, which is relatively heavy, although the power of the wire drawing motor is relatively low as the wire feeding force is small. It can be seen that although the stability of wire feeding is ensured by the wire feeders of this type, the

③推拉丝式。此种类型的送丝机可以增加焊枪操作范围，送丝软管可以加长到10m。除推丝机外，还在焊枪上加装了拉丝机。推丝是主要动力，而拉丝机只是将焊丝拉直，以减小推丝阻力。推力与拉力必须很好地配合，通常拉丝速度应稍快于推丝。这种方式虽有一些优点，但由于结构复杂，调整麻烦，同时焊枪较重，因此实际应用并不多。

(2) 推式送丝机的结构

推式送丝机是应用最广的送丝机，送丝电动机、送丝滚轮和矫直机构等都装在薄铁板压制的机架上。送丝机核心部分的结构如图4-30所示。

③ Wire pushing-drawing type. For wire feeders of this type, the operating range of the welding gun can be increased, and the wire feeding hose can be extended to 10m. In addition to the wire pusher, a wire drawer has also been installed on the welding gun. The pushing force is the main power, while the wire drawer only straightens the welding wire to reduce the pushing resistance. The pushing and drawing forces must match well, with the speed of wire drawing being slightly faster usually than that of wire pushing. Although having some advantages, wire feeders of this type are not widely used due to their complex structure which is inconvenient to adjust, and their relatively heavy welding guns.

(2) Structure of the pushing type wire feeder

The pushing type wire feeders are the most widely used, and their wire feeding motors, wire feeding rollers, and straightening mechanisms are all installed on a frame made of pressed thin iron plates. The structure of the core parts of the wire feeder is shown in Figure 4-30.

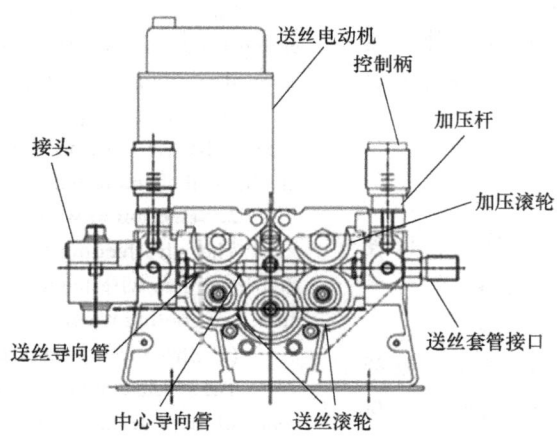

图4-30 送丝机的结构
Figure 4-30 Structure of the Wire Feeder

① 送丝电动机：送丝电动机驱动送丝滚轮，为送丝提供动力。送丝电动机由弧焊焊机电源控制，弧焊焊机电源根据焊接工艺控制送丝速度。

② 加压杆：通过调整预紧力，用于压紧焊丝，控制柄可旋转调节压紧度。

③ 送丝滚轮：电动机带动主动轮旋转，为送丝提供动力。

① Wire feeding motor: it drives the wire feeding roller to provide impetus for wire feeding, and is controlled by the arc welding machine power supply, which controls the wire feeding speed according to the welding technology.

② Pressure rod: it is used to compress the welding wire by adjusting the pre-tightening force, and the control handle can be rotated to adjust the compression degree.

③ Wire feeding roller: the motor drives the driving wheel to rotate, providing impetus for wire feeding.

④ 加压滚轮：加压滚轮将焊丝压入送丝轮上的送丝槽，增大焊丝与送丝轮的摩擦，使焊丝平稳送出。

送丝机以送丝电动机与减速箱为主体，在其上安装送丝滚轮和加压滚轮。加压滚轮通过滚轮架和加压手柄压向送丝轮。根据焊丝直径不同，调节加压手柄可以调节压紧力大小。在它的后面是焊丝校直机构，它由3个滚轮组成，它们之间的相对距离可视焊丝情况进行调整。

在送丝轮的前面是焊丝导向部分，它由导向衬套和出口导向管组成。焊丝从送丝轮的沟槽内送出，由于正对着导向管入口，焊丝顺利地进入送丝软管。为了固定导向衬套，机体上还设有压簧。

送丝滚轮的槽一般有$\phi 0.8mm$、$\phi 1.0mm$、$\phi 1.2mm$三种，应按照焊丝的直径选择相应的输送滚轮。

一般采用他励直流伺服机作为送丝电动机，其机械特征性平硬并可无级调节。

2. 送丝软管的选择

送丝软管是集送丝、导电、输气和通冷却水为一体的输送设备。

(1) 软管结构

软管的结构如图 4-31 所示。软管的中心是一根通焊丝同时也起输送保护气作用的导丝管，外面缠绕导电的多芯电缆，有的电缆中央还有两根冷却水循环的管子，最外面包敷一层绝缘橡胶。

④ Pressure rollers: the pressure rollers press the welding wire into the wire feeding groove on the wire feeding roller, which increases the friction between the welding wire and the wire feeding roller and ensure the smooth feeding of the welding wire.

The wire feeder is mainly composed of the wire feeding motor and gearbox, on which the wire feeding roller and pressure roller are installed. The pressure roller presses toward the wire feeding roller through the roller frame and pressure handle. The compression force can be controlled by adjusting the pressure handle, depending on the diameter of the welding wire. Behind it is the welding wire straightening mechanism, which consists of three rollers whose relative distance can be adjusted based on the conditions of the welding wire.

In front of the wire feeding roller is the welding wire guiding section, which is composed of a guiding sleeve and an outlet guiding tube. The welding wire is sent out from the groove of the wire feeding roller and, as it is directly opposite to the guiding tube inlet, enters smoothly the wire feeding hose. A pressure spring is installed on the feeder body in order to fix the guiding sleeve.

There are three sizes of groove of the wire feeding roller in general: $\phi 0.8mm$, $\phi 1.0mm$, and $\phi 1.2mm$, which shall be selected according to the diameter of the welding wires.

Generally, a separately excited DC servo-motor is used as the wire feeding motor, which has a flat and hard mechanical characteristic and can be adjusted steplessly.

2. Selection of wire feeding hose

The wire feeding hose is a conveying equipment that integrates wire feeding, conductivity, gas transmission, and cooling water circulation.

(1) Structure of the hose

The structure of the hose is shown in Figure 4-31. In the center of the hose there is a wire guiding conduit that carries the welding wire and also serves to transmit the protective gas. It is wrapped around by conductive multi-core cables, some of which have two pipes for cooling water circulation in their center, with a layer of insulating rubber wrapped on the outside.

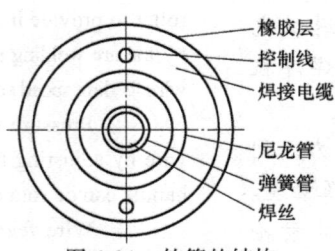

图 4-31　软管的结构
Figure 4-31　Structure of the Hose

焊丝直径与软管内径要配合恰当。软管直径过小，焊丝与软管内壁接触面增大，送丝阻力增大。此时如果软管内有杂质，常常造成焊丝在软管中卡死；软管内径过大，焊丝在软管内呈波浪形前进，在推式送丝过程中将增大送丝阻力。焊丝直径与软管内径匹配见表 4-11。

The diameter of the welding wire should match the inner diameter of the hose appropriately. If the diameter of the hose is too small, the contact surface between the welding wire and the inner wall of the hose will increase, so will the wire feeding resistance. At this time, impurities (if any) in the hose often cause the welding wire to get stuck in the hose; if the inner diameter of the hose is too large, the welding wire will move forward in a wavy shape inside the hose, which will increase the wire feeding resistance during the process of wire feeding of the pushing type. The matching between the welding wire diameter and the hose inner diameter is shown in Table 4-11.

表 4-11 焊丝直径与软管内径匹配
Table 4-11 Matching between the Welding Wire Diameter and the Hose Inner Diameter

焊丝直径 /mm Welding wire diameter/mm	软管内径 /mm Hose inner diameter/mm	焊丝直径 /mm Welding wire diameter/mm	软管内径 /mm Hose inner diameter/mm
0.8～1.0	1.5	1.4～2.0	3.2
1.0～1.4	2.5	2.0～3.5	4.7

(2) 送丝不稳的因素

软管阻力过大是造成弧焊机器人送丝不稳定的重要因素。原因有以下几个方面：

① 用的导丝管内径与焊丝直径不匹配。

② 导丝管内积存由焊丝表面剥落下来的铜屑或铜屑过多。

③ 软管的弯曲程度过大。

目前，越来越多的机器人公司把安装在机器人上臂的送丝机稍微向上翘，有的还使送丝机能做左右小角度自由摆动，目的都是为了减少软管的弯曲，保证送丝速度的稳定性。

3. 焊枪的选择

焊枪的种类很多，根据焊接工艺的不同，选择相应的焊枪。对于机器人弧焊工作站面而言，采用的是熔化极气体保护焊。

(1) 焊枪的选择依据

对于机器人弧焊系统，选择焊枪时，应考虑以下几个方面：

(2) Factors causing unstable wire feeding

The excessive resistance of the hose is an important factor causing unstable wire feeding by arc welding robots. The reasons are as follows:

① Improper matching between the inner diameter of the wire guiding conduit and the welding wire diameter.

② Excessive accumulation of copper or steel particles peeling off the surface of the welding wire in the wire guiding conduit.

③ Excessive bending of the hose.

At present, some wire feeders installed on the upper arm of the robot are tilted upward, and some can even swing freely at small angles from side to side, as more and more robot manufacturers make them this way, to reduce the bending of the hose and ensure the stability of the wire feeding speed.

3. Selection of welding gun

There are many types of welding gun, which shall be selected according to different welding processes. The gas metal arc welding is selected for the robot arc welding workstation.

(1) Selection basis for welding gun

For robot arc welding systems, the following aspects shall be considered when selecting the welding gun:

① 选择自动型焊枪，不要选择半自动型焊枪。半自动型焊枪用于人工焊接，不能用于机器人焊接。

② 根据焊丝的粗细、焊接电流的大小以及负载率等因素选择空冷式或水冷式的结构。

细丝焊时因焊接电流较小，可选用空冷式焊枪结构；粗丝焊时焊接电流较大，应选用水冷式的焊枪结构。

空冷式和水冷式焊枪的技术参数比较见表4-12。

① Choose an automatic welding gun instead of a semi-automatic one, which is used for manual welding rather than robot welding.

② Choose an air-cooled or water-cooled structure based on factors such as the thickness of the welding wire, welding current, and load rate.

An air-cooled structure of welding gun can be selected for the welding of fine wires as the welding current is low, while an water-cooled structure of welding gun can be selected for the welding of coarse wires as the welding current is high.

The comparison of technical parameters between air-cooled and water-cooled welding guns is shown in Table 4-12.

表4-12 空冷式和水冷式焊枪的技术参数比较
Table 4-12 Comparison of Technical Parameters between Air-cooled and Water-cooled Welding Guns

型号 Model	Robo 7G	Robo 7W
冷却方式 Cooling method	空冷 Air-cooled	水冷 Water-cooled
暂载率(10min) Temporary load rate (10min)	60%	100%
焊接电流(Mix) Welding current (Mix)	325A	400A
焊接电流(CO_2) Welding current (CO_2)	360A	450A
焊丝直径 Welding wire diameter	1.0～1.2mm	1.0～1.6mm

根据机器人的结构选择内置式或外置式焊枪。内置式焊枪安装要求机器人末端轴的法兰盘必须是中空的。一般专用焊接机器人如安川MA1400，其末端轴的法兰盘是中空，应选择内置式焊枪；通用型机器人如安川MH6应选择外置式焊枪。

根据焊接电流、焊枪角度选择焊枪。焊接机器人用焊枪大部分和手工半自动焊用的焊枪基本相同。弯曲角

Choose the built-in or externally placed welding guns based on the structure of the robot. For the installation of built-in welding guns, the flange plate of the robot's end axis must be hollow. In general, specialized welding robots such as the Yaskawa MA1400 have a hollow flange plate at the end axis, and built-in welding guns shall be selected for them, while universal robots such as the Yaskawa MH6 should shall have externally placed welding guns.

Select the welding gun based on the welding current and angle of welding gun. The welding guns used for welding robots are mostly the same as the welding

一般都小于45°。根据工件特点选择不同弯曲角，以改善焊枪的可达性。若弯曲角选得过大，送丝阻力会加大，送丝速度容易不稳定，而角度过小，一旦导电嘴稍有磨损，常会出现导电不良的现象。

设备和人身安全方面考虑应选择带防撞传感器的焊枪。

(2) 焊枪的结构

焊枪一般由喷嘴、导电嘴、气体分流环、绝缘套、枪管（枪颈）及防碰撞传感器（可选）等部分组成，如图4-32所示。

guns used for manual semi-automatic welding. The bending angle is generally less than 45°. Different angles can be selected according to the characteristics of the workpiece to improve the reachability of the welding gun. Too large bending angles will lead to an increase in the wire feeding resistance and the possibility of unstable wire feeding speed, while too small ones are easy to cause poor conductivity once the conductive nozzle is even slightly worn.

Welding guns with anti-collision sensors shall be selected for the safety of equipment and personnel.

(2) Structure of the welding gun

The welding gun is generally composed of the nozzle, conductive nozzle, gas shunt ring, insulation sleeve, gun barrel (gun neck), and anti-collision sensor (optional), as shown in Figure 4-32.

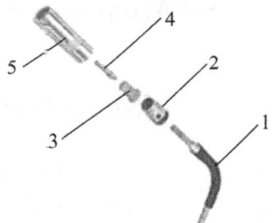

图4-32 焊枪的结构
Figure 4-32 Structure of Welding Gun
1—枪颈 2—绝缘套 3—气体分流环 4—导电嘴 5—喷嘴
1— Gun neck 2— Insulation sleeve 3— Gas shunt ring 4— Conductive nozzle 5— Nozzle

为了更稳定地将电流导向电弧区，在焊枪的出口装一个纯铜导电嘴。导电嘴的孔径和长度因不同直径的焊丝而不同。既要保证导电可靠，又要尽可能减小焊丝在导电嘴的行进路程，以减少送丝阻力，保证送丝的通畅。导电嘴有成锥形、椭圆形、镶套形、锥台形、圆柱形、半圆形和滚轮形七种形式。

喷嘴是焊枪上的重要零件，其作用是向焊接区域输送保护气体防止焊丝末端、电弧和熔池与空气接触。喷嘴的材料、形状和尺寸对气体保护效果和焊接质量有着十分密切的关系。

A red copper nozzle is installed at the outlet of the welding gun in order to guide the current more stably to the arc zone. The aperture and length of the conductive nozzle vary depending on the diameter of the welding wire. It is necessary to ensure reliable conductivity while minimizing the travel distance of the welding wire on the conductive nozzle to reduce the resistance and ensure the smoothness of wire feeding. There are seven forms of conductive nozzle: conical, elliptical, inlaid, pyramidal, cylindrical, and semi-circular shaped, as well as roller type.

The nozzle is an important component on the welding gun, which is used to deliver protective gas to the welding area to prevent the end of the welding wire, arc, and molten pool from coming into contact with air. The material, shape, and size of the nozzle are closely related to the effectiveness of gas protection and weld-

为了减少分流环的附着力，喷嘴应由熔点较高、导热性较好的材料制造，有些表面还需镀铬，以提高其表面光洁点和熔点。

（3）防撞传感器

对于弧焊机器人，除了要选好焊枪以外，还必须在机器人的焊枪把持架上配备防撞传感器。防撞传感器的作用是当机器人在运动时，万一焊枪碰到障碍物，能立即使机器人停止运动，避免损坏焊枪或机器人。

ing quality. In order to reduce the adhesion of the shunt ring, the nozzle shall be made of materials with relatively high melting point and good thermal conductivity, the surface of which needs to be chrome plated also in some cases to improve the surface smoothness and melting point.

(3) Anti-collision sensor

When the welding gun has been selected for the arc welding robot, anti-collision sensors must be equipped on the welding gun holder of the robot. The function of the anti-collision sensor is to immediately stop the robot from moving in case the welding gun encounters an obstacle when the robot is moving, to avoid damaging the welding gun or the robot.

任务 4.3　施工图的设计及建模
Task 4.3　Design and Modeling of Detailed Drawings

【知识目标】

1. 掌握利用 SOLIDWORKS 或者 NX 软件为典型工业机器人弧焊工作站系统施工图建模。

2. 了解工业机器人弧焊工作站系统施工工艺流程。

3. 掌握电气原理图方案设计知识。

【技能目标】

1. 能够根据工业要求选择正确的软件完成建模。

2. 能编写工业机器人弧焊工作站系统施工图的设计选型方案。

【素质目标】

1. 培养学生一丝不苟、精益求精的工匠精神。

2. 使学生树立职业理想，做好人生规划。

[Knowledge Objectives]

1. To master the modeling of detailed drawings for typical industrial robot arc welding workstation systems with the SOLIDWORKS or NX software.

2. To understand the flow of construction process of industrial robot arc welding workstation systems.

3. To master the knowledge of scheme design of electrical schematic diagram.

[Skill Objectives]

1. To be able to select the correct software to complete modeling according to industrial requirements.

2. To be able to compile design selection schemes for detailed drawings of the industrial robot arc welding workstation system.

[Competence Objectives]

1. To cultivate in the student a craftsmanship spirit of being meticulous when constantly striving for perfection.

2. To enable the student to establish their career aspirations and make proper plans for their life.

【任务情景】

通过 SOLIDWORKS 或者 NX 软件为典型工业机器人弧焊工作站系统施工图建模,了解工业机器人弧焊工作站系统施工工艺流程,编写工业机器人弧焊工作站系统施工图的设计选型方案,掌握电气原理图方案设计知识。

【任务分析】

在项目实施的过程中,当完成方案的设计和设备选型以后,在具体生产、安装和调试阶段,往往需要一个团队来完成。负责设计和施工的人员并不一定是同一人或同一小组,因此必须先设计出具体系统的施工图,这样才能便于实施的更好开展。需设计的图样包括设备布局图、系统框图和电气原理图。

【知识准备】

4.3.1 设备布局图

焊接机器人工作站组成如图 4-33 所示。

[Task Scenario]

You are required to perform modelling of detailed drawings of a typical industrial robot arc welding workstation system with the SOLIDWORKS or NX software, understand the flow of construction process of the system, and compile a design selection scheme for detailed drawings of the system, as well as to master the knowledge of scheme design of electrical schematic diagram.

[Task Analysis]

In the process of program implementation, when the design scheme and and equipment selection have been completed, a team is often required to complete production, installation, and commissioning at the corresponding stages. As the person or team responsible for design may not be the same as that of for construction, it is necessary to first complete the detailed drawings of the specific system in order to have better implementation of the program. The drawings to be designed include: equipment layout, system chart, and electrical schematic diagram.

[Assumed Knowledge]

4.3.1 Equipment Layout Diagram

Composition of the welding robot workstation is shown in Figure 4-33.

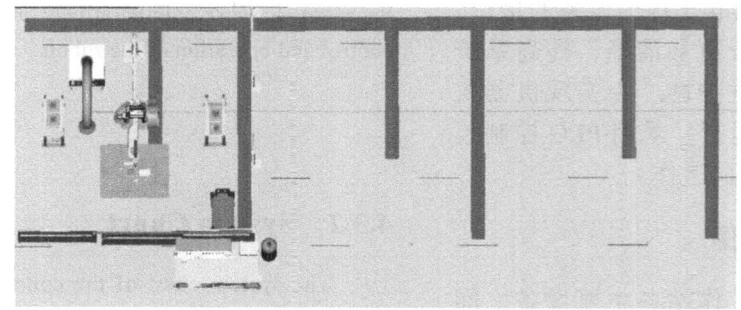

图 4-33 焊接机器人工作站组成

Figure 4-33 Composition of Welding Robot Workstation

1) 一台 FANUC-30iA 弧焊机器人，机器人选型采用德系车生产企业普遍采用的机器人型号，搭建企业真实的工作环境。

2) 一台空气压缩机，空气压缩机为整个工作站供气。

3) 一台周边控制柜(内有 SIMATIC S7-1200 系列的 PLC)和 ET200S 分布式 I/O，它们是整个控制系统的指挥中心。当机器人处于外部控制时，由 PLC 发布指令。

4) 一台带有触摸屏(SIMATIC TP270-10)的操作台，用于检测系统状态。

5) 一台控制柜及其他附属周边设备。

6) 安全光栅及安全门，作为系统重要的安全防护。

周边设备的控制和焊接过程的控制由配套的机器人控制器内在的 SIMATIC 公司生产的 (CP5614) 集成通信处理器、周边柜 (PLC) 和用户焊接示教程序来共同完成。

为了保证系统可靠性和可维护性，采用现场总线连接和 PLC 控制系统控制这些设备按照工艺流程动作完成作业任务。通过这种方式可以总体上节约控制系统硬件成本，使控制硬件模块化，简化和标准化各个设备的接口，使控制任务划分更加清晰，提高系统的可靠性和可维护性。为实现机器人的外部控制和运行，采用 PLC 控制技术，使机器人协同工作。

4.3.2 系统框图

典型焊接工作站各主要设备之间的连接关系及控制关系的系统框图如图 4-34 所示。系统框图根据给定的系统功能要求，进行相应的焊接工作站

1) A FANUC-30iA arc welding robot with a model that is commonly adopted by German car manufacturers; create a real working environment for the enterprise.

2) An air compressor, which supplies air to the entire workstation.

3) A peripheral control cabinet (containing a PLC of SIMATIC S7-1200 series) and ET200S distributed I/O, which are the command center of the entire control system. The PLC issues instructions when the robot is under external control.

4) A console with a touch screen (SIMATIC TP270-10) for detecting the system status.

5) A control cabinet and other auxiliary peripheral equipment.

6) Safety grating and safety door, providing important safety protections for the system.

The control of peripheral equipment and welding process is jointly completed by the (CP5614) integrated communication processor, peripheral cabinet (PLC), and the welding teaching program for the user, produced by SIMATIC Company that also produces the matching robot controller.

In order to ensure the reliability and maintainability of the system, the fieldbus connection and PLC control system are adopted to control these equipment to operate and complete the operation tasks according to the process flow. In this way, the overall cost of control system hardware can be saved, the control hardware modularized, the interfaces of various devices simplified and standardized, and the division of control task clearer, which improves the reliability and maintainability of the system. The PLC control technology is adopted to enable robots to work collaboratively, and achieve external control and operation of the robots.

4.3.2 System Chart

The system chart of the connection and control relationships between the main equipment of a typical welding workstation is shown in Figure 4-34. In the system chart, the welding workstation is designed according to the given functional requirements of the system. At the beginning of design, it is necessary to design a

系统设计。在设计之初，需要设计系统框图，为接下来的电路和程序设计提供一个基础。

system chart which provides a foundation for the subsequent design of circuits and procedures.

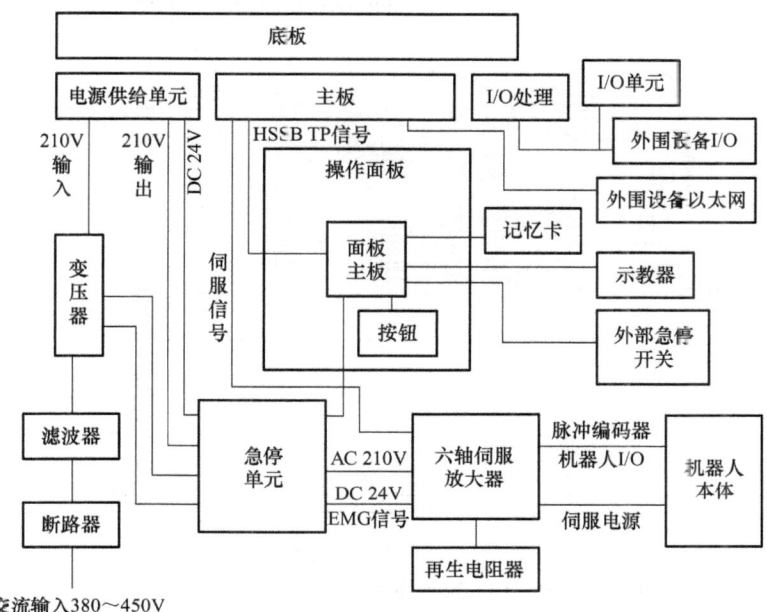

图 4-34 典型焊接工作站各主要设备之间的连接关系及控制关系系统框图

Figure 4-34　System Chart of the Connection and Control Relationship between the Main Equipment of a Typical Welding Workstation

4.3.3　电气原理图

电气原理图是用来表明设备电气的工作原理及各电器元件的作用和相互之间的关系的一种表示方式。运用电气原理图的方法和技巧，对于分析电气线路、排除电路故障、编写程序是十分有益的。电气原理图一般由电气元件分布图、主回路电气原理图、开关电源电路原理图、信号分配电气原理图、接线端子与现场信号接线原理图等部分组成。

这里将部分电气原理图展示给大家，为大家演示如何绘制电气原理图。未绘制的图样请同学们按要求自行绘制。

4.3.3　Electrical Schematic Diagram

The electrical schematic diagram is a representation used to indicate the working principle of equipment electrification and the functions of various electrical components, and their relationships. The use of electrical schematic methods and techniques is very beneficial for analyzing electrical circuits, circuit troubleshooting, and programming. The electrical schematic diagram generally consists of such parts as electrical component distribution diagram, electrical schematic diagram of main circuits, circuit schematic diagram of switch power supply, signal distribution electrical schematic diagram, wiring terminal and on-site signal wiring schematic diagram.

Here, some electrical schematic diagrams are shown to demonstrate how to prepare electrical schematic diagrams. For drawings that have not been prepared, please prepare them by yourself according to the requirements.

1. 主机架配置图

在主机架配置图中详细标注了主机上 PLC 的输入输出位置以及与触摸屏的连接。主机架配置图如图 4-35 所示。

1. Configuration diagram of the host frame

The input and output positions of the PLC on the host and its connection to the touch screen are detailed in the configuration diagram of the host frame. The configuration diagram of the host frame is shown in Figure 4-35.

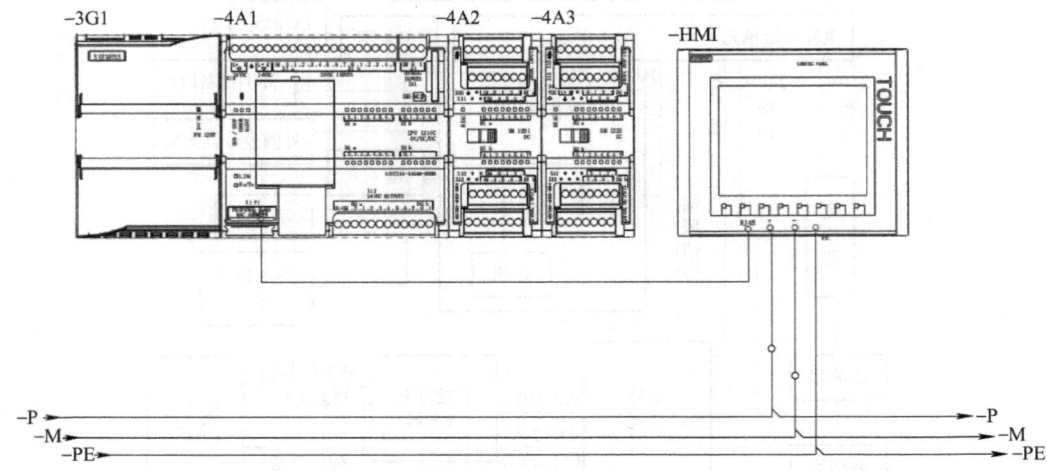

图 4-35 主机架配置图

Figure 4-35 Configuration diagram of the Host Frame

2. 电源回路电气原理图

电气柜进线经过柜门总断路器进入柜内第一层断路器，这是整个电柜的主电源进线。电源回路电气原理图如图 4-36 所示。

柜内断路器分别对变频器机器人、开关电源等分配控制电源。部分柜内断路器电源分配如图 4-37 所示。

3. 信号分配电气原理图

PLC 的信号输入全部由 PLC 模块端子接到电柜第四层对应的接线端子，外围信号接入到接线端子然后完成信号采集。数字量信号输入接线如图 4-38 所示。

PLC 输出信号通过控制继电器控制现场设备，继电器的一路常开触点接到电柜第五层相应的接线端子。数字量信号输出接线如图 4-39 所示。

2. Electrical schematic diagram of power supply circuit

The incoming lines of the electrical cabinet enter the circuit breaker on the first layer inside the cabinet through the main circuit breaker of the cabinet door, which are the incoming lines of the main power supply of the entire electrical cabinet. The electrical schematic diagram of the power supply circuit is shown in Figure 4-36.

The circuit breakers inside the cabinet distribute control power to the frequency converter robots, switch power supplies, etc. Some of the power distribution of the circuit breakers inside the cabinet is shown in Figure 4-37.

3. Electrical schematic diagram of signal distribution

All the signal inputs of the PLC are connected from the PLC module terminals to the corresponding wiring terminals on the fourth layer of the electrical cabinet, and the peripheral signals are connected to the wiring terminals to complete signal acquisition. The digital signal input wiring is shown in Figure 4-38.

The PLC output signal controls the on-site equipment through a control relay, the normally open points of which is connected to the corresponding wiring terminal on the fifth layer of the electrical cabinet. The digital signal output wiring is shown in Figure 4-39.

项目 4　典型工业机器人弧焊工作站系统的设计及应用

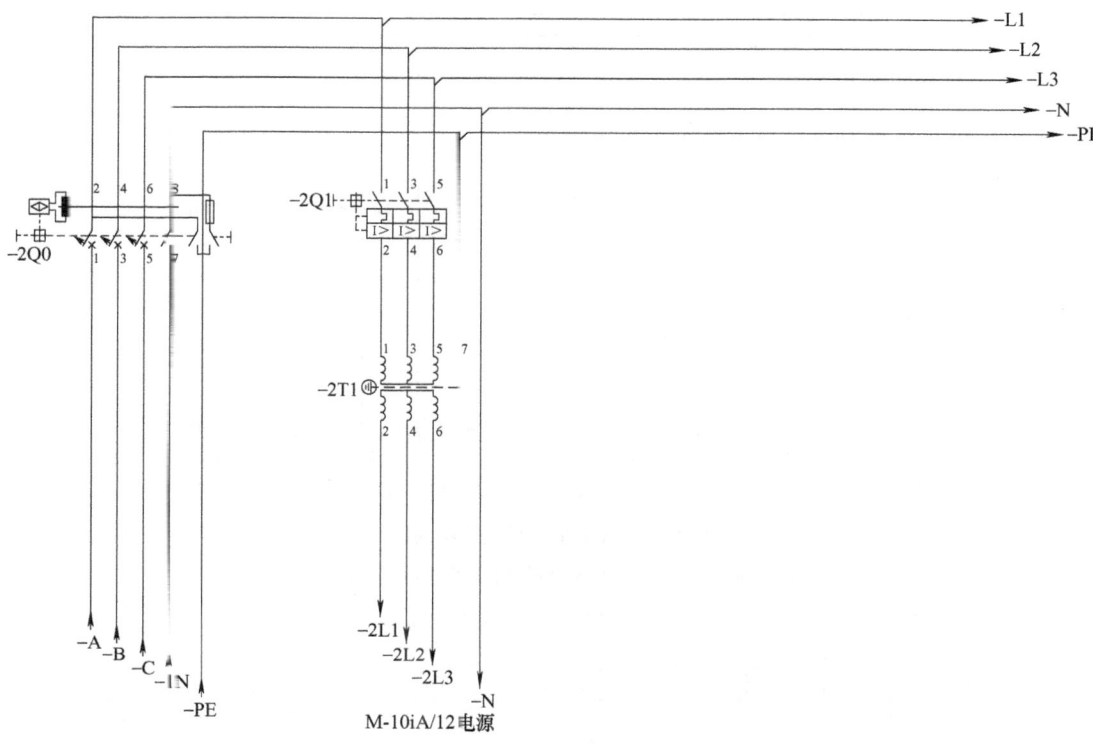

图 4-36　电源回路电气原理图

Figure 4-36　Electrical Schematic Diagram of the Power Supply Circuit

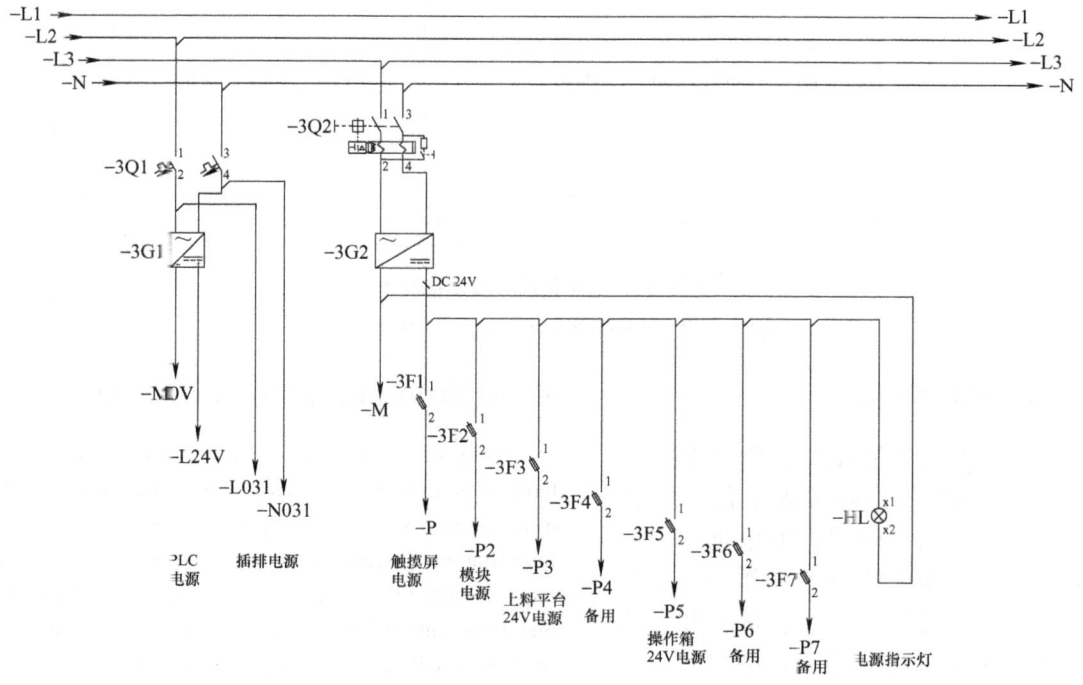

图 4-37　柜内断路器电源分配图

Figure 4-37　Power Distribution Diagram of Circuit Breakers Inside the Cabinet

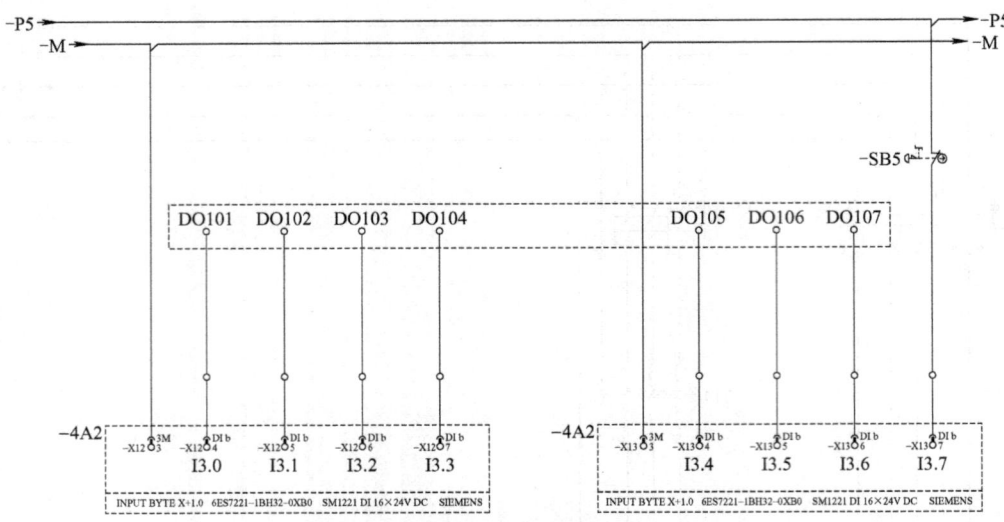

图 4-38 数字量信号输入接线图

Figure 4-38　Digital Signal Input Wiring Diagram

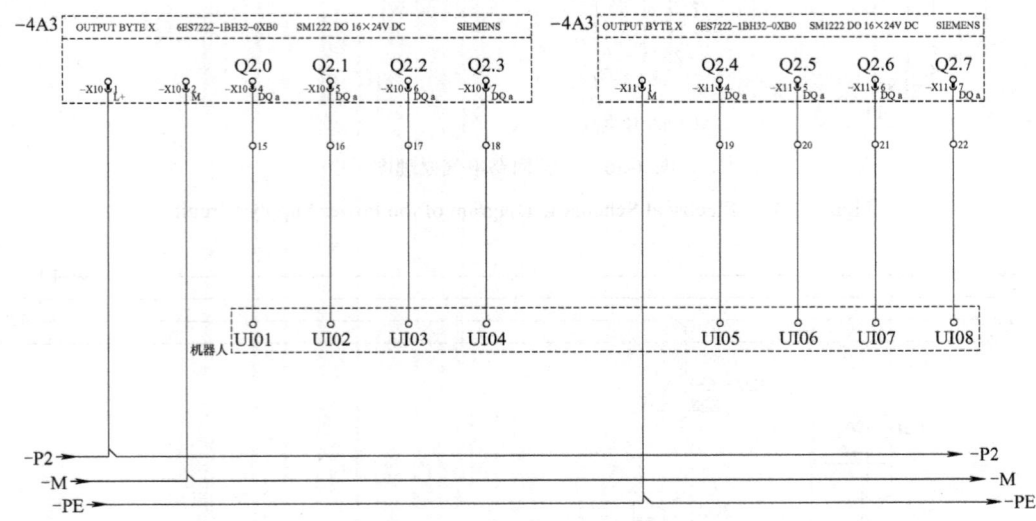

图 4-39 数字量信号输出接线图

Figure 4-39　Digital Signal Output Wiring Diagram

4.3.4　机械零件图

任何机械都是由许多零件组成的，制造机器就必须先制造零件。零件图就是制造和检验零件的依据，它依据零件在机器中的位置和作用，对零件在外形、结构、尺寸、材料和技术要求等方面都提出了一定的要求。一张焊接实训室里面的柔性平台的机械图样如图 4-40 所示。

4.3.4　Drawing of Mechanical Parts

Any machine is composed of many parts, which have to be manufactured first before the manufacturing of machine. The drawing of parts are the basis for manufacturing and inspection of parts, as they specify certain requirements for the appearance, structure, dimensions, materials, and technical requirements of the parts based on their position and function in the machine. A mechanical drawing of the flexible platform in the welding training room is shown in Figure 4-40.

项目 4　典型工业机器人弧焊工作站系统的设计及应用

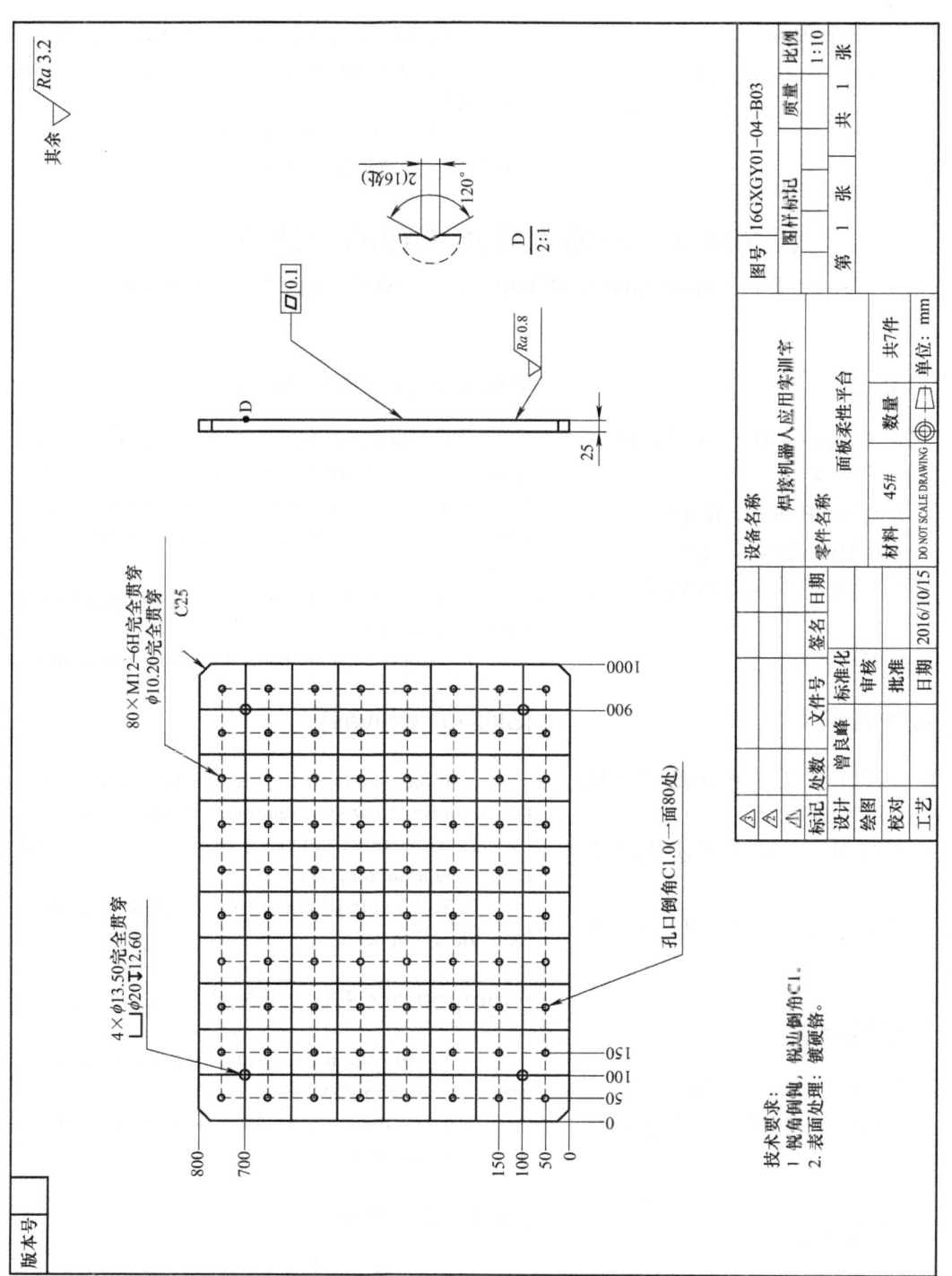

图 4-40　Mechanical Drawing of the Flexible Platform

【课后巩固】

1. 简述弧焊的原理。
2. 简述焊接工作站的布局。
3. 简述焊接工作站的电源电气布局。

[Consolidation after Class]

1. Briefly describe the principle of arc welding.
2. Briefly describe the layout of the welding workstation.
3. Briefly describe the electrical layout of power supply of the welding workstation.

任务 4.4　机器人弧焊工作站的仿真
Task 4.4　Simulation of Robot Arc Welding Workstation

【知识目标】

1. 了解 ROBOGUIDE 仿真软件可以导入的数据模型类型。
2. 掌握在 ROBOGUIDE 仿真软件中导入各种类型的数据模型的方法。
3. 掌握工业机器人仿真程序导入方法；
4. 掌握程序仿真方法。

[Knowledge Objectives]

1. To understand the types of data models that can be imported to the ROBOGUIDE simulation software.
2. To master the method of importing various types of data models into the ROBOGUIDE simulation software.
3. To master the methods of importing industrial robot simulation programs.
4. To master the methods of program simulation.

【技能目标】

1. 能按照工艺要求和规范进行机器人工作站程序编写。
2. 能正确导入/导出工业机器人仿真程序。
3. 能够导入工业机器人搬运工作站仿真布局。

[Skill Objectives]

1. To be able to write programs for robot workstation according to process requirements and codes.
2. To be able to correctly import/export industrial robot simulation programs.
3. To be able to import the simulation layout of industrial robot handling workstations.

【素质目标】

1. 培养学生沟通、协作能力。
2. 培养学生自主探索、善于观察的能力。

[Competence Objectives]

1. To cultivate the students' abilities of communication and collaboration.
2. To cultivate the students' abilities of exploration and observation.

【任务情景】

某压力容器生产企业为了加快生产速度，提高生产效率，准备建立一条工业机器人弧焊工作站，前期完成了生产线的设计方案编写，现在需要在软件

[Task Scenario]

In order to speed up production and improve its efficiency, a pressure vessel manufacturer plans to establish an industrial robot arc welding workstation. As the design scheme of the production line has been completed in the early stage, it is now necessary to simulate

上对生产工艺进行模拟仿真，软件使用 FANUC 仿真软件 ROBOGUIDE，请写出程序并仿真调试。

【任务分析】

当所研究的系统造价昂贵、实验风险性大或需要很长时间才能了解系统参数变化所引起的后果时，仿真是一种特别有效的研究手段。利用计算机实现对于系统的仿真研究不仅方便、灵活，而且也是经济的。本项目通过利用 FANUC 机器人配套的软件 ROBOGUIDE 进行仿真，也可以用 VISIUALONE 和 PROCESS 软件进行仿真，具体选用哪款软件，可根据自身条件进行选用，这里以 ROBOGUIDE 为例进行讲解。

【知识准备】

4.4.1 仿真环境搭建

参照工作站布局图，在仿真软件中导入设备及已设计好的部件模型，仿真环境如图 4-41 所示。

the production process with software, for which purpose the ROBOGUIDE software of FANUC is supposed to be adopted. Please write the program and conduct commissioning of the simulation.

[Task Analysis]

Simulation is a particularly effective research tool for the studied systems which are expensive, or with high experimental risks, or whose parameter changes result in consequences that takes a long time to be understood. Conducting simulation on such systems with computer is a research method that is not only convenient, flexible, but also economical. In this program, simulation can be carried out with the ROBOGUIDE software that matches up with the FANUC robots, or the VISIUALONE or PROCESS software, which can be selected based on one's actual conditions. The ROBOGUIDE software is taken as an example for explanation in this task.

[Assumed Knowledge]

4.4.1 Establishment of the Simulation Environment

Import the equipment and designed component models into the simulation software with reference to the workstation layout, and the simulation environment is shown in Figure 4-41.

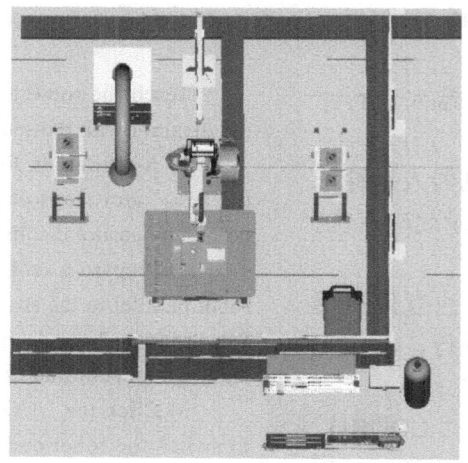

图 4-41 工业机器人弧焊工作站仿真环境

Figure 4-41 Simulation Environment of the Industrial Robot Arc Welding Workstation

4.4.2 工作站仿真

仿真环境建立后，必须在仿真软件中按照工作站工艺流程编写相应的机器人程序才能够进行运动仿真。

要进行真实的焊接，路径规划非常重要，它表征焊接过程中机器人行走的轨迹。机器人行走方式不同焊接质量也会受到影响，因此合理规划路径是进行焊接的前提。根据焊接技术的实际应用发现，再复杂的焊接路径无非是由若干直线或圆弧段所组成。因此，能正确示教直线、圆弧程序非常重要。

1. 仿真示教直线路径程序

在焊接过程中，直线轨迹是应用最为广泛的焊接路径。在操作机器人过程中，要示教机器人走直线的方法很多，最简单的方法是采用两点走直线的方式，如图 4-42 所示。

4.4.2 Workstation Simulation

When the simulation environment is established, the simulation software must be programmed with corresponding robot program according to the process flow of the workstation for the simulation of motion.

To perform real welding, path planning is very important as it represents the walking trajectory of the robot during the welding process. Having reasonable path planning is a prerequisite for welding as the quality of welding can be affected by different walking ways of the robot. It has been found from practical applications of welding technology that welding paths, no matter how complex they are, are composed of several straight or arc segments. Therefore, it is very important to have correct teaching of linear and arc programs.

1. Simulation of linear path teaching program

In terms of welding path, the linear trajectory is the most widely used in the welding process. There are many ways to teach a robot to walk linearly in the process of operating, and the "two-point setting" is the simplest way, as shown in Figure 4-42.

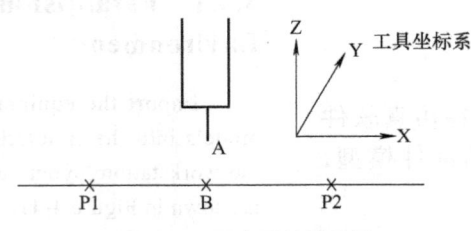

图 4-42 两点示教示意图

Figure 4-42 Schematic Diagram of Teaching of Two-point Setting

按照图示要求，对机器人进行示教，具体步骤如下：

1) 如图 4-42 所示，机器人手臂上的焊枪停于位置 A 处，记录此时点位，注意确保周围无障碍。

2) 制作工具坐标系，注意使机器人在 Y 轴方向上与直线平行，制作工具坐标的具体操作见项目二。

3) 单击示教盒上的"COORD"键切换机器人运行模式，确定运行模式为"关节坐标"，点动机器人运行至图中 B 点位置附近，记录此时点位。

Teach the robot according to the requirements of the diagram with the specific steps as follows:

1) As shown in Figure 4-42, the welding gun on the robot arm stops at position A, record the current point, and ensure that there are no obstacles around.

2) Prepare a tool coordinate system, making the robot parallel to the straight line in the Y-axis direction. See Program 2 for details of the specific procedures of making tool coordinates.

3) Click the "COORD" button on the teach box to switch the robot operation mode, confirm the operation mode as "Joint Coordinate", jog the robot to walk close to point B in the figure, and record the current point.

4) 将机器人运行模式切换为"全局坐标",在全局坐标模式下机器人将沿直线运动到 P1 位置。按下 F2 焊接开始快捷键或直接在指令中选择,调出 "Arc Start[1]" 命令。

5) 以与步骤四相同的方式运动到 P2 点,记录此时的坐标点位。按下 F4 焊接结束快捷键或直接指令选择,调出 "Arc End[1]" 命令。需注意的是方括号中的 1 与焊接开始处的 1 一致。

6) 以步骤三的方式返回到 A 位置,也可通过复制 A 点位置坐标的方式记录该点的坐标。

示教后的程序如图 4-43 所示。

4) Switch the robot operation mode to "Global Coordinates", under which the robot will move in a straight line to the P1 position. Press F2, the shortcut button for Arc Start, or select directly from the command to call the "Arc Start [1]" command.

5) Move to point P2 in the same way as step 4), and record the current coordinate point. Press F4, the shortcut button for Arc End, or select directly from the command to call the "Arc End [1]" command. It shall be noted that the 1 in square brackets is the same as that of the Arc Start[1].

6) Return to position A in the way of step 3), or record the coordinates of point A by copying its position coordinates.

The program after teaching is shown in Figure 4-43.

行	命令	内容说明
1:	JP[1] 100% CNT100	移动到待机 A 位置(关节程序点 1)
2:	JP[2] 100% CNT100	移动到焊接开始 B 位置附近(关节程序点 2)
3:	LP[3] 100% FINE	移动到焊接开始 P1 位置(直线程序点 3)
:	Arc Start[1]	焊接开始(使用焊接条件文件 1)
4:	LP[4] 100% FINE	移动到焊接结束 P2 位置(直线程序点 4)
5:	Arc End[1]	焊接结束(使用焊接条件文件 1)
6:	JP[1] 100% CNT100	返回到待机 A 位置(关节程序点 1)

图 4-43 直线示教焊接程序(调用焊接条件文件)

Figure 4-43 Teaching of Linear Welding Program (Calling the File of Welding Conditions)

图 4-42 所示为示教一个相对较简单直线焊接程序,像这样一个简单的程序,焊接条件设置的过程可以采用直接参数登录的方式,如图 4-44 所示。

Teaching of a linear welding program is relatively easy, as shown in figure 4-42. For a simple program like this, the setting of welding conditions can be achieved by login with direct input of parameters, as shown in Figure 4-44.

行	命令	内容说明
1:	JP[1] 100% CNT100	移动到待机 A 位置(关节程序点 1)
2:	JP[2] 100% CNT100	移动到焊接开始 B 位置附近(关节程序点 2)
3:	LP[3] 100% FINE	移动到焊接开始 P1 位置(直线程序点 3)
:	Arc Start[WP0,5.0V,200A,0.0]	焊接开始(使用焊接条件文件 1)
4:	LP[4] 100% FINE	移动到焊接结束 P2 位置(直线程序点 4)
5:	Arc End[WP0,5.0V,200A,0.0,0.5s]	焊接结束(使用焊接条件文件 1)
6:	JP[1] 100% CNT100	返回到待机 A 位置(关节程序点 1)

图 4-44 直线示教焊接程序(直接数值键入)

Figure 4-44 Teaching of Linear Welding Program (Direct Input of Numbers)

2. 仿真示教圆弧路径程序

相比直线路径，机器人在行走圆弧路径时则要复杂得多。这是因为圆弧的焊接轨迹是由若干个不同的圆弧段构成，需要了解机器人在走圆弧过程中的基本原理。平常画圆或一段圆弧就必须知道圆弧的圆心在何处，半径是多少。FANUC 机器人在设置圆弧路径时采用了 3 点确定一个平面的方法，只要确定 3 点的坐标，机器人自动运算圆弧圆心的坐标，这种方法可完成各种不同圆弧段的路径行走。

使用机器人完成圆弧路径程序是最基本的操作，是绘制各圆弧段的前提。如图 4-45 所示，完成机器人圆弧路径示教程序的编程操作。注意机器人一次只能走一个半圆形，因此整圆需走两个半圆形。

按照图示要求，对机器人进行示教，其具体步骤如下：

1) 如图 4-45 所示，将机器人焊枪移动到待机位置 1（可选择任意的安全位置），记录此时的坐标。需注意的是初学者在使用机器人过程中，经常将机器人的姿态调整到随意位置，这将使机器人在走圆弧路径时经常出现特异点而报警关停机器人，因此应将机器人摆正，可参考机器人零位位置。

2. Simulation of circular arc path teaching program

Compared to linear paths, circular arc paths are much more complex for a robot to walk, as a circular arc path is composed of several different circular arc segments, and it is necessary to understand the basic principles for a robot to walk in circular arc paths. Usually, to draw a circle or a section of a circular arc, it is necessary to know the center and the radius of the circular arc. The method of determining a plane by three points is adopted to set circular arc paths for the FANUC robot, which automatically calculates the coordinates of the arc center once the coordinates of the three points are determined. Walking in various path of different arc segments can be achieved in this way.

Performance of circular arc path program by robot is the most basic operation and a prerequisite for the drawing of various circular arc segments. Complete the programming of teaching of circular arc path program by robot, as shown in Figure 4-45. It should be noted that the robot needs to walk two semi-circles to complete a full circle, as it can walk only a semi-circle at once.

Teach the robot according to the requirements of the diagram with the specific steps as follows:

1) As shown in Figure 4-45, move the robot's welding gun to standby position 1 (any safe position can be selected) and record the coordinates at this point. It shall be noted that the robot is often adjusted to be at a random posture when used by beginners, which will cause it to alarm and stop frequently as it encounters abnormal points when walking on a circular arc path. Therefore, the robot shall be aligned upright, for instance, to its zero position.

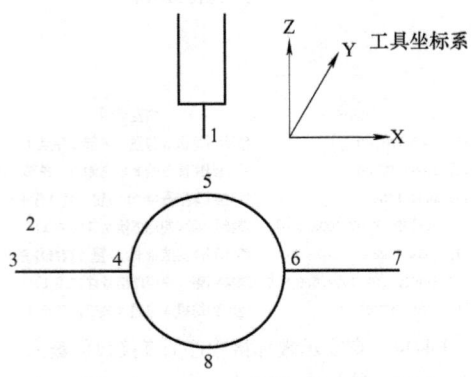

图 4-45 机器人圆弧路径示教程序编程示意图

Figure 4-45 Schematic Diagram of the Programming of Teaching of Circular Arc Path Program by Robot

2) 以任意形式移动机器人到位置 2 焊接开始位置附近，记录此时点位。

3) 以直线方式移动机器人到位置 3 焊接开始位置附近，记录此时点位。

4) 制作工具坐标系，注意使机器人在 X 轴方向上与直线平行。制作工具坐标的具体操作见项目二。

5) 切换机器人为全局坐标系，以直线的方式运动到位置 4 焊接开始位置。按下 F2 焊接开始快捷键或直接在指令中选择，调出"Arc Start[1]"命令。

6) 以任意模式移动到位置 5，记录此时的点位，调节光标到点位上的"J"处，如图 4-46 所示。按下 F4 选择键调出画面如图 4-47 所示。选择"3 圆弧"，此时"J"变为"C"，并同步跳出点位 P[...]，此点位即为圆弧的结束点。只需将机器人移动到位置 6，在下方栏目中会出现位置记忆。按下"SHIFT+F3"完成位置记录，此时 P[...] 将变为有坐标的点位，如图 4-48 所示。

2) Move the robot in any form of path to a place near position 2 of Arc Start as shown in the figure, and record the current point location.

3) Move the robot linearly to a place near position 3 of Arc Start as shown in the figure, and record the current point.

4) Prepare a tool coordinate system, making the robot parallel to the straight line in the X-axis direction. See Program 2 for details of the specific procedures of making tool coordinates.

5) Switch the robot to the global coordinate system, and move the robot linearly to a place near position 4 of Arc Start as shown in the figure. Press F2, the shortcut button for Arc Start, or select directly from the command to call the "Arc Start [1]" command.

6) Under any mode, move to position 5 as shown in the figure, record the current point, and adjust the cursor to position "J" of the point, as shown in Figure 4-46. Press F4, a selection button, to bring up Figure 4-47. Select 3 Circular Arc to change "J" to "C" and pop up simultaneously point P [...], which is the end point of the circular arc. Simply move the robot to position 6, and position memory will appear in the lower column. Press "SHIFT+F3" to complete the position recording. At this time, P [...] will become a point with coordinates as shown in Figure 4-48.

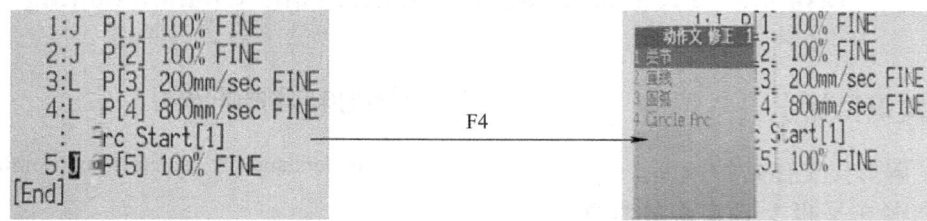

图 4-46　关节坐标画面　　　　　　　图 4-47　关节切换圆弧画面

Figure 4-46　Joint Coordinate Screen　　Figure 4-47　Joint-to-Circular Arc Switching Screen

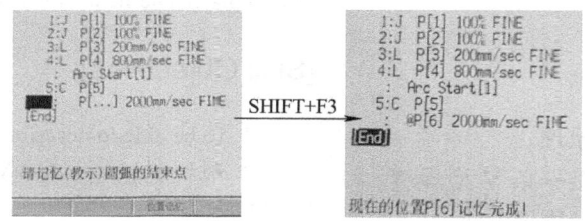

图 4-48　记录圆弧结束点

Figure 4-48　Recording of the End Point of the Circular Arc

7) 以步骤6) 的方式再次进行从位置6经8返回位置4，完成整个圆形示教。也可先返回位置4再经8到6完成整圆路径编制。

8) 按下F4焊接结束快捷键或直接指令选择，调出"Arc End[1]"命令。

9) 以关节的方式返回到图4-45的位置1，完成整个过程的示教操作。

示教后的程序如图4-49所示。

7) Return from position 6 through position 8 to position 4 in the methods specified at step 6) to complete the teaching of the entire circle. Alternatively, it is possible to return to position 4 first, then move through position 8 to position 6 to complete the compilation of the entire circular path.

8) Press F4, the shortcut button for Arc End, or select directly from the command to call the "Arc End [1]" command.

9) Return to position 1 in the manner of joint to complete the teaching of the whole process.

The program after teaching is shown in Figure 4-49.

图 4-49　圆形示教程序

Figure 4-49　Circular Teaching Program

任务 4.5　程序编写、安装与调试
Task 4.5　Programming, Installation and Commissioning

【知识目标】

1. 了解焊接工作站的原理。
2. 掌握在平板上真实焊接程序的编写。
3. 掌握薄板焊接程序编写并调试的流程。

[Knowledge Objectives]

1. To understand the principles of welding workstations.
2. To master the programming for real welding of flat plates.
3. To master the flow of programming for welding of thin plates and its commissioning.

【技能目标】

1. 能描述焊接的原理。
2. 能在平板上焊接一条V型焊缝。
3. 能调试薄板焊接程序作业。

[Skill Objectives]

1. To be able to describe the principles of welding.
2. To be able to weld a V-shaped welding seam on a flat plate.
3. To be able to perform commissioning of the programming for welding of thin plates.

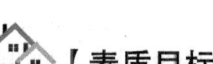 【素质目标】

1. 培养学生自主学习、解决问题能力。

2. 培养学生沟通协作、善于思考的能力。

 【任务情景】

智能制造焊接车间接到一个任务单,需要给实训室焊接一批全由低碳钢板、管组件组装焊接成的全封闭压力容器,而焊接压力容器,需要掌握V型焊接、水平焊接与管和板的焊接,同时还需要去设置焊接参数。本次任务就以V型焊接、水平焊接与管和板的焊接为目标,学习这三种类型的焊接。

 【任务分析】

我们先从了解焊接的原理开始,学习上电开机和操作移动机器人,然后分析压力容器。压力容器主要是由V型焊接、L型焊接和管和板的焊接组成,我们通过分别学习以上三种类型的焊接来完成本次任务需求。

【知识准备】

4.5.1 上电开机和操作移动机器人

1. 开机

若机器人系统连接的是PW455焊接电源,则应先将焊接电源打开。

打开机器人控制柜的断路器,按住"ON"按钮几秒钟,示教盒的开机画面将会显示出来。

手持示教盒,按下并且始终握住

[Competence Objectives]

1. To cultivate students' ability of learning and solving problems independently.

2. To cultivate students' ability to communicate, collaborate, and think effectively.

[Task Scenario]

An intelligent manufacturing welding workshop has received a task order from a training room and Light Industry for the welding of a batch of fully enclosed pressure vessels. The vessels shall be made entirely of low-carbon steel plates and pipes which are to be assembled and welded together. For welding of pressure vessels, one has to master V-shaped welding, horizontal welding, and welding between tubes and plates, as well as the setting of welding parameters. The goal of this task is to master three types of welding: V-shaped welding, horizontal welding, and welding between tubes and plates.

[Task Analysis]

Starting with a basic understanding of the principles of welding, we will learn to power on and operate robots, and then analyze the pressure vessels, assembly of which requires V-shaped welding, L-shaped welding, and welding between tubes and plates. We will complete this task by studying the above mentioned three types of welding separately.

[Assumed Knowledge]

4.5.1 Startup and Operating of the Robot

1. Startup

Turn on the PW455 welding power supply first if it is connected to the robot system.

Turn on the circuit breaker of the robot control cabinet, press and hold the "ON" button for a few seconds to bring up the startup screen of the teach box.

Hold the teach box in hand, press and always hold the "Dead man switch", turn the switch of the teach box to the "ON" position, find the "STEP" button on

"Dead man switch",将示教盒上的开关打到"ON"的位置,在示教盒键盘上找到"STEP"键,按一下并确认左上部的"STEP"状态指示灯亮。如果是新版本示教盒,屏幕顶端的状态显示行将显示"TP off in T1/T2, door open"。按"Reset"键消除报警信息。此时屏幕顶端右面的蓝色状态行应该为 -Joint 10%。

在关节坐标模式下移动机器人。

按下并保持"SHIFT",再配合其他方向键移动机器人。

此时机器人的运动速度可通过示教盒上的"+%"和"-%"键进行调节(或同时配合"SHIFT"进行大范围的调节)。为安全起见,开始时尽量以较低的速度移动机器人,并确认不会发生碰撞时,再适当提高移动速度。

2. 直角坐标模式

松开"SHIFT"键,在键盘上找到并按"COORD"键直到蓝色的状态栏显示"World"。请注意,切换了示教模式之后机器人移动速度会自动降低到10%。

此时再移动机器人时,机器人不再单轴(单关节)转动,而是当按前面三组J1,J2,J3键时,机器人的TCP以直线运动;当按后面三组J4,J5,J6键时,机器人的TCP固定不动,绕相应的直线坐标轴旋转。

3. 轴的软件限位

例如:一直按住"J3,+Z"键,第三轴提升到一定程度将自动停止上升,此时,屏幕顶部信息提示栏中应有限位或位置不可达的报警提示,按"RESET"键消除报警,按住"J3,-Z"键使第三轴往回运动。

the keyboard of teach box, and press and confirm that the "STEP" status indicator on the upper left is on. In addition, the status display line at the top of the screen will display "TP off in T1/T2, door open" should the teach box be of a new version. Press the "Reset" button to cancel the alarm message. The blue status bar on the top right of the screen shall be - Joint 10% at this time.

Move the robot under the Joint Coordinate mode.

Press and hold "SHIFT", and press corresponding direction button(s) to move the robot.

Now, the motion speed of the robot can be adjusted with the "+%" and "-%" buttons on the teach box (or significantly adjusted together with the "SHIFT" button). For safety, move the robot at a lower speed at the beginning and increase the speed appropriately when confirming that there will be no collision.

2. Cartesian coordinate system mode

Release the "SHIFT" button, find and press the "COORD" button on the keyboard until the blue status bar displays "World". Please note that the robot's movement speed will automatically be decreased to 10% when the teaching mode is switched.

If it is moved again at this moment, the robot will not rotate on a single axis (single joint). Instead, its TCP will move linearly when pressing the J1, J2, and J3 buttons in the front, and will not move but rotate around the corresponding linear coordinate axis when pressing the J4, J5, and J6 buttons in the rear.

3. Axis limit by software

For example, press and hold the "J3, +Z" button, the third axis will move up to a certain height and stop automatically. At this moment, there shall be an alarm prompt of limited or unreachable position in the information prompt bar at the top of the screen. Press the "RESET" button to eliminate the alarm, and press and hold the "J3, -Z" button to move the third axis back.

4. Dead-Man/E-Stop 开关

当释放"Dead-Man"开关,状态信息栏中就会有报警信息,要消除报警只要重新按住并保持住,报警信息将自动消失。

新版本机器人的"Dead-Man"开关是个 3 位开关,按压力太大也会导致报警。

紧急或特殊应用情况下,按一下示教盒右上方红色的"E-STOP"急停按钮。屏幕状态信息显示栏同样会有急停报警。要复位该信息,只需顺时针旋转使按钮复位,再按"RESET"键即可。

请注意在进行急停或复位急停操作时,除了可以听到第二轴和第三轴的抱闸声音,还可以听到机器人控制柜内部断路器的跳闸声音。

5. 要点

要求在平板上焊接一条 V 型焊缝。采用直线的、先上后下的焊枪位置,小角度地前向焊。保持干伸长为 5/8in。熄弧行走速度为 40IPM,焊接程序预设置为 25V 和 300IPM。V 型焊缝的两脚长约 2in。

6. 示教程序

生成一个新的程序"Exercise 2"。将焊接试板用夹钳固定,将吸烟尘器的吸尘口放置于离试板 2in 远的地方(如果有的话)。示教 7 个点,在保存起弧点和收弧点时,使用"F2 ARCSTART"和"F4 ARCEND"命令,而焊缝中的其他位置点用"F3 WELD POINT"。

按照焊接程序完整地测试该程序。

1) 设置机器人为焊接模式。

2)按"SHIFT"+"WELD ENBL"键,并确认"weld enable"在点亮状态。如果没有同时操作"SHIFT"和"WELD ENBL",可以

4. Dead-Man/E-Stop switch

When the "Dead Man" switch is released, the status information bar will display an alarm message, which will disappear automatically when the switch is pressed again and held.

The "Dead Man" switch of the robot of the new version is a three-position one, excessive pressure on which will cause an alarm.

In emergency or for special application, press the red "E-STOP" button, the emergency stop button, on the top right of the teach box. Similarly, there will be message of emergency stop alarm in the status information display bar of the screen. To reset this message, simply rotate the button clockwise to reset it, and then press the "RESET" button.

Please note that when performing emergency stop or its reset, one can not only hear the braking sound of the second and third axis, but the tripping sound of the circuit breaker inside the robot control cabinet.

5. Key points

A V-shaped welding seam is required to be welded on a flat plate. The welding gun shall be in a linear, first-up-and-then-down position and performs a small angle forward welding, while maintaining the stick-out of 5/8 inches. The walking speed of arc extinguishing is 40IPM, and the weld procedure pre-set as 25V and 300IPM. The two legs of the V-shaped welding seam are approximately 2 inches long.

6. Teaching program

Generate a new program of "Exercise 2". Fix the welding test panel with pliers and place the suction inlet of cleaner 2 inches (if possible) away from the test panel. Teach 7 points, using the "F2 ARCSTART" and "F4 ARCEND" commands when saving the points of arc start and arc end, and the "F3 Weld Point" for other points in the welding seam.

Test the complete program according to the welding procedures.

1) Set the robot to be in the welding mode.

2) Press the "SHIFT" – "WELD ENBL" buttons and confirm that the "weld enable" is on. If the "SHIFT" and "WELL ENBL" are not operated simultaneously, one can press "EDIT" to return to the mode of program editing. The program can only run when the above three conditions are met!

按"EDIT"回到程序编辑模式。必须满足以上3个条件程序才能运行！

3）将焊丝剪短到接近导电嘴，打开焊机的电源。

4）提醒在该区域内的其他人注意安全。

5）按"SHIFT"+"FWD"运行程序：

1.J P[1] 100% CNT 100
2.J P[2] 100% CNT 100
3.J P[3] 100% FINE
Arc Start [1]
4: L P[4] 40 IPM CNT 100
5: L P[5] 40 IPM FINE
Arc End [1]
6.J P[6] 100% CNT 100
7.J P[1] 100% CNT 100
End

如果发生以下问题，请按照步骤进行排查：

1）不起弧——检查"weld enable"。

2）起弧不正常——松开左手大拇指以暂停程序，操作机器人到合适位置，"weld enable"打到"Off"，"Step"on，复位错误，回到"HOME"位置，将焊丝剪到合适的长度，继续运行程序。如果没有改善，通知老师。

3）如果出现其他任何情况，暂停程序并通知老师。

4.5.2 弧焊参数的选择与设定

1. 弧焊参数的选择

实际中，焊接的种类和焊接的方式很多，应根据工件的材料、尺寸、搭接方式等来确定。通过参考选定材料的焊接条件表选取合理的焊接条件，然后对机器人系统及焊机参数进行设置。在预选电压、电流、焊接速度、气体流量等参数后对工件进行试焊接，

3) Cut the welding wire short enough to approach the conductive nozzle and turn on the power supply of the welding machine.

4) Remind other people in the area to be careful.

5) Press "SHIFT" + " FWD" to run the program:

1.J P[1] 100% CNT 100
2.J P[2] 100% CNT 100
3.J P[3] 100% FINE
Arc Start [1]
4: L P[4] 40 IPM CNT 100
5: L P[5] 40 IPM FINE
Arc End [1]
6.J P[6] 100% CNT 100
7.J P[1] 100% CNT 100
End

Please follow the steps to troubleshoot if the following problems occur:

1) Arc striking failed—Check "weld enable".

2) Abnormal arc striking — release the left thumb to pause the program running, operate to move the robot to an appropriate position, put "weld enable" to "Off", "Step" on, reset the error, return to the "HOME" position, cut the welding wire to have an appropriate length, and continue running the program. Notify the teacher should there be no improvement.

3) Pause the program and notify the teacher should any other situations occur.

4.5.2 Selection and Setting of Arc Welding Parameters

1. Selection of arc welding parameters

In practice, there are many types and methods of welding, which shall be determined based on the material, size, overlapping method, etc. of the workpiece. Select reasonable welding conditions with reference to the table of welding conditions for the selected materials, and then set the parameters of the robot system and welding machine. Conduct trial welding to the work-

根据焊缝的质量对参数进行适当的调整，通过不断的实践来指导焊接质量完全符合技术条件所规定的要求。

(1) 薄板焊接

通常情况下，薄板焊接方式有水平角焊、立向上焊和立向下焊等，其焊接方向如图 4-50 所示。

piece after pre-selection of parameters such as voltage, current, welding speed, and gas flow. Adjusted appropriately the parameters based on the quality of the welding seam until the welding quality fully meets the technical requirements.

(1) Welding of thin plates

Usually, the welding methods for thin plates include horizontal fillet welding, vertical upward welding, and vertical downward welding, with welding directions shown in Figure 4-50.

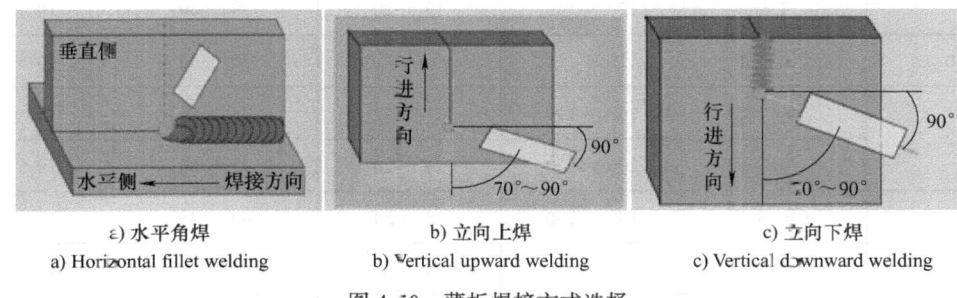

a) 水平角焊　　　　　　　b) 立向上焊　　　　　　　c) 立向下焊
a) Horizontal fillet welding　　b) Vertical upward welding　　c) Vertical downward welding

图 4-50　薄板焊接方式选择

Figure 4-50　Selection of Welding Methods for Thin Plates

现以薄板焊接中的水平角焊为例，介绍弧焊参数的选择。取两块完全对称的薄钢板。焊接工件尺寸如图 4-51 所示，将钢板以水平角接的方式焊接在一起。材质为 20 号钢。

The selection of arc welding parameters is introduced here with the horizontal fillet welding of thin plates as an example. Take two completely symmetrical thin steel plates whose dimensions are shown in Figure 4-51, and weld them together with the method of horizontal fillet welding. The steel is of grade 20.

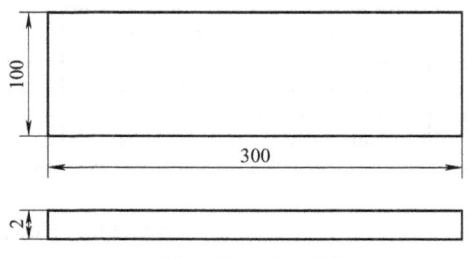

图 4-51　焊接工件尺寸（单位：mm）

Figure 4-51　Dimension of Welding Workpiece (Unit: mm)

根据表 4-13 所示的低碳钢角接焊接工艺参数，选择工艺参数。板厚为 2mm，焊丝直径 1.0mm，焊接电流 115～125A，焊接电压 19.5～20V，焊接速度 50～60cm/min，气体流量 10～15L/min。

Select process parameters with reference to process parameters for fillet welding of low carbon steel shown in Table 4-13. The plate thickness is 2mm, welding wire diameter 1.0mm, welding current 115–125A, welding voltage 19.5–20V, welding speed 50–60cm/min, and the gas flow rate 10–15L/min.

表 4-13 低碳钢角接焊接工艺参数
Table 4-13 Process Parameters for Fillet Welding of Low Carbon Steel

板厚 Plate thickness /mm	焊丝直径 Welding wire diameter /mm	焊接电流 Welding current /A	焊接电压 Welding voltage /V	焊接速度 Welding speed /(cm/min)	气体流量 Gas flow /(L/min)
1.0	0.8	70 ~ 80	17 ~ 18	50 ~ 60	10 ~ 15
1.2	1.0	85 ~ 90	18 ~ 19	50 ~ 60	10 ~ 15
1.6	1.0, 1.2	100 ~ 110	18 ~ 19.5	50 ~ 60	10 ~ 15
	1.2	120 ~ 130	19 ~ 20	40 ~ 50	10 ~ 20
2.0	1.0, 1.2	115 ~ 125	19.5 ~ 20	50 ~ 60	10 ~ 15
3.2	1.0, 1.2	150 ~ 170	21 ~ 22	45 ~ 50	15 ~ 20
	1.2	200 ~ 250	24 ~ 26	45 ~ 60	10 ~ 20
4.5	1.0, 1.2	180 ~ 200	23 ~ 24	40 ~ 45	15 ~ 20
	1.2	200 ~ 250	24 ~ 26	45 ~ 50	15 ~ 20
6	1.2	220 ~ 250	25 ~ 27	35 ~ 45	15 ~ 20
	1.2	270 ~ 300	28 ~ 31	60 ~ 70	15 ~ 20
8	1.2	270 ~ 300	28 ~ 31	55 ~ 60	15 ~ 20
	1.2	260 ~ 300	26 ~ 32	25 ~ 35	15 ~ 20
	1.6	300 ~ 330	25 ~ 26	30 ~ 35	15 ~ 20
12	1.2	260 ~ 300	26 ~ 32	25 ~ 35	15 ~ 20
	1.6	300 ~ 330	25 ~ 26	30 ~ 35	15 ~ 20
16	1.6	340 ~ 350	27 ~ 28	35 ~ 40	15 ~ 20
19	1.6	360 ~ 370	27 ~ 28	30 ~ 35	15 ~ 20

(2) 管和薄板焊接

板与管的焊接方式与薄板之间的焊接方式一样，同样也采用水平角焊方式，因此只需知道工件的厚度即可，如图 4-52 所示。图示中将圆管焊接在钢板上直径为 60mm 孔的位置，因此工件的厚度主要考虑薄板的厚度。根据表 4-13 所示，板厚与上述薄板之间焊接相同，故焊接的基本参数也保持一致。

(2) Welding between tubes and thin plates

The welding method between plates and tubes is horizontal fillet welding, the same as that between thin plates. Therefore, only the thickness of the workpiece needs to be known, as shown in Figure 4-52. Since the circular tube is welded to a hole with a diameter of 60mm on the steel plate in the Figure, the thickness of the workpiece is mainly considered as that of the thin plate. The basic welding parameters remain the same, as the thickness of the plate is the same with that of the welding thin plates mentioned above according to Table 4-13.

2. 焊机参数设置

在焊接时，使用的焊机电源是 FANUC ARC Mate 系列通用的逆变焊接电源 CV350-R，该焊接电源是林肯电气为 FANUC 机器人设计的经济实用型焊接电源。其基本参数见表 4-14。

2. Setting of welding machine parameters

Welding adopts the inverter power supply CV350-R which is universally used for the FANUC ARC Mate series and is an economical and practical welding power supply designed by Lincoln Electric Company for FANUC robots, with its basic parameters as shown in Table 4-14.

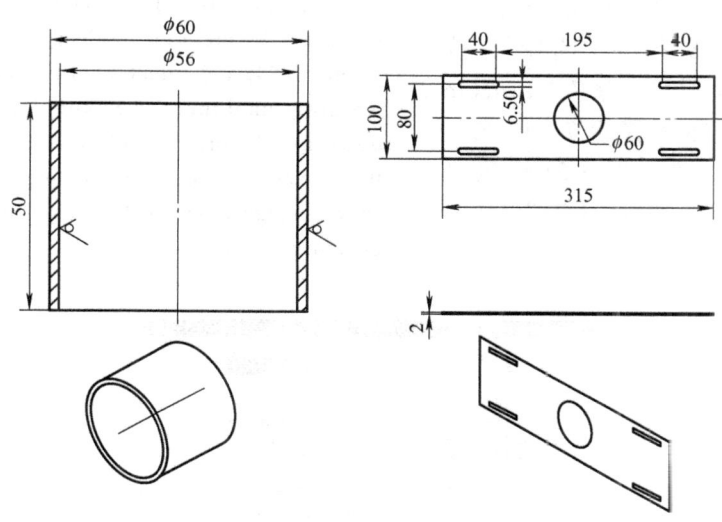

图 4-52 薄板与管的焊接工件尺寸（单位：mm）
Figure 4-52 Dimension of Welding Workpiece of Thin Plates and Tubes (Unit: mm)

表 4-14 CV350-R 基本参数
Table 4-14 Basic Parameters of CV350-R

适用材料 Applicable materials	焊接波形 Welding waveform	电流范围 Current range	通信方式 Communication mode	逆变技术 Inversion technology	输入电源 Input power supply
碳钢 Carbon steel	CV	60～350A	Arc Link	Inverter(30kHz)	380V

由参数表可知，焊接电源主要用于低碳钢的焊接，通过 Arc Link 直接与 FANUC 机器人进行通信，再配合 FANUC 弧焊机器人专用的 LR ARC TOOL 软件，使得焊机的基本参数直接登录到弧焊机器人中。因此，无需在焊机中设置参数，只需在机器人示教盒上直接选择焊机的厂家和焊机电源的形式即可。具体的焊接条件参数

It can be seen from the table of parameters that the welding power supply is mainly used for the welding of low-carbon steel. As the power supply communicates directly with the FANUC robot through Arc Link and is used in conjunction with the LR ARC TOOL software dedicated to the FANUC arc welding robot, the basic parameters of the welding machine are directly logged into the arc welding robot. Therefore, there is no need to set parameters in the welding machine, but directly select the manufacturer of the welding machine and the form of its power supply on the teach box of the robot.

设置可参见本项目任务二中的电弧焊接设定。

如上所述，只需对示教盒上的 ARC TOOL 软件画面进行设置，如图 4-53 所示。画面中各参数的含义见表 4-15。

在图 4-53 中，将光标下移至 6 "制造业者"，按下 F4 "选择"，按图 4-54 选择焊机电源 7 "Lincoln Electric"。将光标下移至 7 "形式"，按下 F4 "选择"，进入图 4-55 焊接电源形式选择画面，一般选择 2 "Power Wave+ENet"。

See the setting of arc welding in Task 2 of this program for the details of the setting of specific parameters for welding conditions.

As mentioned above, only the ARC TOOL software screen on the teach box needs to be set, as shown in Figure 4-53. The meanings of parameters in the screen are shown in Table 4-15.

In Figure 4-53, move the cursor down to 6 " Manufacturers ", press F4 " Select " to enter the screen of Figure 4-54, and select 7 " Lincoln Electric " as the welding machine power supply. Move the cursor down to 7 " Forms " and press F4 " Select " to enter the screen of selection of welding power supply forms shown in Figure 4-55, and select 2 " Power Wave+ENet " generally.

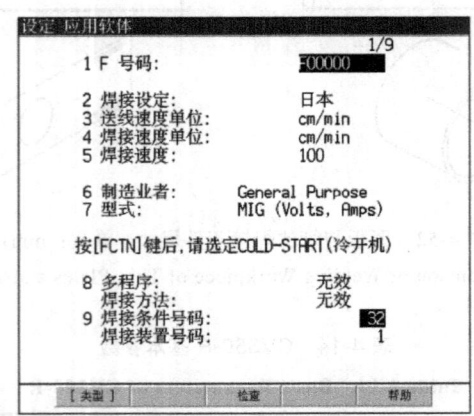

图 4-53 设定 ARC TOOL 软件画面

Figure 4-53　Setting of the ARC TOOL Software Screen

表 4-15　软件画面各参数含义

Table 4-15　Meanings of Parameters in the Software Screen

设定项目 Setting items	说明 Explanations	备注 Remarks
F 号码 F number	软件画面序列号，多种焊机时可对其进行设定 Software screen S/N, which can be set when there are several welding machines	
焊接设定 Settings of welding	可以将送丝速度单位、焊接速度单位自动更改为各国标准的单位 The units of wire feeding speed and welding speed can be automatically changed to standard ones for different countries applicable	单位更改后自动运算 Automatic calculation after unit change
送线速度单位 Wire feeding speed unit	设定送线速度的单位 The unit of wire feeding speed to be set	设定后自动调用 Automatically call after setting

设定项目 Setting items	说明 Explanations	备注 Remarks
焊接速度单位 Welding speed unit	设定焊接速度的单位 The unit of welding speed to be set	设定后自动调用 Automatically call after setting
焊接速度 Welding speed	一旦设定，则"WELD_SPEED"默认为该值 Once set, it will be the default value of "WELD_SPEED" 1:L P[2] WELD_SPEED FINE	直接数值键入 Direct input of values
制造业者 Manufacturers	焊机电源生产厂家选择 Selection of manufacturer of welding machine power supply	
形式 Form	选择焊接电源的形式 Choose the form of welding power supply	电源厂家不同会有差别 It may vary depending on different power supply manufacturers
多程序 Multi-program	多程序执行时的有效/无效选择 Valid/invalid selection during the execution of multi-program	只有林肯焊机才能选择 Selection is available only with Lincoln welding machines
焊接条件号码 Welding condition number	可以设定电弧焊接条件画面中能够使用的焊接条件号码的个数 It is possible to set the quantity of welding condition numbers that can be used in the arc welding condition screen	默认设置为32 Default value is 32
焊接装置号码 Welding machine number	给焊机编号 Number the welding machines	多焊机作业时可设置 Can be set during operations by multiple welding machines

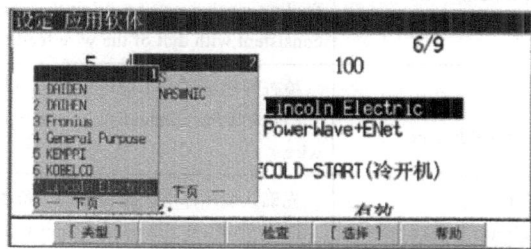

图 4-54 焊机电源厂家选择画面

Figure 4-54 Screen of Selection of Welding Machine Power Supply Manufacturers

图 4-55 焊接电源形式的选择画面

Figure 4-55 Screen of Selection of Welding Power Supply Forms

上述操作设置完毕后，按下示教盒上"FCTN"功能键，从菜单中选择"冷开机"，结束整个焊机电源选择。如果在6"制造业者"中找寻不到焊机电源厂家信息，可在图4-54中选择4"通用模式"，并在7"形式"中选择与焊机电源相符合的焊接控制方式。

When the above settings are completed, press the function button "FCTN" on the teach box and select "Cold start" from the menu to end the entire selection of welding machine power supply. If no information of the welding machine power supply manufacturer can be found in 6 "Manufacturers", you can select 4 "General Purpose" in the Figure 4-54 and select in 7 "Forms" a welding control method that matches the welding machine power supply.

3. 机器人焊接条件设置

根据选定的参数，初设焊接电流为115A，电压为20V。此处选择由焊接命令直接设定：

Arc Start [WP0,20V, 115A,0.0,0.5s]；焊接速度为50cm/min，引弧时间0.5s，再引弧有效。

(1) 机器示教

焊接前准备见表4-16。

3. Setting of robot welding conditions

According to the selected parameters, the initial welding current is 115A and the voltage is 20V. Selections here are set directly by the welding command:

Arc Start [WP0,20V, 115A,0.0,0.5s]；The welding speed is 50cm/min, the arc striking time 0.5s, and the arc re-striking effective.

(1) Teaching of machine

Preparation before welding is shown in Table 4-16.

表4-16 焊接前准备
Table 4-16 Preparation before Welding

序号 S/N	项目 Item	内容 Content
1	焊丝的安装 Installation of welding wire	将适合焊接的焊丝正确安装入送丝机构，确保焊丝的直径与所使用的送丝轮的直径保持一致 Install correctly the welding wire suitable for welding into the wire feeding mechanism, ensuring that the diameter of the welding wire is consistent with that of the wire feeding roller used
2	焊枪的确认 Confirmation of welding gun	检查导电嘴是否与焊丝直径相一致 Check if the conductive nozzle matches the diameter of the welding wire
3	配电柜的断路器闭合 Close of the circuit breaker of the distribution cabinet	先确认变压器电源是否正确，检查无误后闭合断路器 First confirm the power supply of the transformer is correct, and then close the circuit breaker
4	焊接电源的接通 Switch-on of welding power supply	合上焊接电源，背面的冷却风扇开始运转 Switch on the welding power supply, and the cooling fan on the back starts to operate
5	送丝机的设定 Setting of wire feeder	对送丝电动机的种类进行设定，采用机械伺服驱动(4轮：Auto Drive 100)。送丝电动机设定不正确，将不能按照指定的送丝速度送丝，其结果难以保证焊接质量 Set the type of wire feeding motor to be mechanical servo drive (4 rollers: Auto Drive 100). If set incorrectly, the wire feeding motor will not be able to feed wires according to the specified wire feeding speed, which will affect the quality of welding

（续）

序号 S/N	项目 Item	内容 Content
6	机器人侧的设定 Settings on the robot side	对焊机电源和焊接参数进行设定 Set the welding machine power supply and welding parameters
7	示教盒显示值的确认 Confirmation of the displayed value of the teach box	确认示教盒上电压、电流、线速(送丝速度)的设定值及焊接方法的设定 Confirm the setting values of voltage, current, linear speed (wire feeding speed) and the setting of welding method on the teach box
8	点动送丝 Wire feeding by Jogging	机器人发出点动送丝命令(SHIFT+WIRE+)，送出焊丝一直到其从焊枪前端伸出 The robot sends a command of wire feeding by jogging (SHIFT+WIRE+), and feeds the welding wire until it extends from the front end of the welding gun
9	调整保护气的流量 Adjustment of the flow rate of protective gas	在焊接系统画面选择气体喷出"有效"，并选择气体喷出时间，一般设置为5s，5s后自动停止送出气体 In the welding system screen, select "valid" for gas spraying and select the gas spraying time, which is generally set to 5s, thereafter the gas spaying will automatically stop 打开气瓶上的阀门，注意观察流量计，一般流量控制在10～25L/min较适宜。焊接电流越大，所需保护气体流量也应当越大 Open the valve on the gas cylinder, observe the flow meter, and control the flow rate of 10–25L/min which is suitable in general. The higher the welding current, the greater the required protective gas flow rate

在对机器人进行示教焊接之前，需要对焊接系统各部分及参数进行确认，具体内容见表4-16。

(2) 程序点示教

1) 薄板焊接示教。按照弧焊要求对机器人进行编程示教，程序示教步骤如图4-56所示。

Before conduct teaching of welding to the robot, one shall confirm the various parts and parameters of the welding system, as shown in Table 4-16.

(2) Teaching of program points

1) Teaching of welding of thin plates: Perform programming for the teaching of robot according to the requirements of arc welding, with the schematic positions of program points being shown in Figure 4-56.

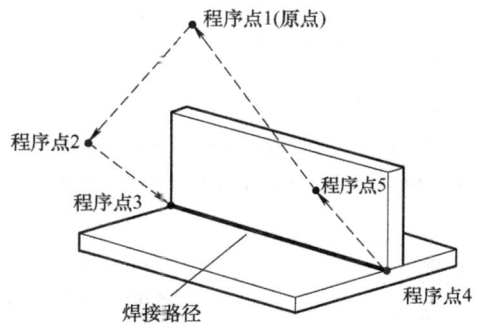

图 4-56 薄板焊接程序点示教步骤

Figure 4-56 Steps for Teaching of Program Points in the Welding of Thin Plates

示教后的程序如图 4-57 所示。
焊接好的成品如图 4-58 所示。

2) 管和板焊接示教。按照弧焊的要求，对机器人进行示教，程序示教步骤如图 4-59 所示。

The program after teaching is shown in Figure 4-57. The finished welded product is shown in Figure 4-58.

2) Teaching of Welding between Tubes and Plates: Teach the robot according to the requirements of arc welding, with the schematic positions of program points being shown in Figure 4-59.

行	命令	内容说明
1:	J P[1] 100% CNT100	移到原点位置（程序点 1）
2:	J P[2] 100% CNT100	移到焊接开始附近（程序点 2）
3:	L P[3] WELD_SPEED FINE	移到焊接开始点（程序点 3）
:	Arc Start[WPO, 20V, 115A, 0.0, 0.5]	焊接开始
4:	L P[4] 50cm/min FINE	移到焊接结束点（程序点 4）
:	Arc End[WPO, 20V, 115A, 0.0, 0.5]	停止焊接
5:	L P[5] WELD_SPEED FINE	离开焊接点（程序点 5）
6:	J P[1] 100% CNT100	返回原点位置（程序点 1）
7:	END	

图 4-57　薄板焊接程序

Figure 4-57　Welding Procedure of Thin Plates

图 4-58　薄板焊接成品

Figure 4-58　Finished Product of Welded Thin Plates

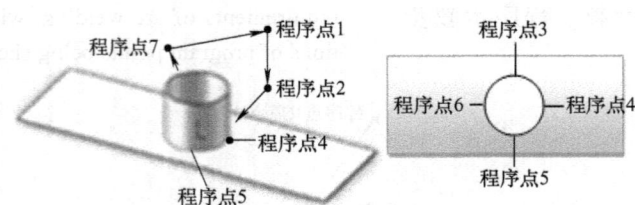

图 4-59　管和板焊接程序点示教步骤

Figure 4-59　Steps for Teaching of Programs Points in the Welding between Tubes and Plates

示教后的程序如图 4-60 所示。
焊接好的成品如图 4-61 所示。

The program after teaching is shown in Figure 4-60. The finished welded product is shown in Figure 4-61.

项目 4　典型工业机器人弧焊工作站系统的设计及应用　　259

```
行          命令                              内容说明
1:      J P[1] 100% CNT100                   移到原点位置（程序点 1）
2:      J P[2] 100% CNT100                   移到焊接开始附近（程序点 2）
3:      L P[3] WELD_SPEED FINE               移到焊接开始点（程序点 3）
 :      Arc Start[WPO, 20V, 115A, 0.0, 0.5]  焊接开始
4:      C P[4]                               前半圆弧中间点（程序点 4）
 :        P[5] 50cm/min FINE                 前半圆弧结束点（程序点 5）
5:      L P[6] 50cm/min FINE                 后半圆弧开始点（程序点 5）
6:      C P[7]                               后半圆弧中间点（程序点 6）
 :        P[8] 50cm/min FINE                 后半圆弧结束点（程序点 3）
 :      Arc End[WPO, 20V, 115A, 0.0, 0.5]    停止焊接
7:      L P[9] WELD_SPEED FINE               离开焊接点（程序点 7）
8:      J P[1] 100% CNT100                   返回原点位置（程序点 1）
9:      END
```

图 4-60　管和板焊接程序

Figure 4-60　Procedure of Welding between Tubes and Plates

图 4-61　管和板焊接成品

Figure 4-61　Finished Product of Welded Tubes and Plates

任务 4.6　技术交底材料的整理和编写
Task 4.6　Organization and Compilation of Technical Disclosure Materials

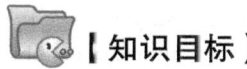

【知识目标】

1. 掌握工业机器人焊接工作站技术交底材料的编写步骤与方法。

2. 掌握工业机器人焊接工作站技术交底材料的清单整理。

【技能目标】

1. 根据工业机器人焊接工作站技术交付材料整理全套说明书。

[Knowledge Objectives]

1. To master the steps and methods for compiling technical disclosure materials for industrial robot welding workstations.

2. To master the organization of list of technical disclosure materials for industrial robot welding workstations.

[Skill Objectives]

1. To be able to organize a complete set of manuals according to the delivered technical materials for industrial robot welding workstations.

2. 能够按照交底材料的要求完成方案整理。

【素质目标】

1. 使用规范的行文格式整理资料。
2. 培养良好的行为习惯。

【任务情景】

某工业机器人系统集成企业为一家汽车生产企业建立了一条工业机器人焊接工作站，已全部完成安装、调试工作，现需要给汽车生产企业提供关于该工作站的技术资料，作为工业机器人系统集成商的技术员，请你完成该工作站技术资料整理与编写工作。

【任务分析】

技术交底是企业极为重要的一项技术管理工作，是施工方案的延续和完善，也是项目质量预控的最后一道关口。其目的是使参与项目施工的技术人员熟悉和了解所承担的项目的特点、设计意图、技术要求、施工工艺及应注意的问题。

通过技术交底，使参与项目施工操作的每一个工人，了解自己所要完成的分项工程的具体工作内容、操作方法、施工工艺、质量标准和安全注意事项等，做到施工操作人员任务明确，心中有数，达到有序地施工，以减少各种质量通病，提高施工质量的目的。

【知识准备】

4.6.1 主要技术资料

工业机器人焊接工作站常见的技术资料如下：

2. To be able to complete the plan organization according to the requirements of the disclosure materials.

[Competence Objectives]

1. To organize documents with normative formats.
2. To cultivate good behavioral habits.

[Task Scenario]

An industrial robot system integrator has built an industrial robot welding workstation for an automobile manufacturer, and completed the tasks of installation and commissioning. Now it is necessary to provide the automobile manufacturer with technical documents about the workstation. As a technician of the industrial robot system integrator, you are required to complete the organization and compilation of the technical documents of the workstation.

[Task Analysis]

Technical disclosure is an extremely important technical management task for enterprises, which is the continuation and improvement of construction schemes, and also the final checkpoint for pre-control of the project quality. The purpose of technical disclosure is to make the technicians involved in the project construction get familiar with and understand the characteristics, design intent, technical requirements, and construction techniques of the project and the precautions of its implementation.

Through technical disclosure, every worker involved in the project construction understands the specific work content, operation methods, construction techniques, quality standards, and safety precautions of the works they need to complete, so that they can carry out the construction in order as they are clear of their tasks and what to be done properly, which reduces various common quality problems and improves construction quality.

[Assumed Knowledge]

4.6.1 Main Technical Documents

The common technical documents of industrial robot welding workstations mainly include the following:

1. 工作站操作说明书

1) 工作站概况和基本软硬件组成。
2) 基本操作流程、关键性的技术及操作中可能会存在的问题。
3) 特殊设备的操作处理细节及其操作须知。
4) 工作站开关机流程及注意事项。
5) 工作站常见故障的现象描述及处理方法。
6) 如果可以,最好能将工作站的程序也列出,无需做出注解。

2. 工作站全套图样

应将方案、设计、施工阶段的所有相关图样均整理好交付给使用方。
1) 工作站布局图及网络拓扑图。
2) 电气原理图及接线图。
3) 工作站系统安装图。
4) 非标准零件图及装配图。

3. 工作站为设备程序

应将调试后的设备程序如工业机器人程序、PLC程序、HMI工程文件以及变频设置文件等全部整理并标注好后交付使用方。

4. 工作站设备说明书

将工作站使用的成品设备说明书整理好后交付给使用方,以确保使用方在使用过程中方便地查阅资料。

4.6.2 操作说明书的编写

工作站的技术资料编写工作,主要是操作说明书的编写,使用户通过工作站操作说明书能够操作、调试与维护设备,并处理简单设备故障。以FANUC工业机器人焊接系统操作说明书为例,了解工作站操作说明书的编写方法。

1. Workstation operating manual

1) Overview of the workstation and composition of basic software and hardware.
2) Basic operating procedures, key technologies, and potential problems in operation.
3) Details and instructions for the operation and handling of special equipment.
4) The startup and shutdown processes of the workstation and their precautions.
5) Description and handling methods of common faults in the workstation.
6) If possible, it would be better to list the programs of the workstation without making annotations.

2. Complete set of workstation drawings

All relevant drawings for the stages of scheme, design, and construction shall be organized and delivered to the user.
1) Workstation layout and network topology diagram.
2) Electrical schematic diagram and wiring diagram.
3) Installation diagram of the workstation system.
4) Drawings of non-standard parts and their assembly.

3. Programs of the workstation equipment

All equipment programs whose commissioning has been done such as industrial robot programs, PLC programs, HMI engineering files, and frequency conversion settings files shall be organized and labeled before being delivered to the user.

4. Workstation equipment manual

The manuals for the finished equipment used in the workstation shall be organized and delivered to the user to ensure easy access to the information during use.

4.6.2 Compilation of Operating Manual

The compilation of technical documents of the workstation mainly involves the compilation of the operating manual, which enables the user to operate equipment, and conduct commissioning, maintenance, and simply troubleshooting of the equipment. The operating manual of the FANUC industrial robot welding workstation is taken as an example to help understand the compilation method of the workstation operating manual.

1. 目录

典型工业机器人焊接系统操作说明书的封面和目录样式如图 4-62 所示。

1. Table of contents

The table of contents of the operating manual for typical industrial robot welding system is shown in Figure 4-62.

图 4-62 焊接系统操作说明书

Figure 4-62 Welding System Operating Manual

2. 正文

这里以 FANUC 机器人 Smart Arc 焊接系统操作说明书第一章的内容编写为例，为大家演示如何进行操作说明书的编写。操作说明书的其他内容请同学们按要求自己编写。

示例：

1. 简介

本章对 Smart Arc 焊接系统的基本构造和各种设备进行说明。

1.1 Smart Arc 焊接系统构成

Smart Arc 焊接系统的主要设备

2. Text

The compilation of the content of the first chapter of the is taken as an example here to demonstrate how to compile the operating manual. Other content of the operating manual shall be compiled by the student as required.

Example：

1. Overview

This chapter explains the basic structure and various equipment of the Smart Arc welding system.

1.1　Composition of the Smart Arc welding system

The main equipment of the Smart Arc welding

由机器人本体、机器人控制柜、焊接电源、送丝机、焊枪、送丝盘架构成，如图 4-63 所示。Smart Arc 焊接系统电缆组成见表 4-17。

system include the robot body, robot control cabinet, welding power supply, wire feeder, welding gun, and wire feeding reel holder, as shown in Figure 4-63. The cable composition of the Smart Arc welding system is shown in Table 4-17.

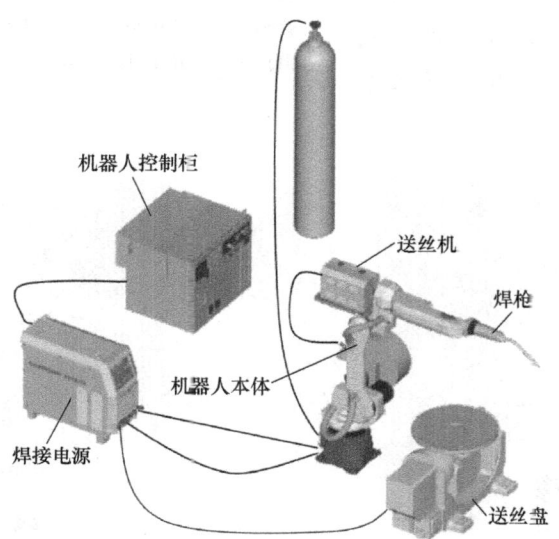

图 4-63 Smart Arc 焊接系统

Figure 4-63 Smart Arc Welding System

表 4-17 Smart Arc 焊接系统电缆组成

Table 4-17 Cable Composition of the Smart Arc Welding System

编号 S/N	名称 Name	作用 Function	备注 Remarks
A	焊机通信电缆 Communication cable for welding machine	机器人与焊机相互通信 Communication between the robot and welding machine	DeviceNet
B	焊机正极电缆 Cable for the positive electrode of welding machine	焊接正极回路 Circuit of the positive electrode of welding	
C	送丝通信电缆 Communication cable for wire feeding	焊机与送丝机相互通信 Communication between welding machine and wire feeder	
D	焊机负极电缆 Cable for the negative electrode of welding machine	焊接负极回路 Circuit of the negative electrode of welding machine	
E	气管 Air tube	输送保护气体 Conveying protective gas	
F	送丝管 Wire feeding tube	输送焊丝 Conveying welding wire	

1.2 Smart Arc 焊接系统设备介绍

1.2.1 机器人设备介绍

机器人设备主要由机器人本体和机器人控制柜组成。机器人本体分为 R-0iB（见图 4-64）和 M-10iA/12（见图 4-65）两个型号。机器人的 J3 及 J6 安装尺寸参见表 4-18。

1.2 Introduction to equipment of the Smart Arc welding system

1.2.1 Introduction to the robot equipment

The robot equipment mainly consists of the robot and its control cabinet. There are robots of two models, i.e. R-0iB (as shown in Figure 4-64) and M-10iA/12 (as shown in Figure 4-65). The installation dimensions of the J3 and J6 axis of robots can be found in Table 4-18.

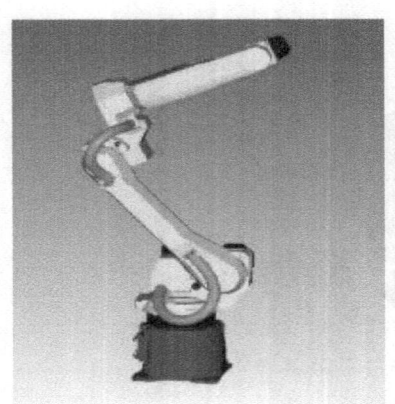

图 4-64 R-0iB 机器人
Figure 4-64 R-0iB Robot

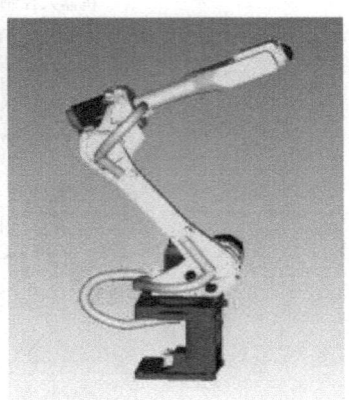

图 4-65 M-10iA/12 机器人
Figure 4-65 M-10iA/12 Robot

表 4-18 机器人本体规格
Table 4-18 Specifications of the Robot

型号 Model	R-0iB			M-10iA/12		
机构 Mechanism	外置式 Externally placed			内置式 Built in		
关节数量 Number of joints	6 轴 6-axis			6 轴 6-axis		
可达半径 Reachable radius	1437mm			1420mm		
安装方式（备注 1） Installation method (Note 1)	地面安装、倒吊安装、倾斜安装 Installed on the ground, installed with upside down, or installed with inclination			地面安装、倒吊安装、倾斜安装 Installed on the ground, installed with upside down, or installed with inclination		
运动范围 Range of motion	J1	J2	J3	J1	J2	J3
	360°	250°	455°	360°	250°	445°
	J4	J5	J6	J4	J5	J6
	380°	280°	720°	380°	380°	720°

项目4　典型工业机器人弧焊工作站系统的设计及应用

（续）

型号 Model	R-0iB			M-10iA/12		
最高速度（备注2） Maximum speed (Note 2)	J1	J2	J3	J1	J2	J3
	225°/s	215°/s	255°/s	230°/s	225°/s	230°/s
	J4	J5	J6	J4	J5	J6
	425°/s	425°/s	625°/s	430°/s	430°/s	630°/s
手腕最高速度 Maximum speed of wrist	2000mm/s			2000mm/s		
手腕最大负载 Maximum load of wrist	3kg			2kg		
J3轴最大负载 （备注3） Maximum load of J3 axis (Note 3)	7kg			12kg		
手腕允许负载转矩 Torque of allowable load of wrist	J4	J5	J6	J4	J5	J6
	8.9N·m	8.9N·m	3.0N·m	22N·m	22N·m	9.8N·m
手腕允许负载惯量 Moment of inertia of allowable load of wrist	J4	J5	J6	J4	J5	J6
	0.28 kg·m²	0.28 kg·m²	0.035 kg·m²	0.65N·m	0.65N·m	0.17N·m
驱动方式 Driving method	交流伺服电动机驱动 Driven by AC servo-motor			交流伺服电动机驱动 Driven by AC servo-motor		
重复定位精度 Repetitive positioning precision	±0.08 mm			±0.08 mm		
机器人质量 （备注4） Robot mass (Note 4)	145kg			130kg		
输入电源功率（平均功耗） Input power kV·A (average power consumption)	2 kV·A (1kW)			2 kV·A (1kW)		

（续）

型号 Model	R-0iB	M-10iA/12
安装条件 Installation conditions	环境温度：0～45℃ Ambient temperature: 0-45℃ 环境湿度：通常在75% RH以下(无结露现象)，短期在90% RH以下(1个月之内) Ambient humidity: below 75% RH usually (without condensation), below 90%RH for short-term (within 1 month) 振动加速度：4.9 m/s²(0.5g)以下 Vibration acceleration: below 4.9 m/s² (0.5g)	环境温度：0～45℃ Ambient temperature: 0-45℃ 环境湿度：通常在75% RH以下(无结露现象)，短期在90% RH以下(1个月之内) Ambient humidity: below 75% RH usually (without condensation), below 90%RH for short-term (within 1 month) 振动加速度：4.9 m/s²(0.5g)以下 Vibration acceleration: below 4.9 m/s² (0.5g)
注： Notes:	1. 如采用倾斜安装方式，机器人J1轴和J2轴的运动范围将受到限制。 The range of motion of the J1 and J2 axis of the robot will be limited if installation with inclination is adopted. 2. 短距离运动时，可能达不到各轴的最高标称速度。 The maximum nominal speed of each axis may not be reached during short-distance movement. 3. 根据手腕部负载重量的不同而受到不同的限制。 Limitations vary depending on the carrying load of the wrist. 4. 不含机器人控制器的质量。 The mass of the robot controller is excluded.	

M-10iA/12机器人内部装有内置管线包(包括：送丝机通信电缆、焊接正极电缆、保护气管、压缩气管)。内置管线一头安装于机器人J1轴底座上，另一头裸露在机器人J3轴。

The M-10iA/12 robot is internally equipped with a built-in pipeline package (including: the communication cable for wire feeder, cable for positive electrode of welding, protective gas tube, and compressed air tube). One end of the built-in pipelines is installed on the based on the J1 axis of the robot, and the other end is exposed on the J3 axis of the robot.

1.2.2 焊接设备介绍

焊接设备包括焊接电源和送丝机。焊机及送丝机接口示意图如图4-66和图4-67所示。焊接电源技术规格和送丝机技术规格见表4-19和表4-20。

1.2.2 Introduction to the Welding Equipment

The welding equipment includes the welding power supply and wire feeder. The interfaces of welding machine and wire feeder are shown respectively in Figure 4-66 and Figure 4-67.The technical specifications of the welding power supply and the wire feeder are shown in Table 4-19 and Table 4-20.

图4-66 焊机
Figure 4-66　Welding Machine

图4-67 送丝机
Figure 4-67　Wire Feeder

表 4-19 焊接电源技术规格
Table 4-19 Technical Specifications of the Welding Power Supply

焊接电源型号 Welding power supply model	C350iA	P400iA	C500iA/P500iA
控制方式 Control mode	数字控制 Digital control		
额定输入电压/相数 Rated input voltage/ phases	3 相 AC 380V(±25%) 3-phase AC 380V(±25%)		
输入电源频率 Input power frequency	45～65Hz		
额定输入容量 Rated input capacity	15kV·A/12.7kW	19.7kV·A/18kW	24kV·A/22.3kW
功率因数 Power factor	0.94	0.94	0.93
输出特性 Output characteristic	CV		
额定输出电流 Rated output current	350A	400A	500A
额定输出电压 Rated output voltage	31.5V	34V	39V
额定负载持续率 Rated load duration	直流 100% DC 100%	直流 100% DC 100%	直流 60% DC 60%
额定输出空载电压 Rated output no-load voltage	73.3V	73.3V	73.3V
输出电流范围 Output current range	30～400A	30～400A	30～500A
输出电压范围 Output voltage range	12～45V	12～45V	12～45V
外壳防护等级 Protection level of the shell	IP23S		
环境温度 Ambient temperature	-10～40℃（主机 -39℃可启机） -10-40℃ (the main machine can be started at-39℃)		
绝缘等级 Insulation level	H		

表 4-20 送丝机技术规格
Table 4-20 Technical Specifications of the Wire Feeder

项目 Item	参数 Parameter
送丝传动控制方式 Control mode of wire feeding transmission	光电编码器反馈+独立芯片高速环路控制 Optoelectronic encoder feedback+ high-speed loop control with independent chip
送丝机额定电流 Rated current of wire feeder	3.5A

（续）

项目 Item	参数 Parameter
送丝机额定电压 Rated voltage of wire feeder	36V
送丝速度 Wire feeding speed	1.4～24 m/min
送丝轮直径 Wire feeding roller diameter	0.8～1.6mm
焊丝盘类型 Type of welding wire reel	所有标准化的焊丝盘 All are standardized welding wire reels
驱动装置 Driving device	四轮送丝驱动装置 Four-roller driving device for wire feeding
焊枪接口 Welding gun interface	欧式接口 European interface

项目 5　数字化生产线的构架及技术特点

Program 5　Architecture and Technical Characteristics of Digital Production Lines

【项目场景】

某铝壳电机关键构件现有生产线的加工工艺为：毛坯件入库→上料→粗加工→精加工→成品检测→入库。每道工序花费的时间较长，再加上生产线自动化、智能化的程度普遍偏低，对于不同工序之间的上下料多依靠人工，增加了劳动负荷，生产安全性降低，生产效率降低，并且工件在机床装夹过程中定位精度不高，生产质量也会受到影响，更有可能损坏刀具，造成一定的经济损失。而对于工件粗、精加工完成后工件尺寸的检测，大多是人工通过普通的测量仪进行检测，由于仪器的精度和操作者的熟练程度都会影响工件的质量检测，并且检测数据并不能及时反馈给机床系统，对于加工不合格的工件不能及时地修正机床的相关参数，增加了生产线加工的废品率。除此以外，对于仓储模块并不能对每个仓位的产品规格信息进行实时统计与数据反馈。

数字化生产线基于先进控制技术及工业机器人、视觉检测、传感和RFID等其他技术，集成了多功能控制

[Program Scenario]

A key component of a motor with an aluminum shell is machined on the existing production line as follows: blanking storage → loading → rough machining → fine machining → finished product detection → storage. The long time taken by each procedure, low level of automation and intelligence of the production line, manual labor on which most of the loading and unloading between different procedures rely, all these have led to an increase in labor load and a reduction in production safety and efficiency. Moreover, the positioning precision of the workpiece during clamping by the machine tool is not high, which will affect the production quality and be likely to damage the tool, causing certain economic losses. For the detection of the size of workpieces after rough and fine machining, most of them are manually detected with ordinary instruments for detection. The machining rejection rate of the production line increases due to the fact that the quality detection of the workpiece can be affected by the precision of instruments and the proficiency of operators, and that the detection data cannot be timely fed back to the machine tool system and, for unqualified workpieces, relevant parameters of the machine tool cannot be corrected timely. In addition, for the storage module, it is not possible to have real-time statistics and data feedback on the information of product specifications for each bin.

The digital production line is based on advanced control technology and other technologies such as industrial robot, visual detection, sensing, and RFID. It integrates multi-functional control systems and top-

系统和顶尖检索设备，能够进行工序内容多且复杂的作业，可以实现产品多样化定制、批量生产。基于新一代信息通信技术与先进制造技术的深度融合，在生产过程中也可同步优化整个生产流程，如图 5-1 所示。

notch retrieval equipment, and can perform tasks with procedures that have multiple and complex contents, to have diversified customized products and products that are produced in batches. Based on the deep integration of the information and communication technology of a new generation and the advanced manufacturing technology, the entire production flow can be synchronously optimized during the production process as shown in Figure 5-1.

图 5-1 数字化生产线
Figure 5-1 Digital Production Line

本项目以某铝壳电机关键构件生产为背景，根据数字化生产线智能化、信息化的发展要求，通过分析试件的生产工艺和生产线的功能需求，本项目设定完成以下任务：对生产线总体布局、工艺流程进行仿真设计且有需求时进行优化，对工业机器人进行示教与轨迹规划，对控制系统进行设计。本项目通过完成以上所有任务实现整条生产线从生产调度、自动化上下料、数控加工、智能检测、自动化仓储到数据管理与远程监控的高自动化生产过程，提高该铝壳电机关键构件的生产效率与质量，提高生产线的数据管理水平，节约人力成本，降低安全风险。因此，本项目具有一定的工程实践价值。

In the context of the production of a key component of a motor with an aluminum shell, and according to the development requirements of an intelligent and information-based digital production line, and the analysis of the production technology of specimens and function requirements of the production line, the following tasks are set for this program: conduct a simulated design of the overall layout and process flow of the production line, and optimize the design when necessary; conduct teaching of the industrial robots, planning of their paths, and design of the control systems. By doing all of these, this program aims to realize a highly automated production for the entire production line from production scheduling, automated loading and unloading, CNC machining, intelligent detection, automated storage to data management and remote monitoring, improving the productivity and quality of the key component of the motor with an aluminum shell, and the data management of the production line, while saving labor costs, and reducing safety risks. Therefore, this program is of a certain practical value in engineering.

【项目描述】

认知数字化生产线的基本构架，了解数字化生产线的基本技术特点与调试方法。

【知识目标】

1. 了解数字化生产线技术的基本系统构架。
2. 熟悉数字化生产线的关键技术及特点。
3. 熟悉数字化生产线调试方法。

【技能目标】

1. 能说出数字化生产线技术的基本系统构架。
2. 能说出数字化生产线的关键技术及特点。
3. 能说出数字化生产线调试方法。

[Program Description]

Recognize the basic architecture of digital production lines, and understand their basic technical characteristics and commissioning methods.

[Knowledge Objectives]

1. To understand the basic system architecture of digital production line technology.
2. Familiar with the key technologies and characteristics of digital production lines.
3. To be familiar with commissioning methods for digital production lines.

[Skill Objectives]

1. To be able to describe the basic system architecture of the technologies of digital production lines.
2. To be able to describe the key technologies and characteristics of digital production lines.
3. To be able to describe the commissioning methods for digital production lines.

任务 5.1　数字化生产线的系统构架
Task 5.1　System Architecture of Digital Production Lines

【知识目标】

了解数字化生产线的基本系统构架。

【技能目标】

能说出数字化生产线技术的基本系统构架。

【素质目标】

1. 了解推进工业数字化转型的发展趋势。
2. 提高学生的创新和发展思维能力。

[Knowledge Objectives]

To understand the basic system architecture of digital production lines.

[Skill Objectives]

Can describe the basic system architecture of digital production line technology.

[Competence Objectives]

1. Understand the development trend of promoting industrial digital transformation
2. Improve students' innovative and developmental thinking abilities.

【任务情景】

原有的铝壳电机关键构件工艺流程生产线结构单一，主要为机加工生产线，工人劳动强度大，如图 5-2 所示。

[Task Scenario]

The existing production line for the key component of the motor with an aluminum shell is mainly a machining production line with a single structure, for which the workers have a high labor intensity, as shown in Figure 5-2.

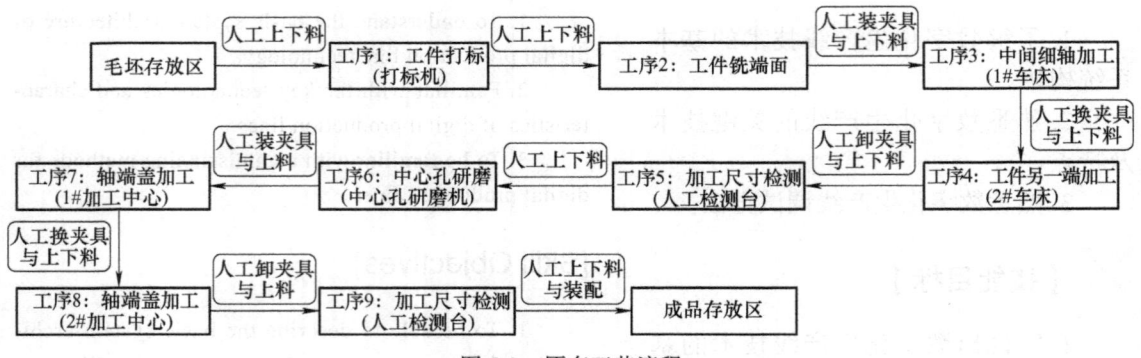

图 5-2　原有工艺流程
Figure 5-2　Existing Process Flow

【任务分析】

本次任务以某铝壳电机关键构件生产线自动化升级改造项目为背景。经调研，该厂现有生产线存在自动化、智能化程度较低，生产线柔性较差，生产线数据缺乏有效管理，生产效率较低下，工人劳动强度较大等缺点。改造后的数字化生产线共有 5 台机器人，分属 5 大制造环节：OP1～OP5（如原料仓、车铣中心、传送机构、加工中心、组装与成品仓）。每个制造环节都有工业机器人，能执行多种不同任务，如机床上下料、搬运、装配等，可大幅提高生产效率。为适应现代制造业的升级与转型，提高生产线自动化、智能化、信息化水平，提出如下需求：智能仓储，自动化上下料，智能化检测，生产线信息化管理，柔性化生产。

[Task Analysis]

This task is based on the program of automation upgrade and renovation of the production line for a key component of a motor with an aluminum shell. It was found from investigation that the existing production line in the factory has shortcomings such as low automation and intelligence, poor flexibility, ineffective data management for the production line, low production efficiency, and high labor intensity among workers. The renovated digital production line has a total of 5 robots, which belong to five major manufacturing stages: OP1-OP5 (i.e. raw material store, milling center, conveying mechanism, machining center, assembly and finished product store). In every manufacturing stage, there are industrial robots that can perform various tasks, such as machine tool loading and unloading, handling, and assembly, which leads to a significant improvement in production efficiency. In order to adapt to the upgrade and transformation of modern manufacturing industry, and improve the level of automation, intelligence, and informatization, the following requirements are proposed for the production line: intelligent storage, automated loading and unloading, intelligent detection, information-based management for the production line, flexible production.

【知识准备】

5.1.1 功能模块

根据生产线的改造需求,在现有机加工工艺不变的前提下,对生产线进行自动化与智能化改造,增加如下几个功能模块:

(1) 智能仓储模块

该模块主要包括立体仓库、出入库中转平台、RFID 检测系统、产品视觉检测系统等。该模块中,立体仓库满足产品自动化出入库的硬件需要,并实时监控库位工件信息;出入库中转平台满足零件出入库过程中定位及放置姿态的要求;RFID 检测系统满足产品自动化、数字化生产的通信技术要求;产品视觉检测系统识别并检测产品的形状及尺寸。

(2) 机器人自动上下料模块

该模块主要包括 6 轴工业机器人、地轨、机器人夹具快换平台。该模块主要完成零件在加工、检测、存储过程中自动化上下料和夹具自动快换的任务。

(3) 智能检测模块

该模块主要包括试件外径尺寸和残留切屑等外观检测装置。

(4) MES 生产线管理与监控模块

该模块主要包括监控显示屏、MES 生产管理系统和 MES 系统服务器。其中,监控显示屏用于对整个制造单元的信息及设备状态的可视化监控与上位操作。MES 生产管理系统实

[Assumed Knowledge]

5.1.1 Function Modules

According to its transformation requirements, while maintaining the existing machining process unchanged, the production line is proposed to be transformed with automation and intelligence by the addition of the following functional modules:

(1) Intelligent storage module

This module mainly includes the warehouse, transfer platform for store-input and store-exit, RFID detection system, and visual detection systems for products, the warehouse meets the hardware requirements for automated store-input and store-exit of products and real-time monitoring of workpiece information in relevant storage positions, the transfer platform for store-input and store-exit meets the requirements for the positioning and placement posture of parts during the store-input and store-exit process, the RFID detection system meets the communication technology requirements for automated and digital production of products, while visual detection systems for products for store-input and store-exit meet the requirement of recognition and detection of the shape and size of the product.

(2) Module of automated loading and unloading by robot

This module mainly includes six-axis industrial robots, ground rails, and quick change platform of robot fixtures, and is mainly for the tasks of automated loading and unloading of parts during machining, detection, and storage, as well as quick change of fixtures.

(3) Intelligent detection module

This module mainly includes devices for the detection of outer diameter size of specimens and appearance of residual chips.

(4) Module of MES production line management and monitoring

This module mainly includes the monitoring display screen, MES production management system, and MES system server, wherein the monitoring display screen is used for visual monitoring and upper level operating of the information and equipment status of the entire manufacturing unit, the MES production manage-

现自动任务排产、单元生产过程的自动控制、生产过程质量管理、单元设备履历管理、产品单件质量追溯、各种看板监控、各种报表和统计等需求；MES 服务器满足 MES 生产管理软件的硬件支持。

铝壳电机关键构件数字化生产线系统构架图如图 5-3 所示。

ment system for the realization of automated task scheduling, automated control of the production process of the unit, quality management in the production process, record management of the equipment in the unit, quality traceability of individual products, and various Kanban monitoring, as well as various reports and statistics, while the MES server is used to meet the requirement of hardware support for the MES production management software.

The system architecture diagram of the digital production line for the key component of the motor with an aluminum shell is shown in Figure 5-3.

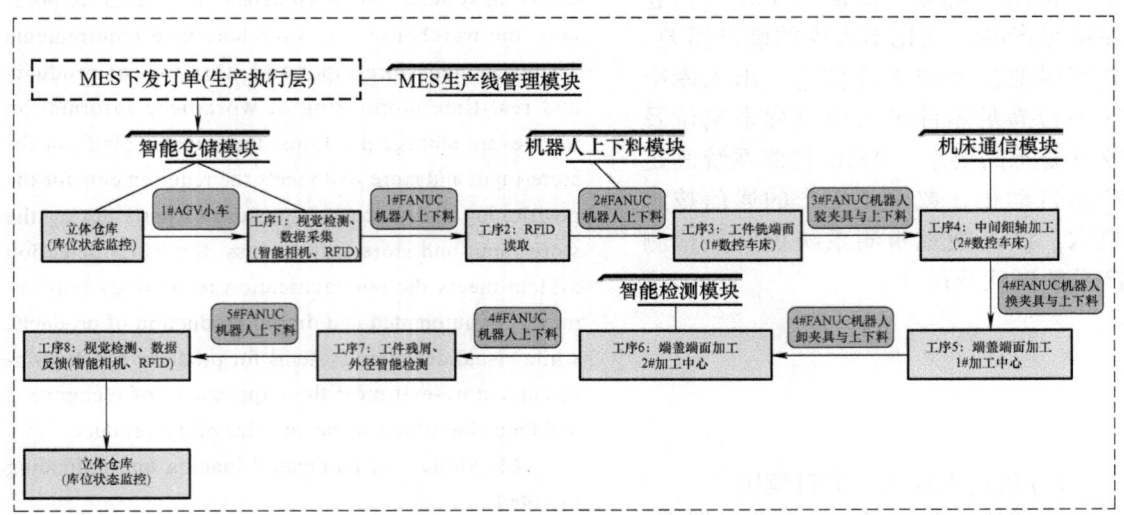

图 5-3 数字化生产线系统构架图

Figure 5-3　System Architecture Diagram of the Digital Production Line

5.1.2　关键技术与特点

1. 数字化生产线关键技术

数字化生产线一般是以机器人为中心，以信息技术和网络技术为纽带，将所有设备联系在一起的大型自动化生产线。在全球市场中，日益加剧的竞争压力对生产线提出了更高的要求，数字化生产线将越来越多地应用于汽车制造、电子电器生产、物流仓储等

5.1.2　Key Technologies and Characteristics

1. Key technologies of digital production lines

Digital production lines are generally large-scale automated production lines centered around robots, linked by information and network technologies to connect all equipment together. In the global market, increasing competitive pressure has put forward higher requirements for production lines, and digital production lines will be increasingly applied in industries such as automobile manufacturing, electronic and electrical

行业。

在数字制造技术发展与应用研究方面,美国处于国际研究的前沿,许多大学和科研机构都在从事虚拟制造的研究工作。当前,新一轮的工业革命正在深化,以数字化技术为基础,在互联网、物联网、云计算、大数据等技术的强力支持下催生的产业模式创新,也会使制造业的产业模式发生根本性变化。中国依靠科技创新,以智能制造为核心,抢占国际竞争制高点,提高经济发展核心竞争力,谋求未来发展的主动权,在智能制造方面已经走在前列。

数字化生产线使用成套自动化设备替代传统设备或人工作业组合,设备具有自动识别、检测、传感等功能,能够实现物料上下料、传送和储存等工序的自动化。采用 RFID(射频识别技术)、条形码、二维码等技术,实现对生产线的制造、刀具、设备、质量等进行控制与数据采集。生产采用单独的控制系统,实现关键工序设备自动控制。采用 DCS(分散控制系统)、DNC(分布式数控)等生产过程控制与调度自动化系统,各装备之间能够实现连续运转。信息监视控制采用 HMI、SCADA 等在生产线内实现生产数据的采集、监控和传递。

MES 生产管理系统通过控制包括物料、设备、人员、流程指令和设施在内的所有工厂资源来提高制造竞争力,提供了一种系统地在统一平台上集成诸如质量控制、文档管理、生产调度等功能的方式。随着 MES 在车间

production, logistics and warehousing.

In terms of the development and application research of digital manufacturing technology, the United States is at the forefront of international research as many universities and research institutions there are engaged in the research on virtual manufacturing. Currently, with the on-going development of a new round of industrial revolution, the innovation of industrial models that is based on digital technology and strongly supported by technologies such as the Internet, the Internet of Things, cloud computing, and big data will fundamentally change the industrial model of the manufacturing industry. China has taken the lead in the intelligent manufacturing as it relies on technological innovation that is centered around the intelligent manufacturing, and take the initiative to seize the commanding heights in the international competition and to improve its core competitiveness in the economic and future development.

The digital production line uses automated equipment in complete sets that takes the place of traditional equipment or manual operations and has functions such as automatic identification, detection, and sensing to achieve automation in procedures such as material loading and unloading, handling, and storage. Technologies such as RFID (Radio Frequency Identification Technology), barcode, QR code are adopted for the control of and data collection on the manufacturing, machining tools, equipment, quality, etc. of the production lines. For production, a separate control system is adopted to achieve automatic control of the equipment in key procedures, and automation systems for production process control and scheduling such as DCS (Distributed Control System) and DNC (Distributed Numerical Control) are used for continuous operation of each equipment, while HMI, SCADA, etc. are used for Information monitoring and control to achieve the collection, monitoring, and transmission of production data within the production line.

The MES production management system improves manufacturing competitiveness by controlling all factory resources, including materials, equipment, personnel, flow instructions, and facilities, which provides a systematic way to integrate functions such as quality control, document management, and production scheduling on a unified platform. With the implementation

执行层面的引入，大大改善了车间生产流程的调度和管理效率。ERP系统是一个基于客户机/服务机架构的开放的、集成的企业资源计划系统。其功能覆盖与PLC生产制造和销售相关的供应链管理、订单管理、生产计划、库存管理等方面。

2. 数字化生产线的特点

1) 产品智能化。产品智能化是指把传感器、处理器、存储器、通信模块和传输系统融入各种产品，使产品具备动态存储、感知和通信的能力，从而实现产品的可追溯、可识别、可定位。

2) 装备智能化。通过先进制造、信息处理、人工智能等技术的集成和融合，可以形成具有感知、分析、推理、决策、执行、自主学习及维护等自组织、自适应功能的智能生产系统以及网络化、协同化生产设施，这些都属于智能装备。

3) 生产方式智能化。个性化定制、极少量生产、服务型制造以及云制造等新业态、新模式，其本质是在重组客户、供应商、销售商以及企业内部组织的关系，重构生产体系中信息流、产品流、资金流的运行模式，重建新的产业价值链、生态系统和竞争格局。

4) 管理智能化。随着纵向集成、横向集成和端到端集成的不断深入，企业数据的及时性、完整性、准确性不断地得到提高，必然使管理更加准确、更加高效、更加科学。

5) 服务智能化。智能服务是智能制造的核心内容，越来越多的制造企业已经意识到从生产型制造向生产服务型制造转型的重要性。

of MES at the level of plant execution, the scheduling and management efficiency of the production flow of the plant has been greatly improved. ERP system is an open and integrated enterprise resource planning system based on the client/server architecture, with its functions covering supply chain management, order management, production planning, inventory management, etc. related to PLC production and sales.

2. Characteristics of digital production lines

1) Intelligent products. Intelligent products refer to the products that integrate sensors, processors, memories, communication modules, and transmission systems, and therefore, have the ability of dynamical memory, perception, and communication, thereby can be traced, identified, and located.

2) Intelligent equipment. Intelligent equipment includes intelligent production systems and facilities that are formed by the integration and fusion of technologies of advanced manufacturing, information processing, artificial intelligence, etc. The intelligent production systems have self-organizing and adaptive functions such as perception, analysis, reasoning, decision-making, and execution, as well as autonomous learning and maintenance, while the intelligent production facilities are network based and work collaboratively.

3) Intelligent production methods. The essence of new business forms and models such as personalized customization, minimal production, service-oriented manufacturing and cloud manufacturing is to restructure the relationship between the customers, suppliers, distributors and the enterprises themselves, reshape the operation modes of information flow, product flow and capital flow in the production system, and rebuild the industrial value chains, ecosystems and competition patterns.

4) Intelligent management. With the continuous development of vertical integration, horizontal integration, and end-to-end integration, the data timeliness, integrity, and accuracy of enterprises are constantly improved, which will inevitably make management more accurate, efficient, and scientific.

5) Intelligent services. As intelligent services are the core content of intelligent manufacturing, more and more manufacturing enterprises have realized the importance of transforming from the production-oriented manufacturing to the production-and-service-oriented one.

5.1.3 工作站调试

数字化生产线设备在安装完成之后，首先要进行单站调试，单站调试无误后，才能保证联调工作的顺利完成。此外，在单站调试的过程中，一些调试助手也能够快速地帮我们验证。

上电前的检查工作是非常重要的，通常分为短路检查、断路检查和对地绝缘检查。推荐方法：用万能表一根一根地检查。尽管这样花费的时间很长，但是检查是比较完整的。为了减少不必要的损失，一定要在通电前进行输入电源的电压检查，确认是否与原理图所要求的电压一致。对于有PLC、变频器等价格昂贵的电气元件一定要认真地执行这一步骤，避免电源的输入输出反接对元件的损害。推荐方法：打开电源总开关之前，先进行一次电压的测量，并记录。

检查PLC的输入输出。下载程序包括PLC程序、触摸屏程序和显示文本程序。将写好的程序下载到相应的系统内，并检查系统的报警。调试工作不会总是很顺利的，总会出现一些系统报警，一般是因为内部参数没设定或是外部条件构成了系统报警的条件。这就要根据调试者的经验进行判断，对配线再次检查确保正确。如果还不能解决故障报警，就要对PLC等的内部程序进行详细的分析，逐步分析确保正确。

设备功能的调试。排除上电后的报警后就要对设备功能进行调试了。首先要了解设备的工艺流程，然后进行手动空载调试。手动工作动作无误

5.1.3 Workstation Commissioning

When equipment installation of a digital production line is completed, a single station commissioning must be carried out first. Only when the single station commissioning is conducted and proved correct can the smooth implementation of joint commissioning be ensured. In addition, some commissioning assistants during the single station commissioning can help us obtain a quick verification.

The checks before power on are very important, and usually consist of short circuit check, open circuit check, and ground insulation check, with the recommended method as: check one by one with a universal meter, which can ensure a relative completeness of checks although taking a long time. In order to reduce unnecessary losses, the voltage of the input power supply must be checked before powering on to confirm whether it is consistent with the voltage required by the schematic. For expensive electrical elements such as PLCs and frequency converters, the above mentioned step must be strictly followed to avoid reverse connection of the input and output of the power supply which may cause damages to the elements, with the recommended checking method as: measure and record the primary voltage before turning on the main power switch.

The check of the PLC input and output. Download programs that include PLC program, touch screen program, and display text program. Download the written programs to the corresponding systems and check for system alarms. The commissioning will not always be smooth, as some system alarms are most likely to occur due to the internal parameters that have not been set or external conditions that constitute the conditions for system alarms. The wiring shall be rechecked to prove correct, which requires a judgement based on the experience of the person who conducts commissioning. If the fault alarm cannot be resolved yet, it is necessary to analyze in details step by step the internal programs, such as that of the PLC, to ensure their accuracy.

The commissioning of equipment functions. The commissioning of equipment functions shall be conducted when the alarms after powering on are eliminated. Firstly, one shall understand the process flow of the equipment; then perform manually the commissioning

再进行自动的空载调试。最后，空载调试完毕后，进行带载的调试。

without load, which shall be proved correct, before performing the automated commissioning without load. Finally, perform the commissioning with load when the commissioning without load is completed.

【课后巩固】

1. 简述数字化生产线的系统构架。
2. 简述数字化生产线的关键技术与特点。
3. 简述数字化生产线的调试步骤。

[Consolidation after Class]

1. Briefly describe the system architecture of digital production lines.
2. Briefly describe the key technologies and characteristics of digital production lines.
3. Briefly describe the commissioning steps of the digital production lines.

任务 5.2　数字化生产线各模块的设计与仿真
Task 5.2　Design and Simulation of Various Modules of Digital Production Lines

【知识目标】

1. 掌握数字化生产线的 Process Simulate 仿真建模方法。
2. 掌握 Process Simulate 的运动学创建方法。
3. 掌握 Process Simulate 的装配仿真。

[Knowledge Objectives]

1. To master the simulation modeling method with the Process Simulate for digital production lines.
2. To master the kinematics creation method with the Process Simulate.
3. To master the assembly simulation with the Process Simulate.

【技能目标】

1. 能够对数字化生产线各模块进行仿真和建模。
2. 能够对模型进行运动学定义。
3. 能够对数字化生产线进行装配仿真。

[Skill Objectives]

1. To be able to conduct simulation and modeling for various modules of digital production lines.
2. To be able to define the model in terms of kinematics.
3. To be able to perform assembly simulation of digital production lines.

【素质目标】

1. 培养学生发现问题、解决问题的能力。
2. 培养学生合作、交流的能力，培养学生的团队意识。

[Competence Objectives]

1. To cultivate students' ability to discover and solve problems.
2. To cultivate students' ability to cooperate and communicate in the spirit of teamwork.

【任务情景】

Process Simulate 软件是西门子公司 Tecnomatix 数字孪生解决方案中的一款仿真软件，用于人机工程仿真、装配过程仿真和机器人离线仿真。通过 Process Simulate 软件，进行数字化生产线的建模与仿真。

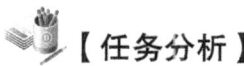【任务分析】

Process Simulate 软件支持多款机器人控制器，如 FANUC、ABB、YASKWA、KUKA、SCARA(三菱)、NC、Panasonic、UR 等知名品牌。Process Simulate 软件可进行多机器人、多工位的过程仿真验证，根据任务搭建合适的仿真场景。

【知识准备】

5.2.1 基于 Process Simulation 的仿真建模

Process Simulation 软件仿真数据类型为 .cojt 格式，可通过外部插件进行快速批量转换常用软件 Catia、UG、Pro/E 等 3D 设计软件数据，也可通过 Process Simulate 软件自带的数据转换功能进行转换，但其只能转换 .STP 格式的 3D 数据。其操作步骤如下：使用"设置建模范围"命令激活选定对象建模的范围，如图 5-4 所示；这时会加载所选组件的 COJT 文件并打开它以进行建模，同时将其设置为活动组件。

[Task Scenario]

The Process Simulate is a simulation software in Siemens' Tecnomatix digital twin solution, and is used for the simulation of human-machine engineering, and assembly process, as well as the offline simulation of robots. Conduct the modeling and simulation of digital production lines with the Process Simulate software.

[Task Analysis]

The Process Simulate software supports multiple robot controllers, such as FANUC, ABB, YASKWA, KUKA, SCARA (Mitsubishi), NC, Panasonic, and UR which are well-known brands. It can perform process simulation verification for multiple robots and stations, and build appropriate simulation scenarios based on specific tasks.

[Assumed Knowledge]

5.2.1 Simulation and Modeling Based on the Process Simulation

The simulation data type of Process Simulation software is .cojt format, which can either be quickly converted in batches through external plugins into 3D design software data (with common software such as Catia, UG and Pro/E), or be converted through the built-in data conversion function of the Process Simulate software which, however, can convert only the 3D data in the .STP format. The operating steps are as follows: use the "Set Modeling Scope" command to activate the modeling scope of the selected object, as shown in Figure 5-4. The COJT file of the component will be loaded and opened for modeling at this moment, and set as the active component.

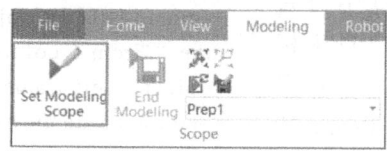

图 5-4 选定对象建模的范围

Figure 5-4 Modeling Scope of the Selected Object

对于处于建模状态的组件，在其名称前有一个图标叠加层，如图 5-5 所示。默认情况下，在建模结束时，编辑原型零件或资源的单个实例会将更改传输到该原型的所有实例。但是，如果希望修改单个实例而不影响其他实例，则可以执行 Save Component As 命令，系统会创建一个新的原型并加载它的一个新实例。

For a component in the modeling state, there is an icon overlay layer before its name, as shown in Figure 5-5. By default, at the end of modeling, editing a single instance of a part or resource of a prototype will transfer the changes to all instances of that prototype. However, if one wants to modify a single instance without affecting other instances, execute the Save Component As command, and the system will create a new prototype and load a new instance of it.

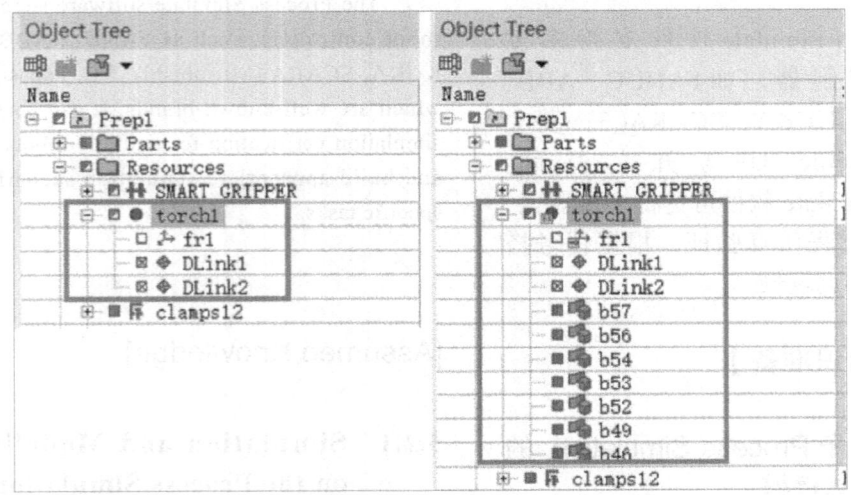

图 5-5　模型图层
Figure 5-5　Model Layer

在系统根目录下选择要保存的路径并输入文件名，单击"保存"按钮，可以看到新增的另存组件会显示在对象树中，如图 5-6 所示。

用户可以使用 Set New Working Frame 创建新的坐标系的方式来指定新的工作坐标系，或单击 Reset to Origin 将工作坐标系恢复成系统默认位置。

在选取了对象之后单击，可以在弹出的如图 5-7 所示的对话框中设置对象的自身坐标系。在设置自身坐标系对话框中，Objects 栏中列出了将要被设置自身坐标系的对象，From 栏是对象目前的自身坐标系位置，在 To frame 栏中可以定位希望设置新自身坐

Select the path to save in the system root directory and enter the file name. Click the "Save" button to see the newly saved component displayed in the object tree, as shown in Figure 5-6.

Users can create a new frame through the "Set New Working Frame" to specify a new working frame, or click "Reset to Origin" to restore the working frame to the default position of the system.

When the object is selected, click it to set its self frame in the pop-up dialog box as shown in Figure 5-7. In the dialog box of "Set Self Frame", the "Objects" column lists the objects whose self frame will be set, the "From" column is the current position of the object's self frame, and in the "To frame" column, one can locate the position of the new Self Frame of the object.

标系对象的坐示位置。

完成建模后，使用 End Modeling 命令可结束对组件的建模，Process Simulate 会将该组件保存在系统根目录下。在关闭 Process Simulate 之前，如果不结束建模会话，下次打开 Process Simulate 时，对象仍然会显示处在建模状态。使用 End Modeling 命令，系统会更新链接到该组件的所有实例。要保存 Process Simulate 的 Study，可以单击 File → Disconnected Study → Save，如图 5-8 所示。

When modeling is completed, execute the "End Modeling" command to end the modeling of the component, and the Process Simulate will save the component in the system root directory. If the modeling session is not ended before closing the Process Simulate, it will appear that the object is still in the modeling state the next time the Process Simulate is opened. The system will update all instances linked to the component when using the "End Modeling" command. To save the "Study" of the Process Simulate, one can click File → Disconnected Study → Save, as shown in Figure 5-8.

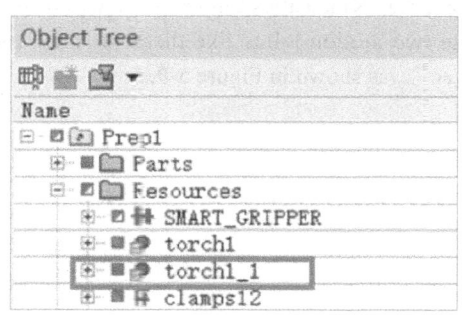

图 5-6 新的实例保存
Figure 5-6 Saving a New Instance

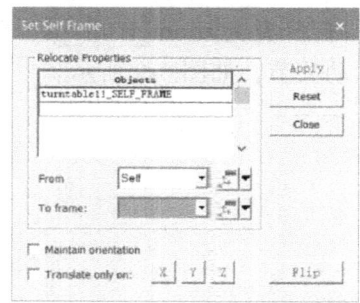

图 5-7 设置对象的自身坐标系
Figure 5-7 Setting the Object's Self Frame

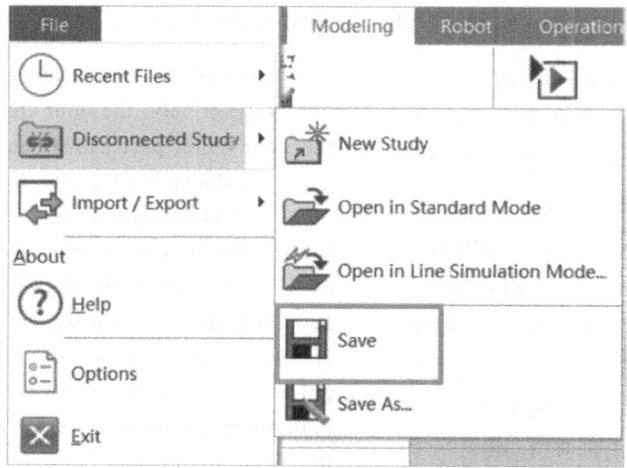

图 5-8 保存页面
Figure 5-8 the Page of Saving

5.2.2 基于 Process Simulation 的运动学创建

1) 在 Process Simulate 标准模式下新建一个 Robcad Study, 并使用 Insert-Component 命令将 room_door_geo_user2.cojt 作为 ToolProtype 类型插入。

2) 在对象树中, 选中 room_door_geo, 将其设置为建模状态。

3) 对于本例中的简单结构对象——门 (room_door_geo), 它有两个运动关节, 一个是门板绕着门铰链相对于门框的转动, 另一个是门把手绕着门板的转动。所以, KinematicsEditor 页面将打开并为 room_door_geo 创建两个运动关节, 如图 5-9 所示。

5.2.2 Creation of Kinematics Based on the Process Simulation

1) Create a new " Robcad Study" under the standard mode of the Process Simulate, and then insert the " room_ door_ geo_ user2. cojt " as a ToolType using the "InsertComponent" command.

2) In the object tree, select " room_ door_ geo ", and set it to the modeling state.

3) In this example, the object is a simple structure—a door (room_door_geo) that has two motion joints: one motion is the rotation of the door panel around the door hinge relative to the door frame, and the other is the rotation of the door handle around the door panel. Therefore, the "Kinematics Editor" page will be opened to create two motion joints like these for the " room_ door_ geo ", as shown in Figure 5-9.

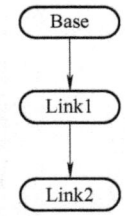

图 5-9 创建运动关节
Figure 5-9 Creating Motion Joints

4) 打开 Kinematics Editor, 分别创建 3 个 Link: 门框部分命名为 BASE, 门板部分命名为 Link1, 门把手部分命名为 Link2, 如图 5-10 所示。

5) 添加 BASE 和 Link1 之间的运动关节 j1: 绕着沿铰链 Z 向的轴转动。选中 BASE 和 Link1, 单击 Create Joint, 在 Joint Properties 对话框中, 设定运动轴的起点是铰链 Z 向的顶部中心点, 终点是铰链 Z 向的底部中心点。一般情况下, 门向里的方向为 90°, 所以也要给 j1 的运动范围设定限制。如图 5-11 所示, 设定 j1 的 Joint Properties, 完成后单击 OK 按钮。

4) Open the " Kinematics Editor" and create three links respectively: the door frame part is named Base, the door panel part Link1, and the door handle part Link2, as shown in Figure 5-10.

5) Add the motion joint j1 between the BASE and Link1: that rotates around the axis in the Z-direction of the hinge. Select the BASE and Link1, click Create Joint and, in the dialog box of Joint Properties, set the starting point of the motion axis to be the top center point in the Z-direction of the hinge, and the ending point to be the bottom center point in the Z-direction of the hinge. As the door is usually turned inward by 90 °, a limit should be set for the range of motion of j1. Set the Joint Properties of j1 as shown in Figure 5-11, and click the OK button when finished.

项目 5　数字化生产线的构架及技术特点　283

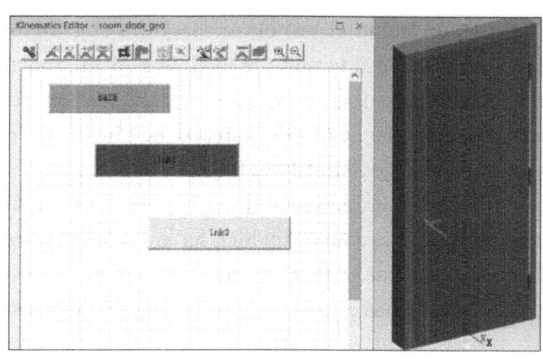

图 5-10　创建 3 个 Link
Figure 5-10　Creating Three Links

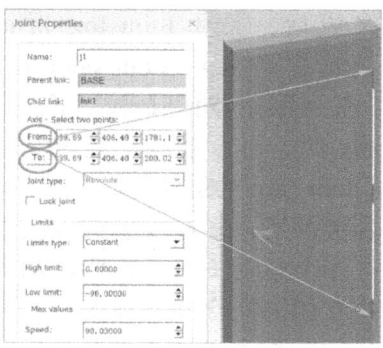

图 5-11　设定门的运动范围
Figure 5-11　Setting the Motion Range of the Door

6) 添加 Link1 和 Link2 之间的运动关节 j2：门把手绕着 X 方向的轴转动，如图 5-12 所示。设定 j1 的 Joint Properties，完成后单击 OK 按钮，如图 5-12 所示。

7) 打开 room_door_geo 的 Joint Jog 页面，拖动 Steering/Poses 滑动条，如图 5-13 所示，可以在图形查看器中看到相应部件随运动关节的运动，这有助于检验对象运动学定义的正确性。

6) Add the motion joint j2 between the Link1 and Link2: the door handle rotates around the axis in the X-direction, as shown in Figure 5-12. Set the Joint Properties of j1, and click the OK button when finished, as shown in Figure 5-12.

7) Open the Joint Joint page of the room_ door_ geo, and drag the Steering/Poses slider, as shown in Figure 5-13. The movement of the corresponding parts with the motion joints can be seen in the graphic viewer, which helps to verify the correctness of the kinematic definition of the object.

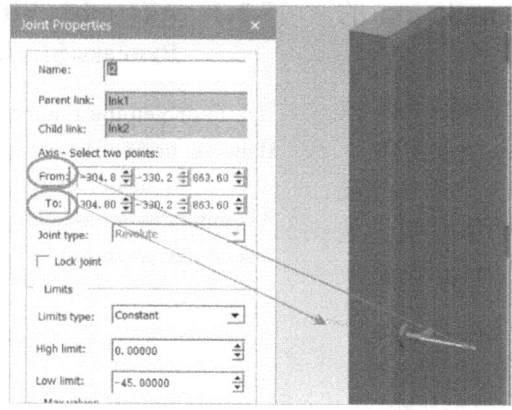

图 5-12　设定门把手的运动范围
Figure 5-12　Setting the Motion Range of the Door Handle

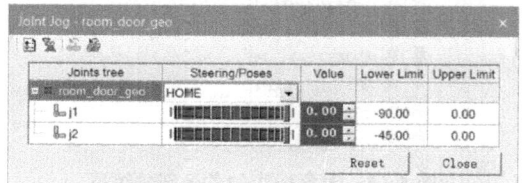

图 5-13　Joint Jog 页面
Figure 5-13　Joint Jog Page

8) 完成后，关闭 Joint Jog 对话框。可以选择结束对 room_door_geo 的建模，然后保存当前 Study。下次打开该 Study 时，可以看到已经定义好的 room_door_geo 的运动学。

很多情况下，某些设备或者机器人中存在多个运动关节，这些运动关节中有些互相之间存在从属关系，如图 5-14 所示的机器人，该机器人后背有一个四连杆机构。

8) Upon completion, close the Joint Jog dialog box. One can choose to end the modeling of the room_ door_ geo and then save the current study. when this Study is opened the next time, can see the kinematics of the room_ door_ geo already defined.

In many cases, certain devices or robots have multiple motion joints, some of which have subordinate relationships with each other, as shown in Figure 5-14 of the robot, which has a four-rod mechanism on the back .

Joint parameters of j2
Leading joint j1
Follow factor of–1
Z
U
X

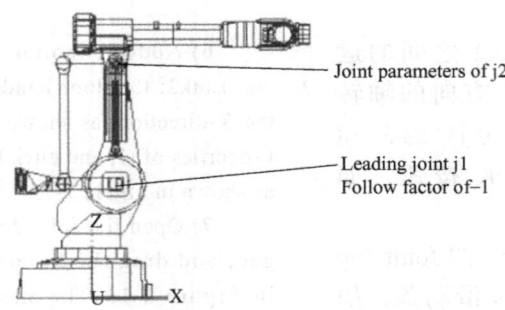

图 5-14 互相之间存在从属关系的运动关节

Figure 5-14 Motion Joints Having Subordinate Relationships with Each Other

5.2.3 基于 Process Simulation 的装配仿真

1) 在 Process Simulate 标准模式下，打开 Engine Assembly. psz 文件。

2) 在对象树中，可以看到 Engine Assembly，作为一个装配体总成，如图 5-15 所示，它一共有 6 个分总成部件。

5.2.3 Assembly Simulation Based on the Process Simulation

1) Open the Engine Assembly. psz file under the standard mode of Process Simulate.

2) In the object tree, it can be seen that the Engine Assembly, as an assembly, has six components as sub assemblies, as shown in Figure 5-15.

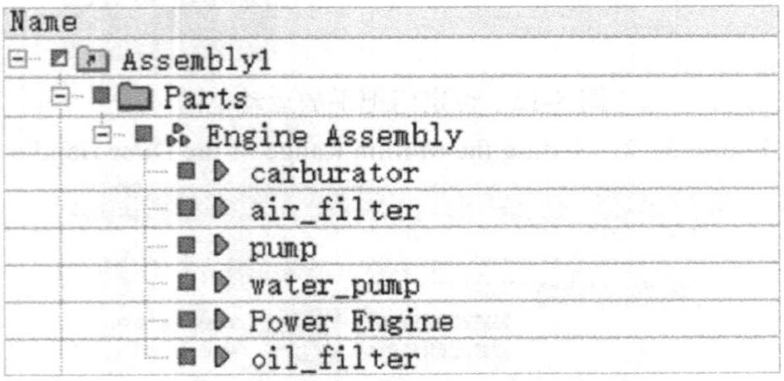

Name
Assembly1
Parts
Engine Assembly
carburator
air_filter
pump
water_pump
Power Engine
oil_filter

图 5-15 对象树中的各个部件组成

Figure 5-15 Composition of the Components in the Object Tree

3) 右击操作树中的空白区域,选择并新建一个复合操作,并将其命名为 Engine Assembly。

4) 一般情况下装配仿真过程是一个相对装配总成的拆卸过程,所以将从 EngineAssembly 最外侧的部件入手,由外而内地对其部件结构进行分解,最后再用 Sequence Editor 进行排序和链接。

5) 设置 Pick Level 为 Component,选择 Engine Assembly 复合操作,按住 <Ctrl> 键,在图形查看器中选择发动机总成上部的 air_filter 部件,右击并在弹出的快捷菜单中选择;在弹出的对话框中的终止点 End Point 位置选择 air_filter 部件上方(Z 向)的某个点,将持续时间 Duration 设置为 3.5s,如图 5-16 所示,完成后单击 OK 按钮。

6) 在操作树中,可以看到已经创建的复合操作和它包含的子操作 air_filter_Op。在 air_filter_Op 操作下,包含了两个路径点,这两个路径点就是刚才在创建 air_filter 的 Object Flow-Operati/On 时输入的 loc 和 loc1,如图 5-17 所示。

3) Right click on the blank area in the action tree, select, create a new composite operation and name it Engine Assembly.

4) In general, the process of assembly simulation is a disassembly process relative to the assembly. Therefore, starting from the outermost part of the EngineAssembly, the structure of the components will be decomposed from the outside to the inside, and finally sorted and linked using the Sequence Editor.

5) Set the Pick Level as Component, select the composite operation of Engine Assembly, hold the<Ctrl>button, and select the component of air_filter on the upper part of the engine assembly in the graphic viewer, right-click and select from the pop-up shortcut menu. At the End Point in the pop-up dialog box, select a certain point above the component of air_filter (Z-direction), set the duration of Duration to 3.5s, as shown in Figure 5-16, and click the OK button when finished.

6) In the operation tree, one can see the created composite operation and its contained sub operations of air_filter_Op. Under the air_filter_Op operation, two path points are included, i.e. the Start Point (loc) and End Point (loc1) entered during the creation of Object FlowOperati/On of the air_filter, as shown in Figure 5-17.

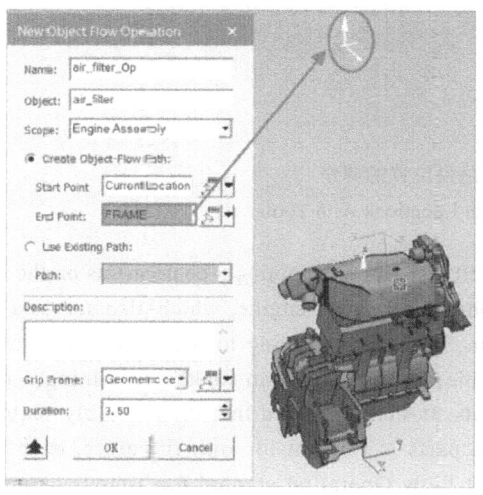

图 5-16 参数设置
Figure 5-16 Parameter Settings

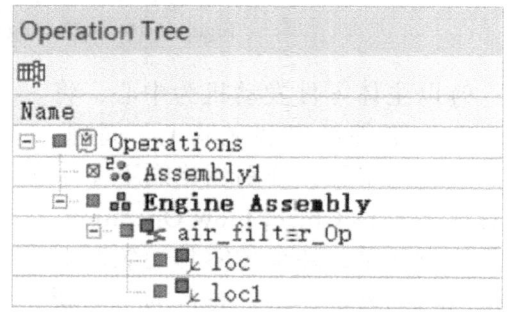

图 5-17 在操作树中显示创建对象
Figure 5-17 Displaying the Created Object in the Operation Tree

7) 在操作树中，选中 air_filter 的 Object Flow Operation，然后在路径编辑器中单击工具栏上的按钮，可以看到 air_filter_Op 被添加到路径编辑器中，如图 5-18 所示。

7) From the operation tree, select the Object Flow Operation of the air_filter, and then click the button on the toolbar of the path editor, it can be seen that the air_filter_op has been added to the path editor, as shown in Figure 5-18.

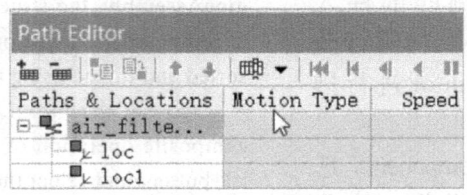

图 5-18 创建对象被添加到路径编辑器
Figure 5-18 Addition of the Created Object to the Path Editor

8) 因为要做的装配仿真是发动机各部件从分散状态到组装为总成的过程，所以需要将 air_filter_Op 中起始、终点的位置互换一下。可以在路径编辑器中直接用鼠标拖动对应的路径进行操作，或者使用工具进行路径位置的调整。完成后如图 5-19 所示，可以看到对象树中对应操作下的路径顺序也发生了相应的变化。

8) Since the assembly simulation to be done is the process of assembling various components of the engine which are in a dispersed state to form an assembly, it is necessary to swap the positions of the start and end points in the air_filter_Op. One can directly drag the corresponding path with the mouse in the path editor, or adjust the path locations with tools. Upon completion, as shown in Figure 5-19, it can be seen that the path order under the corresponding operation in the object tree has changed accordingly.

图 5-19 使用工具进行路径位置的调整
Figure 5-19 Adjusting the Path Locations with Tools

9) 以主体零件发动机为中心，将其余 5 个部件都安装在其上。同理，参照上述的步骤，分别创建剩下 4 个分总成零件的 Object Flow Operation。需要注意的是，装配体零件的装配顺序是从内到外的，所以每次为一个新的部件创建的 New Object Flow Operation 都应该排列在复合操作下的最前

9) Install the remaining 5 components on the main component of power engine, which also serves as the center. Similarly, referring to the above steps, create the Object Flow Operation for the remaining 4 components. It shall be noted that the assembly sequence of the parts is from inside out; therefore, each New Object Flow Operation created for a new component shall be listed at the first place in the composite operation. When all are completed, all operations under the operation tree can be seen, as shown in Figure 5-20.

项目 5　数字化生产线的构架及技术特点　287

面。全部完成后，可以看到操作树下所有操作，如图 5-20 所示。

10) 在序列编辑器中，可以看到右侧的甘特图区域中所有的子操作都是对齐的，那是因为所有操作设置的时间 (Duration) 都是一样的，如 3.5s。单击序列编辑器上的"播放仿真"按钮，可以看到仿真运行的过程是 5 个组件同时往发动机移动的过程。如果希望看到部件一个接一个地安装，可以在序列编辑器里将 5 个组件的 Flow Operation 使用 Link 功能连起来。选中序列编辑器树状区域中复合操作下的所有子操作，单击按钮，可以看到操作序列的变化，如图 5-21 所示。

10) In the sequence editor, it can be seen that all sub operations in the Gantt chart area on the right are aligned, for the reason that the time (Duration) set for all operations is the same, i.e. 3.5s. Click the "Play Simulation" button on the sequence editor, it can be seen that the simulation process is the process of 5 components moving to the power engine simultaneously. If one wants to see the installation of components one by one, it is possible to use the Link function to link the Flow Operation of the 5 components in the sequence editor. Select all sub operations under the composite operation in the tree area of the sequence editor and click the button to see the changes in the operation sequence, as shown in Figure 5-21.

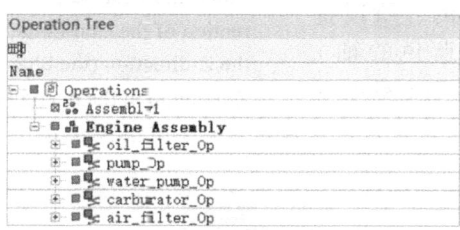

图 5-20　操作树下显示所有操作

Figure 5-20　Displaying all Operations under the Operation Tree

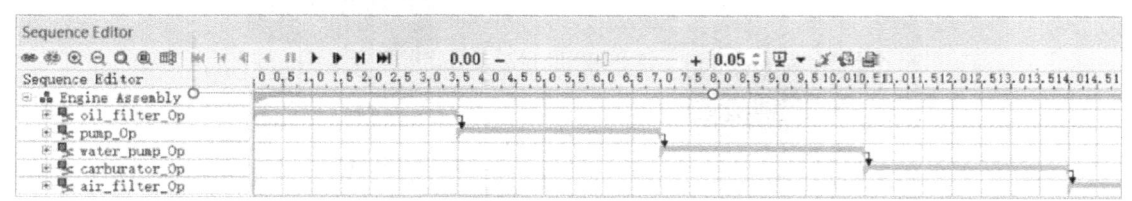

图 5-21　操作序列的变化

Figure 5-21　Changes in the Operation Sequence

11) 再次单击序列编辑器上的"播放仿真"按钮，这次可以看到 5 个组件依次安装在发动机上。其中，在安装 air_filter 部件时，安装的最佳路径应该是沿着发动机的正上方垂直向下安装。由于之前在定义 Object Flow Operation 的起始点时，并没有精确定位，所以需要修正一下安装 air_filter 部件的路径。

11) Click the "Play Simulation" button on the sequence editor again. It can be seen this time that the 5 components are installed on the power engine in sequence. Among them, when installing the component of air_filter, the best path should be the one from the top of the power engine vertically downwards to the bottom. Due to the lack of precise positioning when defining the starting point of the Object Flow Operation, it is necessary to correct the installation path of the air_filter.

12) 在操作树中，单击 air_filter_Op 操作，然后在路径编辑器中单击最后一个路径 loc，最后单击 Operation 选项卡 → Add Location → Add Location Before，如图 5-22 所示。

12) In the action tree, click the air_filter_Op, then click the last path, loc, in the path editor, and finally click the Operation tab → Add Location → Add Location Before, as shown in Figure 5-22.

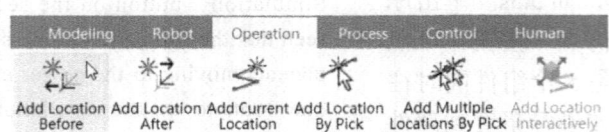

图 5-22 启用 Add Location Before 功能

Figure 5-22 Enabling the Add Location Before Function

13) 可以看到 Placement robotic arm 对话框弹出，并在图形查看器中出现位置的操纵器坐标系。由于希望路径是沿着终止点的 Z 向移动的，所以拖动操纵器坐标系的 Z 向（蓝色的轴），或者直接使用对话框中的 Translate Z 功能，如图 5-23 所示。

13) It can be seen that the dialog box of Placement robotic arm pops up and the Placement Manipulator frame of the location appears in the graphics viewer. Since the desired path to is to translate along the Z-direction of the End Point, drag the manipulator frame along the Z-direction (the blue axis), or directly use the Translate Z function in the dialog box, as shown in Figure 5-23.

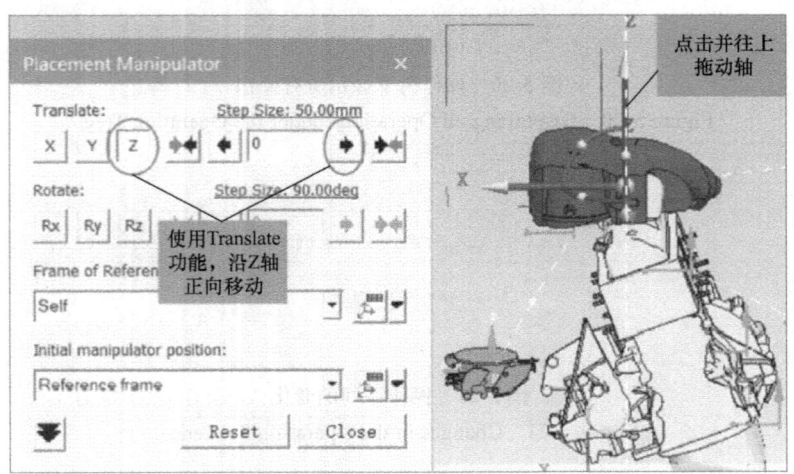

图 5-23 路径沿 Z 向移动

Figure 5-23 Path Translating along the Z-direction

14) 完成后单击 Placement robotic arm 对话框中的 Close 按钮，可以看到路径编辑器中，air_filter_Op 操作的路径位置增加为了 3 个，包括刚刚插入的新位置，如图 5-24 所示。

14) When finished, click the Close button in the dialog box of Placement robotic arm, and it can be seen that 3 path locations appear under the operation of air_filter_Op in the path editor, including the newly added location, as shown in Figure 5-24.

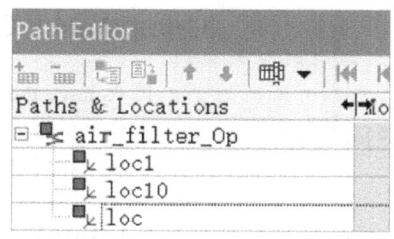

图 5-24 路径编辑器中显示的新增路径位置

Figure 5-24 Displaying the New Path Location in the Path Editor

15) 单击序列编辑器，展开其中的 air_filter_Op，可以看到它下面包含的路径位置也相应地更新了。再次单击序列编辑器上的"播放仿真"按钮，可以看到 air_filter_Op 操作的变化。对于其他的子操作，也可以采用相同的方法在路径中添加新的位置。

16) 完成后，可以保存或者另存相关的 Study 文件。

15) Click the sequence editor and expand its air_filter_Op, it can be seen that the path locations included have also been updated accordingly. Click the "Play Simulation" button on the sequence editor again, and the changes of the operation of air_filter_Op can be seen. For other sub operations, the same method can be used to add new locations in the path.

16) When finished, the Study file can be saved, or saved as other relevant files.

【课后巩固】

抓取与放置操作作为机器人最基本的运动操作，该如何让零件通过传输带来到机器人旁边，又该如何让机器人准确地抓取零件并放到合适的位置上呢？

[Consolidation after Class]

Picking and dropping are the most basic motion operations of a robot. How can we make the parts brought through transmission to a place beside the robot, and enable the robot to accurately pick the parts and drop them to appropriate positions?

任务 5.3　识读数字化生产线的设计图样
Task 5.3　Reading of the Design Drawings of Digital Production Lines

【知识目标】

1. 了解识读机械原理图的一般方法。
2. 了解识读电气原理图的一般方法。

[Knowledge Objectives]

1. To understand the general methods for reading mechanical schematic diagrams
2. To understand the general methods for reading electrical schematic diagrams.

【技能目标】

能正确识读机械原理图和电气原理图等图样。

[Skill Objectives]

To be able to correctly read drawings including mechanical and electrical schematic diagrams.

【素质目标】

1. 具有分析与决策能力。
2. 具有发现问题、解决问题的能力。
3. 具有团体协作能力。

【任务情景】

基于面向工作过程的铝壳电机关键构件数字化生产线的安装与调试，探讨对机械原理图和电气原理图领域的课程任务学习，使课程学习内容与工作内容良好对接。

【任务分析】

在前序课程中，同学们已经有了机械制图、三维计算机辅助设计与电路分析等课程基础，掌握了一些制图与识图的基本知识与技能，对于以后要走上对生产线进行安装、调试、维护与维修岗位的同学来说，将所学知识运用到相对复杂的数字化生产线，也是一项必备的技能。

【知识准备】

5.3.1 识读机械图样

对于典型数字化生产线系统，本次任务是要求学生识读中等复杂程度的零件图和装配图。

同学们已经学过组成机器的最小单元是零件，读零件图的目的就是根据零件图想象出零件的结构形状、了解零件的尺寸和技术要求，以便指导生产和解决技术问题，这是作为机械工程技术人员和生产制造企业应具备的素质。本次任务中出现的一些通用零件，如螺栓、螺母、轴承、键和销

[Competence Objectives]

1. To have the ability to conduct analysis and decision-making.
2. To have the ability to discover and solve problems.
3. To have the ability for teamwork.

[Task Scenario]

Based on the installation and commissioning of the digital production line for a key component of a motor with an aluminum shell which are work process oriented, explore the study of course tasks in the fields of mechanical and electrical schematic diagrams, so as to ensure a good connection between the study and work contents.

[Task Analysis]

Previously, students have taken such courses as mechanical drawing preparation, computer-aided 3D design, and circuit analysis, and have mastered some basic knowledge and skills in drawing preparation and reading. For students who will engage in the installation, commissioning, maintenance and repair of production lines in the future, it is a necessary skill for them to apply their knowledge to digital production lines that are relatively complex.

[Assumed Knowledge]

5.3.1 Reading of Mechanical Drawings

For typical digital production line systems, students are required to complete this task of reading part and assembly drawings that are moderately complex.

Students have learned that a part is the smallest unit that makes up a machine. The purpose of reading part drawings is to imagine the structural shapes of the parts, understand their dimensions and technical requirements based on the part drawings, in order to guide production and solve technical problems, which is also a competence that mechanical engineering technicians and manufacturing enterprises should possess. Some common part that appeared in this task, such as the drawing of bolts, nuts, bearings, keys, and pins, have been uni-

等，在国家标准中都做了统一的规定。读零件图的方法与步骤如下：

1) 了解视图名称、材料、比例等。从名称大致了解零件的用途；从材料可知其大概的制造方法；从图样比例可估计零件的大小。

2) 分析零件各个视图，弄清它们的视图名称、剖切位置、投影关系以及其所表达的内容。用形体分析法和线面分析法分析结构的相对位置，然后相互联系，想出零件的整体结构形状。

3) 分析零件的总体尺寸、定形尺寸、定位尺寸、尺寸基准以及零件的主要尺寸，明确零件各部分的大小。

4) 分析零件的尺寸极限要求与几何公差、表面结构等技术要求和质量指标。

图 5-25 所示为铝壳电动机轴承固定件零件图。一般情况下，轴承内径用轴承内径代号 (基本代号的后两位数) × 5 = 内径 (mm)，例：轴承 6204 的内径是 04 × 5 = 20mm。

特殊情况：当轴承内径小于 20mm，如轴承内径尺寸为 10、12、15、17(mm)，对应内径代号为 00、01、02、03，如图 5-25 所示。当轴承内径小于 10mm，直接用基本代号的最后一位表示轴承内径尺寸。

装配图主要用于表达机械部件的形状结构、传动关系、工作原理以及零件间的装配关系。在生产过程中，一般根据装配图组织生产，将零件装配成部件和机器。在日常的生产与技术交流过程中，看懂装配图是组装技术人员必备的能力，在设计、装配、安装、调试及技术交流时都需要装配图。

formly specified in national standards. The methods and steps for reading part drawings are as follows:

1) Understand the names, materials, scales, etc. of the views. Understand roughly the purpose of the parts from their names, the manufacturing methods from their materials, and the sizes from the scales of their drawings.

2) Analyze various views of the parts to understand their view names, cutting positions, projection relationships, and the content they represent. Analyze the relative position of the structure by the methods of shape analysis and line-surface analysis, and then connect them with each other to come up with the overall structural shape of the parts.

3) Analyze the overall size, shaping size, positioning size, size reference, and main dimensions of the parts, and be clear of the size of each section of the parts.

4) Analyze the technical requirements and quality indicators of the parts including the dimensional limit requirements, geometric tolerances, and surface structure.

Figure 5-25 is a part drawing of a bearing fixing part of a motor with an aluminum shell. Generally, the bearing inner diameter is designated by the bearing inner diameter code (the last two digits of the basic code) × 5= inner diameter (mm). For example, the inner diameter of bearing 6204 is 04 × 5=20mm。

Special situations: when the inner diameter of the bearing is less than 20mm, e.g. 10, 12, 15, and 17 (mm), its corresponding inner diameter code is 00, 01, 02, and 03 respectively, as shown in Figure 5-25. When the inner diameter of the bearing is less than 10mm, the last digit of the basic code can be used directly to represent the inner diameter of the bearing.

Assembly drawings are mainly used to express the shape, structure, transmission relationship, and working principle of the mechanical parts and their assembly relationship. In the production process, production is generally organized based on assembly drawings, where parts are assembled into components and machines. In daily production and technical communication, assembly technicians shall be able to read assembly drawings, which are required for design, assembly, installation, commissioning, and technical communication.

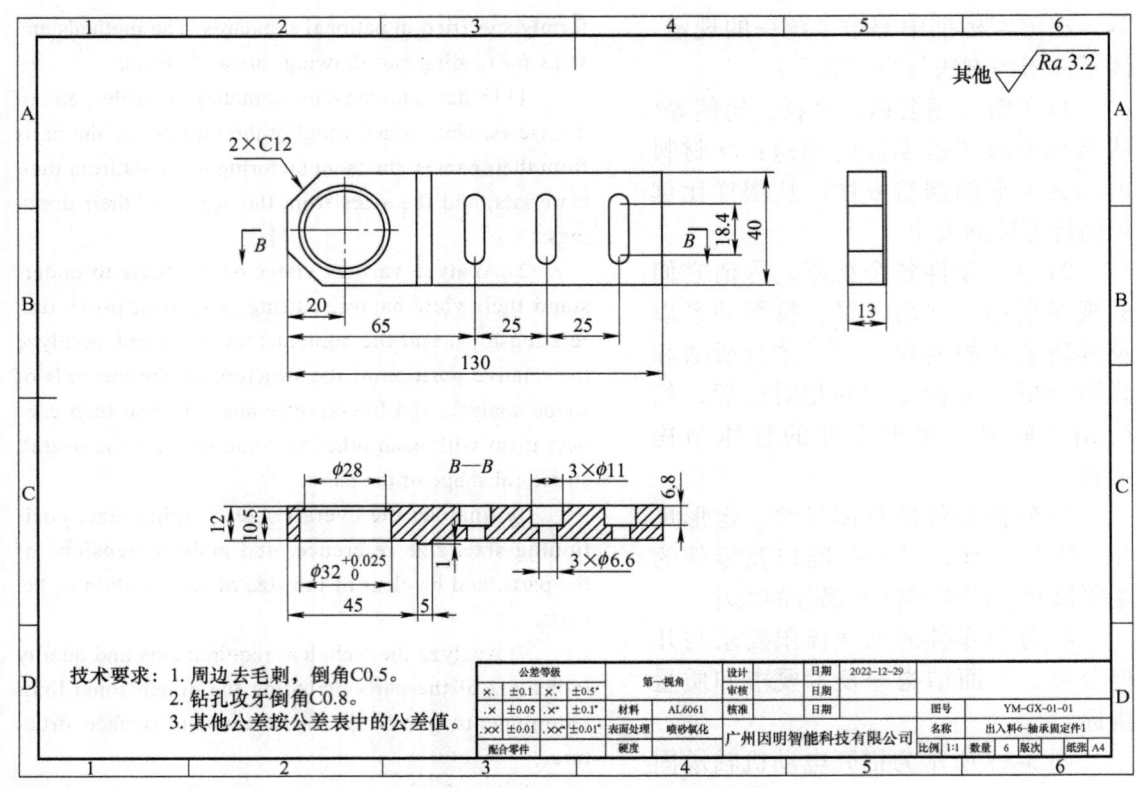

图 5-25　铝壳电动机轴承固定件零件图
Figure 5-25　Part Drawing of a Bearing Fixing Part of a Motor with an Aluminum Shell

在学习数字化生产线系统时，装配图的识读更是一项必须具备的技能。数字化生产线的使用和维修的过程中，装配图是了解生产线各个部件工作原理、性能，从而决定拆装、生产、维护和维修方法的重要依据。在机械制图中可表达部件或整台机器的结构、工作原理、装配关系等技术图样，如图 5-26 所示。在生产中依据它来装配、检验、安装和维修机器。一张完整的装配图应包括下列内容：视图，选用一组视图并采用各种表达方法正确、完整地表示出来。

When learning digital production line systems, reading assembly drawings is a necessary skill. For the use and repair of digital production lines, assembly drawings are an important basis for understanding the working principles and performance of various components of the production lines, thus determining the methods for disassembly, production, maintenance, and repair. Mechanical drawings can express patterns of technical information such as the structure, working principle, and assembly relationship of a component or the entire machine, as shown in Figure 5-26. Assembly, inspection, installation, and repair of machines during production are based on assembly drawings. A complete assembly drawing shall include the following content: views, and a set of views are selected and expressed accurately and completely using various expression methods.

项目 5　数字化生产线的构架及技术特点　293

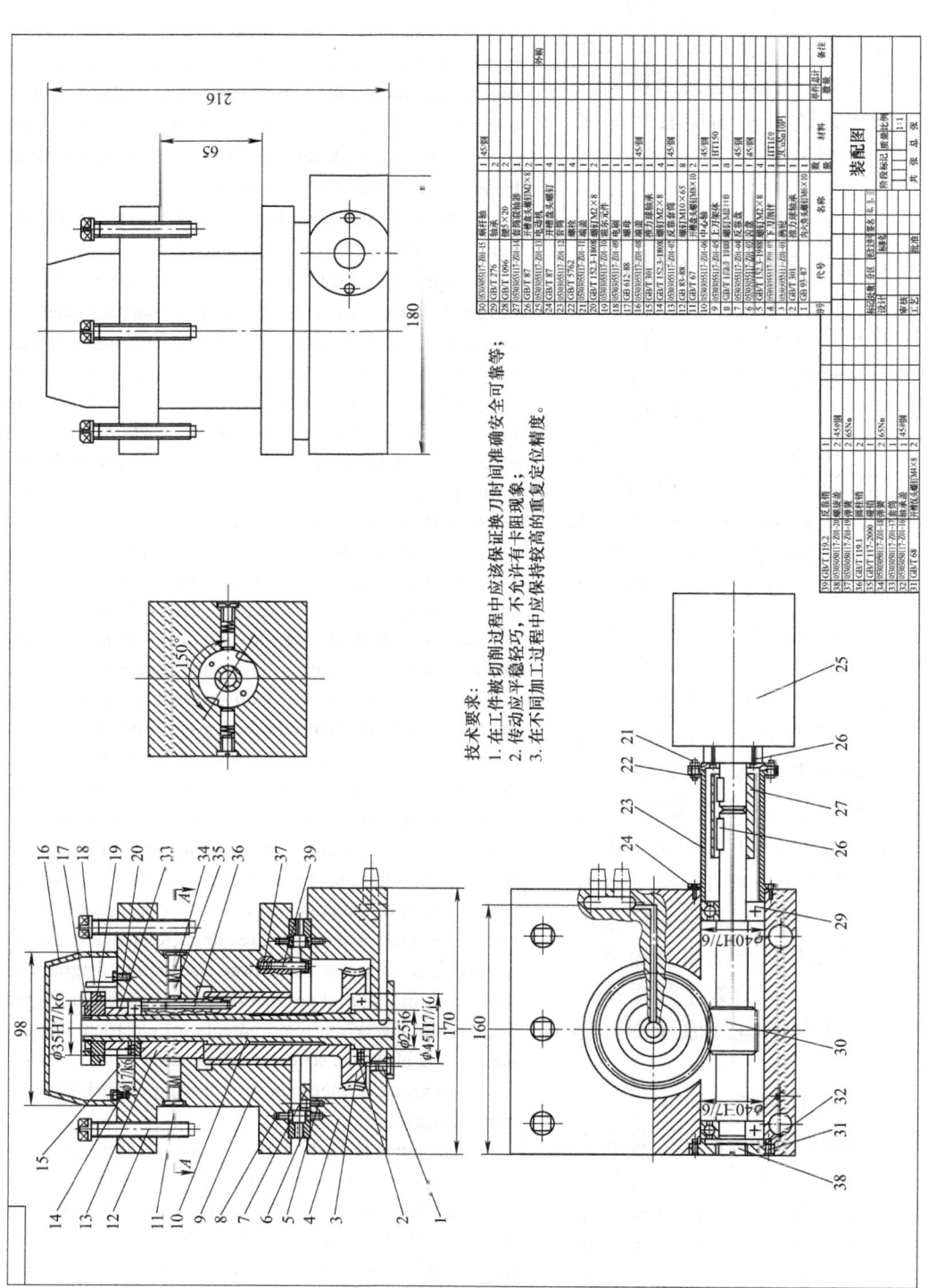

图 5-26　数控车床四工位刀架装配图

Figure 5-26　Assembly Drawing of a Four-station Tool Holder for CNC Lathe

5.3.2 识读电气原理图

电气原理图的识读是非常重要的，因为它是数字化生产线安装、调试和维护的理论依据。识读电气原理图的方法和步骤和试读机械图样有相同之处，也有其自身特点。

(1) 了解电气原理图中电气元件的布局规则

电气原理图中电气元件的布局，应根据便于阅读的原则安排。电气原理图尽可能按功能布置，按动作顺序从上到下、从左到右排列。当同一电气元件的不同部件分散在不同位置时，为了表示是同一元器件，要在电气元件的不同部件处标注统一的文字符号。对于同类元器件，要在其文字符号后加数字序号来区别。所有电器的可动部分均按没有通电或没有外力作用时的状态画出。应尽量减少线条和避免线条交叉。各导线之间有电联系时，在导线交点处画实心圆点。根据图面布置需要，可以将图形符号旋转绘制，一般逆时针方向旋转 90°，但文字不可倒置。另外，图样上方的 1、2、3 等数字是图区编号，它是为了便于检索电气电路，方便阅读分析从而避免遗漏设置的。图 5-27 所示的数字化生产线系统配电图就满足了电气原理图元器件的布局要求。

(2) 看电气原理图的一般方法

看电气原理图的一般方法是先看主电路，明确主电路控制目标与控制要求，再看辅助电路，并通过辅助电路的回路研究主电路的运行状态。电气原理图中所有电气元件都应采用国家标准中统一规定的图形文字符号表示。

5.3.2 Reading of Electrical Schematic Diagrams

The reading of electrical schematic diagrams is very important because it serves as the theoretical basis for the installation, commissioning, and maintenance of digital production lines. Except their own characteristics, the methods and steps for reading electrical schematic diagrams are similar to those for reading mechanical drawings.

(1) Understand the layout rules of electrical components in electrical schematic diagrams

A principle for the layout of electrical components in the electrical schematic diagram is that they shall be arranged for easy reading. The electrical schematic diagram shall be arranged according to functions as much as possible, from top to bottom and from left to right in the order of actions. When different parts of the same electrical component are scattered in different positions, unified textual symbols shall be marked at the said different parts of the electrical component in order to indicate that they belong to the same component. For components of the same kind, a numerical sequence number shall be added after their textual symbol to so that they can be distinguished. The movable parts of all electrical appliances are drawn according to the state when they are not powered on or exerted with external force. Minimize the lines and avoid their intersections as much as possible. When there is an electrical connection between the wires, draw a solid dot at the intersection. When required by the layout of drawings, the graphic symbols can be rotated, usually by 90° counterclockwise, but the text cannot be inverted. In addition, the numbers of 1, 2, 3, etc. are the serial number of the drawing areas, which are set for easier retrieval of electrical circuits, reading and analysis, and the avoidance of missing. The power distribution diagram for the system of a digital production line, as shown in Figure 5-27, meets the layout requirements for electrical components in the electrical schematic diagram.

(2) General methods for viewing electrical schematic diagrams

The general method of viewing electrical schematic diagrams is to view the main circuit first to be clear of its control objectives and requirements, then view the auxiliary circuit, and study the operating status of the main circuit based on the loops of the auxiliary circuit. All electrical components in the electrical schematic diagram shall be represented with graphic and textual symbols uniformly specified in national standards.

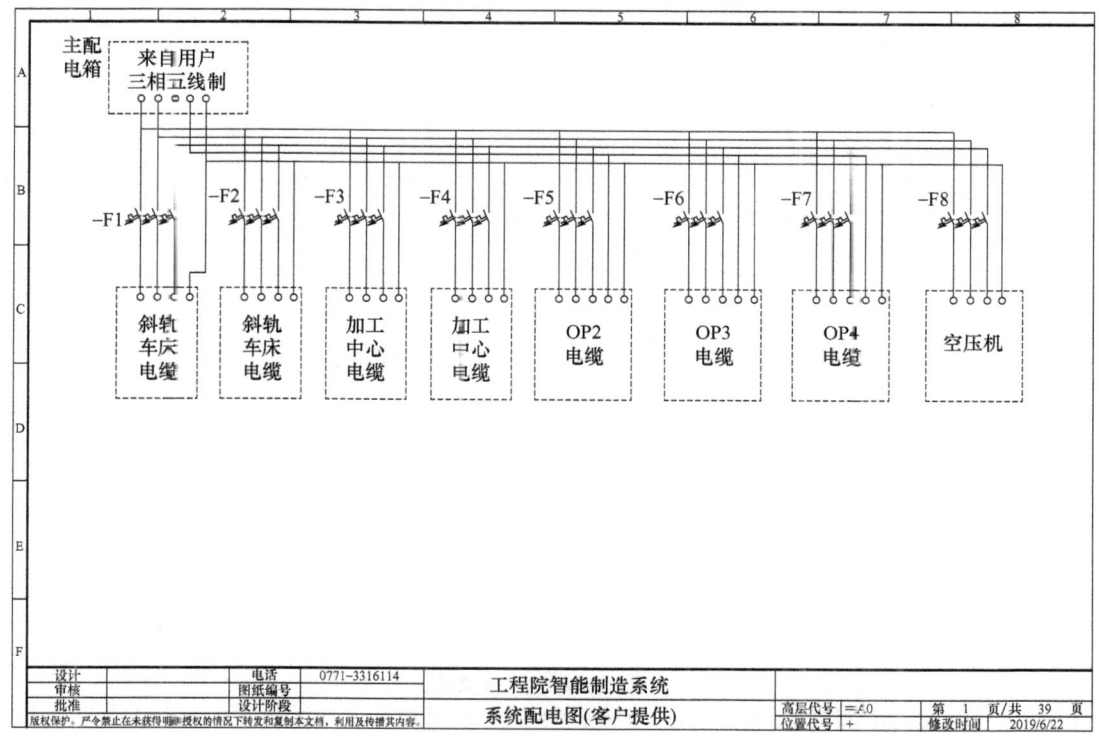

图 5-27 数字化生产线的系统配电图

Figure 5-27 Power Distribution Diagram for the System of a Digital Production Line

主电路一般是电路中的动力设备，它将电能转变为机械运动的机械能，典型的主电路就是从电源开始到电动机结束的那条电路。辅助电路包括控制电路、保护电路、照明电路。通常来说，除了主电路以外的电路都可以称之为辅助电路。图 5-28 所示为斜轨车床操作台电路，B1-B4、C1-C4、D1-D4 属于电路中的主控电路，其他辅助电路则包括废料仓电源控制、PLC 电源控制、触摸屏电源控制、自动注油系统电源控制。

(3) 识读主电路的步骤

1) 看清主电路中的用电设备。用电设备指消耗电能的用电器或电气设备，要搞清用电设备是怎样从电源取电的，了解主回路中所用的控制电气及保护装置，如短路保护的熔断器、过载保护的热继电器等。

The main circuit is generally the one that includes dynamic equipment which converts electrical energy into mechanical energy for mechanical movement. A typical main circuit is the one that begins with the power supply and ends with the motor. Auxiliary circuits include control circuits, protection circuits, and lighting circuits. Generally speaking, circuits other than the main circuit can be referred to as auxiliary circuits. In the circuit of the operating console of an inclined rail lathe, as shown in Figure 5-28, B1–B4, C1–C4, and D1–D4 belong to the main control circuit, while other auxiliary circuits include power supply control for the waste bin, PLC, touch screen, and automatic oil injection system.

(3) Steps for reading of main circuits

1) Identify the power consuming equipment in the main circuit. Power consuming equipment refers to the appliances or equipment that consume power (electrical energy). It is necessary to understand how the power consuming equipment takes power from the power supply, and to understand the electrical and protective devices for control used in the main circuit, such as fuses for short circuit protection, and thermal relays for overload protection.

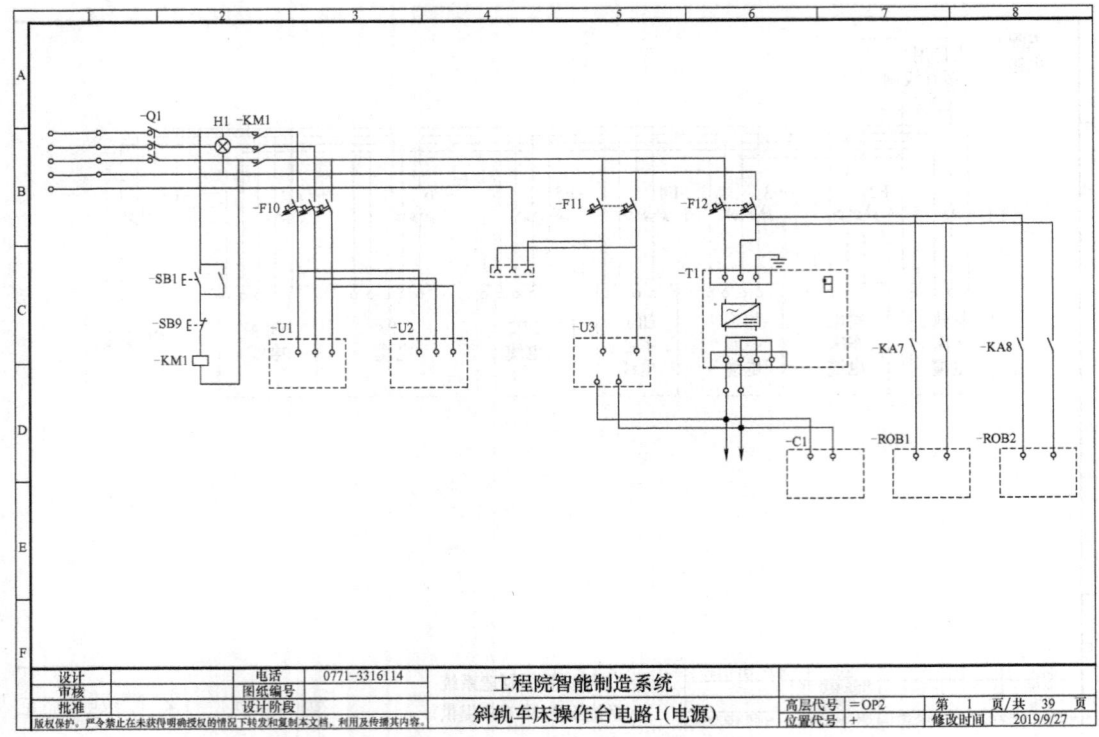

图 5-28 斜轨车床操作台电路

Figure 5-28　Circuit of the Operating Console of an Inclined Rail Lathe

2) 看清楚用电设备是用什么元器件控制的。控制用电设备的方法很多，有的直接用开关控制，有的用启动器控制，有的用接触器控制。

3) 了解主电路所用的控制器及保护电器。前者是指常规的接触器以外的其他控制元器件，如电源开关。

4) 了解电源电压的等级，如是380V还是220V。

如图 5-27 所示，数字化生产线的系统配电图就标明了所有车床及控制操作台的电源电压等级都是三相五线制380V 电压。从图 5-28 所示的主控电路也可以看得出来使用的是三相五线制的380V 的电压，并且主控电路中都具有断路器和熔断器的保护。

(4) 识读辅助控制电路的步骤

1) 分析辅助控制电路。根据主电

2) Be clear of what electrical components are used for the control of the power consuming equipment. There are many ways to control power consuming equipment, some of which are directly controlled by switches, some are controlled by starters, and some are controlled by contactors.

3) Understand the controllers and protective appliances used in the main circuit. The former refers to the control components other than conventional contactors, such as power switches.

4) Understand the level of power supply voltage, e.g. 380V or 220V.

The power distribution diagram for the system of a digital production line shown in Figure 5-27 indicates that the power supply voltage for all lathes and consoles is 380V in a three-phase and five-wire system, which can also be seen from the circuit of the operating console of an inclined rail lathe as shown in Figure 5-28. In addition, the main control circuit is protected by circuit breakers and fuses.

(4) Steps for reading of auxiliary control circuits

1) Analyze the auxiliary control circuit. Find

路中各电动机和执行电器的控制要求，逐步找出辅助控制电路中的其他控制环节。

2) 看电源：首先，看清电源的种类，是直流还是交流；其次，要看清控制电路的电源是从什么地方接来的，以及电压等级。电源一般是从主电路的两条相线上接来的，其电压为380V。也有从主电路的一条相线和一条零线上接来的，电压为单相220V。

3) 了解控制电路中所采用的各种继电器、接触器的用途，如采用了一些特殊的继电器，还应了解它们的动作原理。

4) 根据辅助电路来研究主电路的动作情况。

图5-29所示为滚筒线操作台电路，1-5图区为主控电路，6-8图区为辅助控制电路。主控电路主要控制的是滚筒运行的三个电动机，辅助控制电路主要是通过触摸屏来控制滚筒的启停以及滚筒的正反转、轴流风机的启停。主控电路的电源为380V，辅助控制电路的电压为220V。主控电路采用了按钮+接触器启动的方案，辅助控制电路采用的是触摸屏+PLC的控制方案。

5.3.3 识读其他图样

加工工序卡：是规定某一工序内具体加工要求的文件。除工艺守则已作出规定的之外，一切与工序有关的工艺内容都集中在工序卡片上，如机加工工序卡、装配工序卡、操作指导卡等。

加工工序卡的识读是生产过程与工艺过程必须掌握的一项基本技能。目前组织产品加工的工艺文件主要以"加工工序卡"为主，如图5-30所示。而加工工序卡与工艺过程卡的区别在

gradually other control links in the auxiliary control circuit according to the control requirements of each motor and actuator in the main circuit.

2) Check the power supply: firstly, be clear of the type of power supply, whether it is DC or AC; secondly, be clear of where the power supply of the control circuit is coming from and what the voltage is. The power supply comes generally from two phase lines of the main circuit with a voltage of 380V, or from one phase line and one neutral line of the main circuit with a voltage of single-phase 220V.

3) Understand the purpose of various relays and contactors used in the control circuits, and the operation principles of special relays (if any).

4) Study the operation of the main circuit based on the auxiliary circuits.

Figure 5-29 shows the circuits of the console of a drum line, with areas 1-5 as the main control circuits and areas 6-8 the auxiliary control circuits. The main control circuits mainly control the three motors for the operation of the drum, while the auxiliary control circuits mainly control, through the touch screen, the start and stop, the forward and backward rotation of the drum, as well as the start and stop of the axial flow fan. The power supply of the main control circuits is 380V, while that of the auxiliary control circuits is 220V. The main control circuits adopt a start scheme of button+contactor, while the auxiliary control circuits adopt a control scheme of touch screen+PLC.

5.3.3 Reading of Other Drawings

Machining procedure card: a document that specifies specific machining requirements for a certain procedure. Except those already specified in the process code, all content related to process is concentrated on procedure cards, such as machining procedure cards, assembly procedure cards, operating instruction cards.

The reading of machining procedure cards is a basic skill that must be mastered in the production and technological processes. Currently, the technological documents for organizing product machining are mainly "machining procedure cards", as shown in Figure 5-30. The difference between the machining procedure card and the technological process card is that the latter only reflects the steps that the parts have undergone during machining

于：工艺过程卡仅反映了零件加工所经过的步骤，不能直接用于指导工人的操作，如图 5-31 所示。按照工艺过程卡的顺序，一旦零件流转到的机床，作为生产岗位的操作者必须由自己来确定零件在本机床上的安装方法、工序尺寸、切削用量等。作为生产和施工岗位的技术指导，要不断地去解决生产中出现的各类问题。学生不但要能正确理解零件图及工艺过程卡的设计意图，而且还要懂得工艺原理，以及具有丰富的实践经验。职业院校学生在学校学习阶段必须加强识读零件图及工艺过程卡的思维能力。

and cannot be directly used to guide workers' operations, as shown in Figure 5-31. According to the sequence of the technological process card, once the parts are transferred to the machine tool, the operator in the corresponding post must determine by himself the installation method, procedure size, cutting amount, etc. of the parts on the machine tool. As technical guides for the production and construction posts, they should solve various problems that arise during production. Not only should students be able to understand correctly the design intent of part drawings and technological process cards, but need to understand the process principles and have rich practical experience. Vocational college students must improve their ability to read part drawings and technological procedure cards when learning in the college.

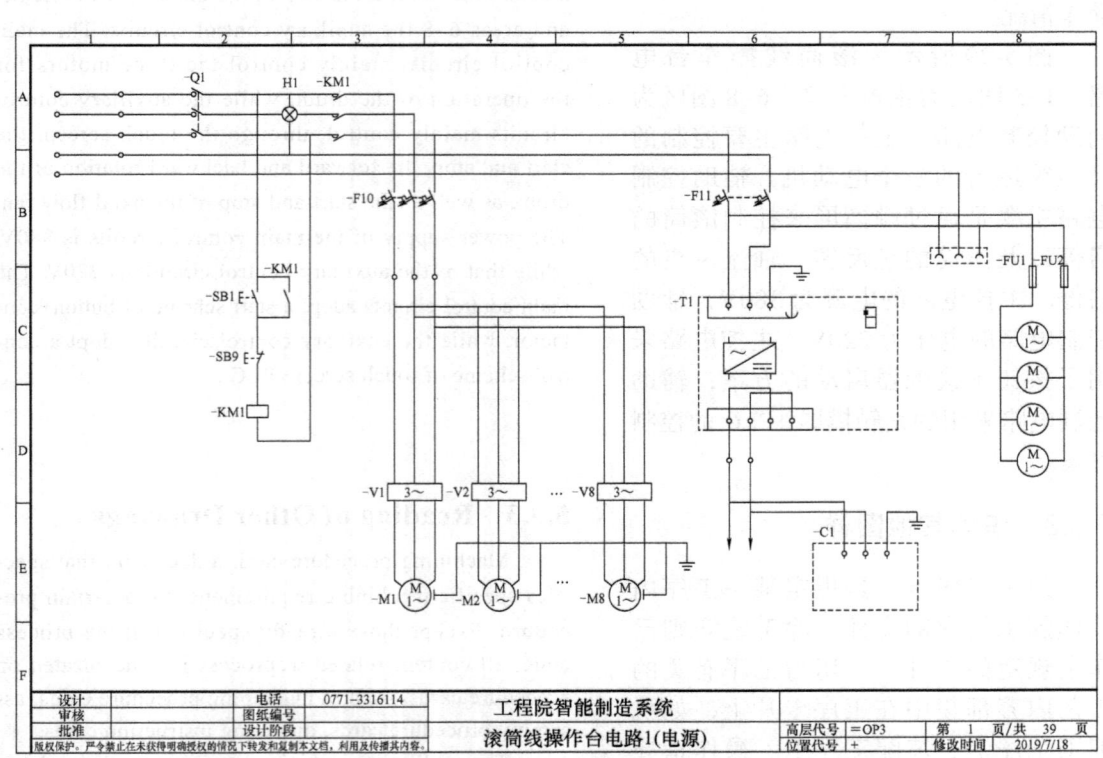

图 5-29 滚筒线操作台电路

Figure 5-29 Circuits of the Console of a Drum Line

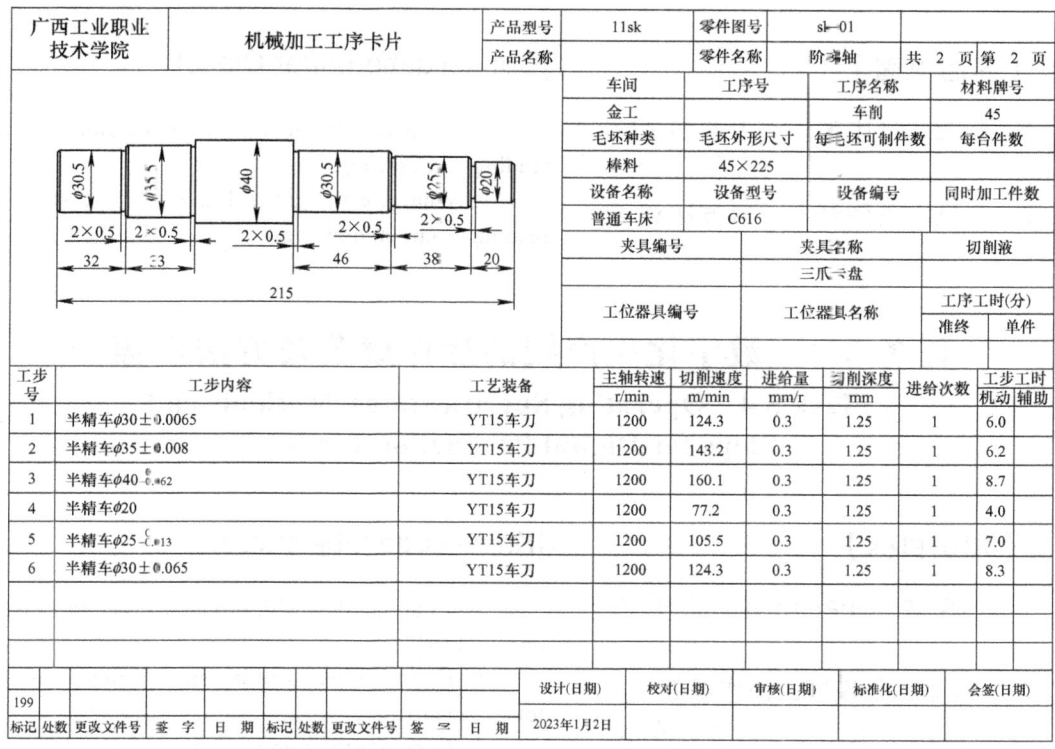

广西工业职业技术学院	机械加工工序卡片	产品型号	11sk	零件图号	sl—01		
		产品名称		零件名称	阶梯轴	共 2 页	第 2 页

车间	工序号	工序名称	材料牌号
金工		车削	45
毛坯种类	毛坯外形尺寸	每毛坯可制件数	每台件数
棒料	45×225		
设备名称	设备型号	设备编号	同时加工件数
普通车床	C616		

夹具编号	夹具名称	切削液	
	三爪卡盘		
工位器具编号	工位器具名称	工序工时(分)	
		准终	单件

工步号	工步内容	工艺装备	主轴转速 r/min	切削速度 m/min	进给量 mm/r	切削深度 mm	进给次数	工步工时 机动	工步工时 辅助
1	半精车$\phi30\pm0.0065$	YT15车刀	1200	124.3	0.3	1.25	1	6.0	
2	半精车$\phi35\pm0.008$	YT15车刀	1200	143.2	0.3	1.25	1	6.2	
3	半精车$\phi40_{-0.062}^{0}$	YT15车刀	1200	160.1	0.3	1.25	1	8.7	
4	半精车$\phi20$	YT15车刀	1200	77.2	0.3	1.25	1	4.0	
5	半精车$\phi25_{-0.013}^{0}$	YT15车刀	1200	105.5	0.3	1.25	1	7.0	
6	半精车$\phi30\pm0.065$	YT15车刀	1200	124.3	0.3	1.25	1	8.3	

										设计(日期)	校对(日期)	审核(日期)	标准化(日期)	会签(日期)
199										2023年1月2日				
标记	处数	更改文件号	签字	日期	标记	处数	更改文件号	签字	日期					

图 5-30　铝壳电机轴承加工工序卡

Figure 5-30　Machining Procedure Card for a Bearings of Motors with an Aluminum Shell

微电机壳	机械加工工艺过程卡片	产品型号		零件图号			
		产品名称		零件名称	微电机壳	共 10 页	第 1 页

材料牌号	HT200	毛坯种类	铸铁	毛坯外形尺寸	114*122*126mm	每毛坯件数	1	每台件数		备注	

工步号	工序名称	工序内容	车间	工段	设备	工艺装备	工时 准终	工时 单件
1	备料							
2	铸造	锻造毛坯	铸工					
3	清砂	清砂	金工					
4	热处理	退火(消除内应力)人工时效	金工					
5	铣	粗铣底座底面	金工		XA6132卧式铣床	专用夹具　高速钢圆柱铣刀		
6	钻孔	划线，钻削底座上ϕ7.8mm的孔	金工		Z535立式钻床	专用夹具　高速钢复合钻头		
7	扩孔	按孔端面找正压紧扩孔到ϕ8mm	金工		Z535立式钻床	专用夹具		
8	镗孔	从底座底面镗ϕ8mm到ϕ12mm锪头沉孔2mm	金工		Z535立式钻床	专用夹具		
9	车	粗车ϕ122mm两个端面	金工		CA6140车床	专用夹具		
10	铣	粗铣车内圆中的6个肋板表面和外圆表面及底座上表面	金工		XA6132卧式铣床	专用夹具		
11	钻孔	划线，钻外表面上2×M4—7H的底孔，ϕ10mm通孔	金工		Z535立式钻床	专用夹具		
12	攻螺纹	用丝锥攻2×M4—7H螺纹	金工		Z535立式钻床	专用夹具		
13	攻螺纹	攻两端面3×M5—7H	金工		Z535立式钻床	专用夹具		
14	去毛刺	去除全部毛刺	钳工			钳工台		
15	终检	按零件图样要求全面检查	质检					

										设计(日期)	校对(日期)	审核(日期)	标准化(日期)	会签(日期)
										何锦				
标记	处数	更改文件号	签字	日期	标记	处数	更改文件号	签字	日期					

图 5-31　电机端盖工艺过程卡

Figure 5-31　Technological Process Card for an End Cover of Motor

【课后巩固】

1. 识读零件图的一般方法和步骤是什么？

2. 识读电气图纸的一般方法和步骤是什么？

[Consolidation after Class]

1. What are the general methods and steps for reading part drawings?

2. What are the general methods and steps for reading electrical drawings?

任务 5.4　数字化生产线的操作规范及方法步骤
Task 5.4　Operating Specifications, Methods and Steps for Digital Production Lines

【知识目标】

1. 熟悉基于 PROFINET 的生产网络。
2. 掌握基于触摸屏的操作规范。
3. 掌握模块化生产单元的调试方法。

[Knowledge Objectives]

1. To be familiar with production networks based on PROFINET.
2. To master the operating specifications based on touch screens.
3. To master the commissioning methods for modular production units.

【技能目标】

1. 能够说出数字化生产线基于 PROFINET 生产网络的基本构架。
2. 能够按照操作规范对触摸屏进行操作。
3. 能够选用正确的方法对模块化的生产单元进行调试。

[Skill Objectives]

1. To be able to describe the basic architecture of digital production lines based on the PROFINET production networks.
2. To be able to operate the touch screen according to operating specifications.
3. To be able to select correct methods for the commissioning of modular production units.

【素质目标】

1. 具有分析与决策能力。
2. 具有发现问题、解决问题的能力。
3. 具有精益求精的工匠精神。
4. 培养奉献社会的职业道德情操。

[Competence Objectives]

1. To have the ability to conduct analysis and decision-making.
2. To have the ability to discover and solve problems.
3. To cultivate a craftsmanship spirit of striving for excellence as a craftsman.
4. To cultivate professional ethics and sentiments that contribute to society.

【任务情景】

铝壳电机关键构件数字化生产线利用立体仓库设备可实现仓库高层合理化、存取自动化、操作简便化，包括几大模块：立体仓库、出入库平台、FANUC 机器人、AGV 磁条导航小车、工位操作台。工位配有机器人编程、PLC 编程、WINCC 上位机组态、AGV（磁条导航小车）调度、RFID（无线射频）、变频器驱动、伺服驱动、传感器技术、气动技术和故障模块排除技术等。

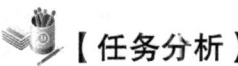【任务分析】

PROFINET 支持多种数据传输介质，多种网络拓扑结构，多种技术协议（如 HTTP、SNMP、ARP、TCP/IP 等），多种有线、无线、非实时、实时（RT、IRT）通信。通过建立 PROFINET 生产网络可实现数据一网到底跨越现场层、控制层和管理层的实时控制与传输，将数字化生产线各部分联为一个整体。各个单元互联具备即插即用的兼容性，现场增加设备操作方便，不仅与内部元件、外接元件、工业物联网、IT 因特网互联互通，而且用户还可以通过计算机 IE 浏览器，甚至手机、平板等移动设备进行网络在线监控。

【知识准备】

5.4.1 基于 PROFINET 的生产网络

PROFINET 基于工业以太网技术，使用 TCP/IP 和 IT 标准，是一种实时以太网技术。它无缝集成了所

[Task Scenario]

The digital production line for a key component of a motor with an aluminum shell can achieve rational high stories, automated access, and easy operating by the use of warehouse equipment, including several major modules: warehouse, platform for store-input and store-exit, FANUC robots, AGV magnetic strip automatic guided vehicle, and operating platform at the station; the stations are equipped with such technologies as robot programming, PLC programming, WINCC upper machine configuration, AGV (magnetic strip automatic guided vehicle) scheduling, RFID (Radio Frequency Identification), frequency converter driving, servo driving, sensor technology, pneumatic technology, and fault module troubleshooting.

[Task Analysis]

The PROFINET supports multiple data transmission media, network topologies, technical protocols (such as HTTP, SNMP, ARP, and TCP/IP), and various wired, wireless, non-real-time, and real-time (RT, IRT) communications. By establishing a PROFINET production network, real-time control and transmission of data from the field layer, to the control and management layers can be achieved within a single network, connecting various parts of the digital production line as a whole. The units are interconnected, and therefore, compatible for plug-and-play, which makes it easy to add equipment on the site. They are not only interconnected with internal components, external components, industrial Internet of Things, and IT Internet, but also allow users to conduct online monitoring through IE browsers of computer, even mobile devices such as mobile phones and tablets.

[Assumed Knowledge]

5.4.1 Production Network Based on the PROFINET

The PROFINET is a real-time Ethernet technology based on the industrial Ethernet technology, adopting TCP/IP and IT standards. It integrates

有的现场总线，实现了工业以太网和实时以太网的技术功能。PROFINET 是自动化领域处于领先地位的工业以太网标准，它包括全厂范围的现场总线通信以及车间与管理室之间的通信。PROFINET 可以同时进行标准的以太网传输和毫秒级的实时数据传输。

本课程的数字化生产线所需设备自身都集成了 PROFINET 接口，只需要分别设置 IP 地址，保证设备间的 IP 地址在同一网段不重复即可。S7-1500、S7-1200、G120、V90、TP900 精致面板等设备通过工业以太网交换机连接，如图 5-32 所示。交换机不需要单独设置和硬件组态，只需要提供 24V 直流供电即可，即插即用。

seamlessly all fieldbuses, achieving the technical functions of the industrial Ethernet and real-time Ethernet. PROFINET is a leading industrial Ethernet standard in the field of automation, and includes the factory-wide fieldbus communication and the communication between plants and administrative offices. PROFINET can perform simultaneously both the standard Ethernet transmission and real-time data transmission at the millisecond level.

The devices required for the digital production line of this course are integrated with PROFINET interfaces, and only the IP addresses need to be set separately to ensure that they are not duplicated in the same network segment for different devices. Devices such as S7-1500, S7-1200, G120, V90, and TP900 delicate panels are connected through industrial Ethernet switches, as shown in Figure 5-32. The switches do not require separate settings and hardware configuration, but only a 24V DC power supply for their plug-and-play.

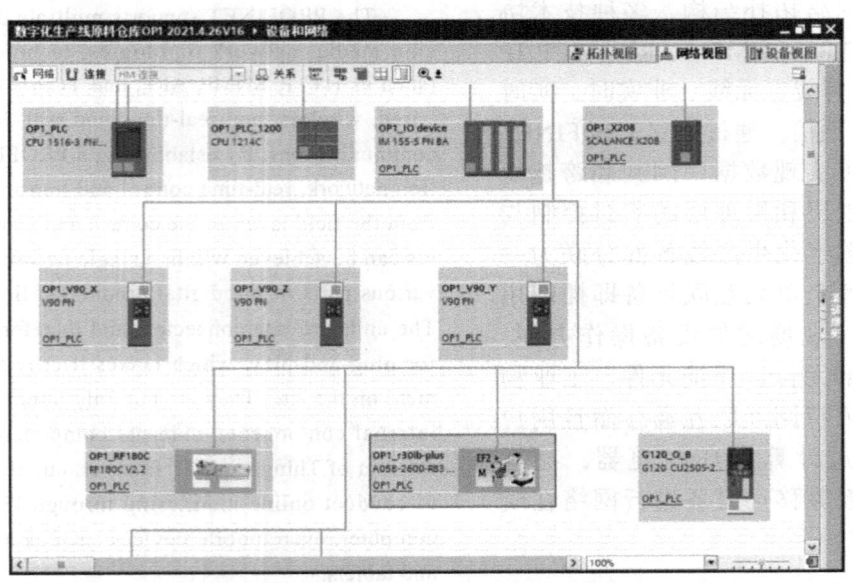

图 5-32 基于 PROFINET 的生产网络

Figure 5-32 Production Network Based on the PROFINET

5.4.2 基于触摸屏的操作规范

WINCC 是用来组态 SIMATIC 面板的可视化软件，也是 SIMATIC 工业 PC 的工程组态软件。WINCC 可以与多种工业化现场设备以及控制软件有

5.4.2 Operating Specifications Based on Touch Screens

WINCC is a visualization software used to configure SIMATIC panels, and an engineering configuration software for SIMATIC industrial PCs. WINCC

效兼容。它具有优良的操作界面，用户可以方便地通过菜单窗口进行可视化操作。在WINCC中，用户通过创建画面形成对实际现场设备的控制、监视，运行数据的采集和管理，也可以形成对控制过程中的报警画面、数据报表等。本系统通过WINCC实现柔性自动化实训生产线的人机界面(HMI)，满足系统的运行要求。本系统所使用的HMI为TP900 Comfort精致系列面板。参考WINCC使用教程，根据系统各个组成单元模块、设备、位置、监控对象，进行HMI的连接和WINCC画面的设计。

PROTAL中触摸屏的设计要遵循以下原则：

1. 以用户为中心的基本设计原则

在系统设计过程中，设计人员要抓住用户特征，发现用户需求。在整个开发过程中要不断征求用户的意见，向用户咨询。系统设计决策要结合用户的工作和应用环境，必须理解用户对系统的要求。最好的方法就是让真实的用户参与开发，这样开发人员就能正确了解用户的需求和目标，系统就会更加成功。

2. 顺序原则

按照处理事件顺序、访问查看顺序（如由整体到单项、由大到小、由上层到下层等）与控制工艺流程等设计监控管理和人机对话主界面及其二级界面。

3. 功能原则

按照对象应用环境及场合具体使用功能要求、各种子系统控制类型、不同管理对象的同一界面并行处理要求，设计分功能区、多级菜单、分层提示信息和多项对话栏并举的窗口的人机交互界面，从而使用户易于分辨

can be effectively compatible with various industrial on-site devices and control software. Its good operating interface enables users to easily perform visualized operations through the menu window. In WINCC, users can create screens to control and monitor the on-site devices, collect and manage their running data, and form alarm screens and data reports, etc. during the control process. With WINCC, this system can achieve the human-machine interface (HMI) of the flexibly automated training production line, meeting the running requirements of the system. The HMI used in this system is the TP900 Comfort exquisite series panel. The HMI connections and WINCC screens are designed with reference to the WINCC usage tutorial, various unit modules forming the system, devices, locations, and monitored objects.

The design of touch screens in the PROTAL should follow the principles below.

1. User-centered design principle

In the process of system design, designers need to focus on the characteristics of users, find their needs, continously solicit their opinions by consulting with them throughout the entire process of system development. The decisions on the system design shall be made with reference to the user's working environment and the environment of applications, based on the understanding of the user's requirements for the system. The best way to have more successful systems is to let real users participate in the development, so that developers can correctly understand their needs and goals.

2. Principle of sequence

The main interface and its secondary interface for monitoring, management, and human-machine dialogue should be designed according to the order of handling events, the order of access and viewing (e.g. from overall to individual, from large to small, and from upper to lower layers), as well as the control process flow.

3. Principle of function

A human-machine interaction interface should be designed to have sub function zones, multi-level menus, hierarchical prompt information, and windows with multiple dialogue bars according to the specific functional requirements of the application environment and occasion of the objects, control types of various subsys-

和掌握交互界面的使用规律和特点，提高其友好性和易操作性。

4．一致性原则

这一原则包括色彩的一致、操作区域的一致和文字的一致。一方面，界面的颜色、形状、字体与国家、国际或行业通用标准相一致；另一方面，界面的颜色、形状、字体自成一体，不同设备的相同设计状态的颜色应保持一致。界面细节美工设计的一致性使运行人员看界面时感到舒适，从而不分散他的注意力。对于新运行人员，或紧急情况下处理问题的运行人员来说，一致性还能减少操作失误。

5．频率原则

按照管理对象的对话交互频率高低设计人机界面的层次顺序和对话窗口菜单的显示位置等，提高监控和访问对话频率。

6．重要性原则

按照管理对象在控制系统中的重要性和全局性水平，设计人机界面的主次菜单和对话窗口的位置和突显性，从而有助于管理人员把握好控制系统的主次，实施好控制决策的顺序，实现最优调度和管理。

7．面向对象原则

按照操作人员的身份特征和工作性质，设计与之相适应和友好的人机界面。根据其工作需要，宜以弹出式窗口显示提示、引导和帮助信息，从而提高用户的交互水平和效率。

以上七点就是工业触摸屏的设计原则，本着简单易懂好操作的基本原则，充分体现了以人为本的理念。

本次任务以数字化生产线原料仓

tems, and the requirement that different management objects can be processed in parallel in the same interface. Therefore, it will be easier for users to distinguish and master the usage patterns and characteristics of the interaction interface which becomes more use-friendly and easier to be operated.

4. Principle of consistency

This principle includes consistency in color, operating area, and text. On the one hand, the color, shape, and font of the interface shall be consistent with that of the national, international, or industrial standards. On the other hand, the color, shape, and font of the interface should have their own characteristics, and the color under the same design state for different devices shall be the same. The consistency of artistic design of interface details makes the operator feel comfortable when viewing the interface, without being distracted. For new operators or operators handling problems in emergency situations, consistency can reduce operational errors.

5. Principle of frequency

This principle requires that the hierarchical order and display position of the dialogue window menu of the human-machine interface should be designed according to the dialogue interaction frequency of the management object to improve the monitoring and access dialogue frequency.

6. Principle of importance

This principle requires that the position and prominence of the main and secondary menus and dialogue box of the human-machine interface should be designed according to the importance and global level of the management object in the control system, which helps the administrator understand the priority of the control system, and implement the order of making control decisions, achieving the optimal scheduling and management.

7. Object-oriented principle

This principle requires that a suitable and operator-friendly human-machine interface should be designed according to the identity characteristics and job nature of the operators. Pop-up windows are recommended to be used to display information of prompts, guidance, and help, as required by the work of the operators, in order to improve their interaction and efficiency.

The above seven points are the design principles of industrial touch screens, which are centered around

触摸屏操作为例,掌握规范操作触摸屏的基本方法。

在操作触摸屏前,应具备以下安全条件:

1) 检查设备是否异常。

2) 电气设备电源、开关及插座等设备良好,接地正常。

3) 机械设备结构无损坏、设备气路气压正常。

4) 操作人员具备相关的设备操作培训。

5) 运行产线时操作人员必须在监控产线状况。

6) 发生紧急故障时,按下工作台上的紧急按钮停止产线运行。

具备以上安全条件后,触摸屏的基本操作规范如下:

1) 进入开始页面,查看使用注意事项,如图 5-33 所示。

2) 进入系统控制页面,查看原料仓各设备当前状态是否正常,如图 5-34 所示。

the basic principle of easy to understand and operate, fully reflecting the people-oriented idea.

In this task, the operating of the touch screen for the raw material bin of the digital production line is taken as an example to demonstrate the basic methods for normative operation of the touch screen.

The following safety conditions shall be met before operating the touch screen:

1) Check if the equipment is normal.

2) The power supply, switches, sockets, etc. for the electrical equipment are in good condition and grounded normally.

3) The structure of mechanical equipment is undamaged and the air pressure of the air circuit for the equipment is normal.

4) Operators have been trained for operating relevant equipment.

5) When the production line is running, its status must be monitored by operators.

6) When an emergency malfunction occurs, press the emergency button on the workbench to stop the production line from running.

When the above safety conditions are met, the basic operating specifications of the touch screen are as follows:

1) Enter the start page to view the precautions for use, as shown in Figure 5-33.

2) Enter the system control page and check whether the equipment for the raw material bin is under a normal status, as shown in Figure 5-34.

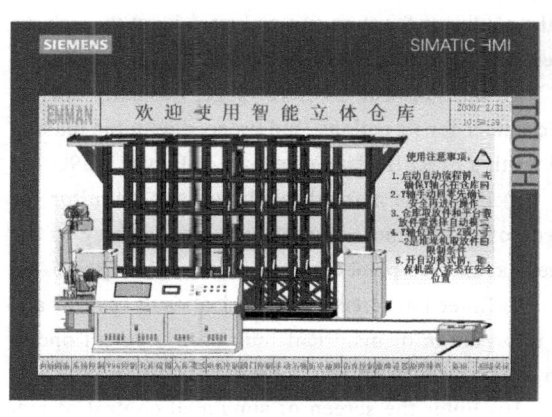

图 5-33 开始页面

Figure 5-33 Start Page

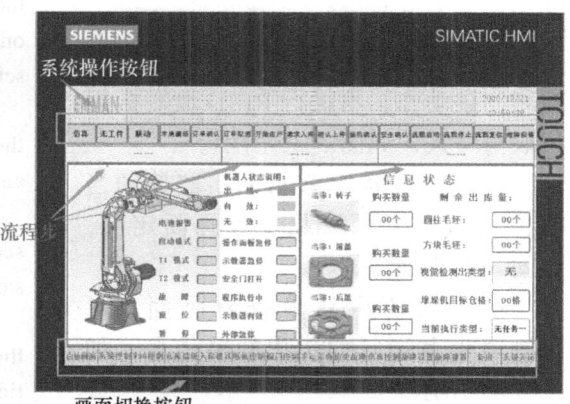

图 5-34 控制页面

Figure 5-34 Control Page

3) 进入V90伺服控制页面，对V90进行回零点动、绝对位置与相对位置的移动检查，如图5-35所示。

4) 进入仓库信息页面，查看立体仓库料盘与工件信息，如图5-36所示。

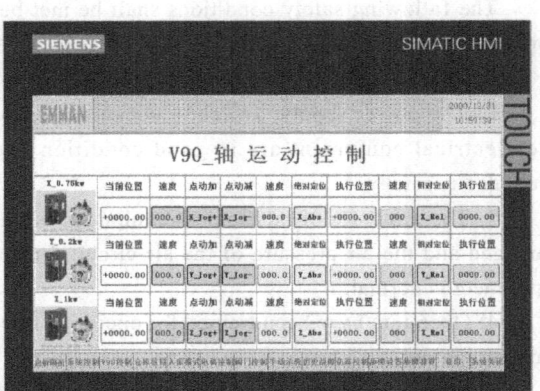

图5-35　V90伺服控制页面
Figure 5-35　V90 Servo Control Page

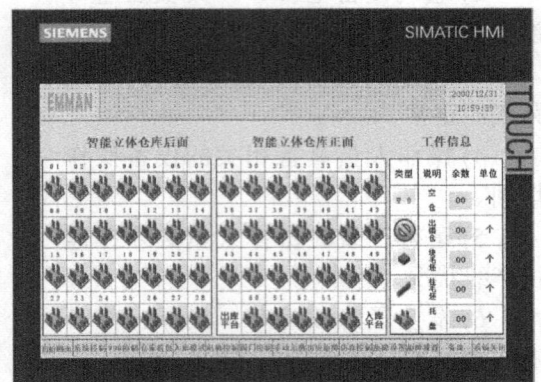

图5-36　仓库信息页面
Figure 5-36　Store Information Page

3) Enter the V90 servo control page and check the following: jog for back-to-zero, motion of absolute and relative positions of the V90, as shown in Figure 5-35.

4) Enter the store information page and view the information of the material trays and workpieces in the warehouse, as shown in Figure 5-36.

5) 进入入库模式页面，选择一个空仓格进行手动入库，入库成功之后查看仓库信息变化，之后再进行手动出库操作，如图5-37所示。

6) 进入电动机控制页面，手动控制出库平台、入库平台与出件平台的电动机正反转、电动机复位等功能，再使用AGV调度功能对出入库平台进行原料的出入库，如图5-38所示。

7) 进入阀门控制页面，手动操作控制气缸动作，再进行RFID读写数据读取与写入，如图5-39所示。

8) 进入手动示教页面，监控轴当前位置、仓库示教及零点画面定义，如图5-40所示。

9) 进入历史故障查看画面，可以查看设备运行的历史故障信息，如图5-41所示。

10) 进入仿真控制上件画面，先选择立体仓库空料盘，再选择原材料，可对空料盘进行一件上料操作，如图5-42所示。

5) Enter the page of store-input mode, select an empty bin for the operation of manual store-input and, when finished, check the changes in the store information, and then proceed with the operation of manual store-exit, as shown in Figure 5-37.

6) Enter the motor control page, manually operate to check the functions of forward and backward rotations, reset, etc. of the motors for the platforms of store-exit, store-input, and workpiece delivery; then, proceed with the operation of store-exit and store-input of raw materials on the platforms for store-exit and store-input through the scheduling function of AGV, as shows in Figure 5-38.

7) Enter the valve control page, manually operate the cylinder for its actions, and proceed with the reading and writing of RFID data, as shown in Figure 5-39.

8) Enter the manual teaching page, monitor the screen of current positions of the axis, teaching of the store, and the zero-point defining, as shown in Figure 5-40.

9) Enter the screen of historical fault review, and the information of historical fault of equipment operation can be seen, as shown in Figure 5-41.

10) Enter the screen of simulated control of materials loading, first select the empty material tray of the warehouse, then the raw materials to proceed with the operation of loading of one piece of materials to the empty material tray, as shown in Figure 5-42.

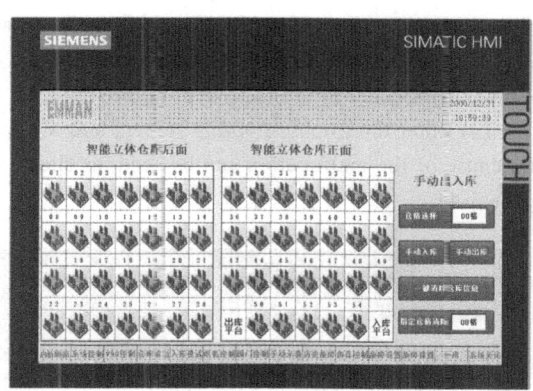

图 5-37　手动出入库

Figure 5-37　Manual Store-input and Store-exit

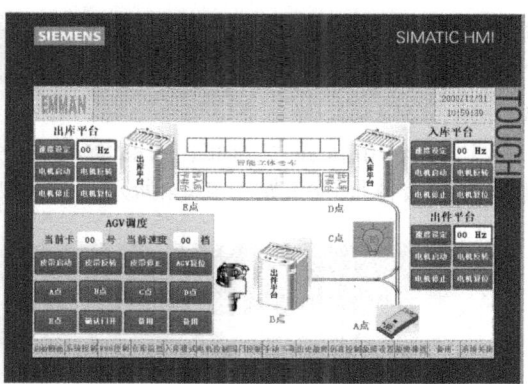

图 5-38　电动机控制页面

Figure 5-38　Motor Control Page

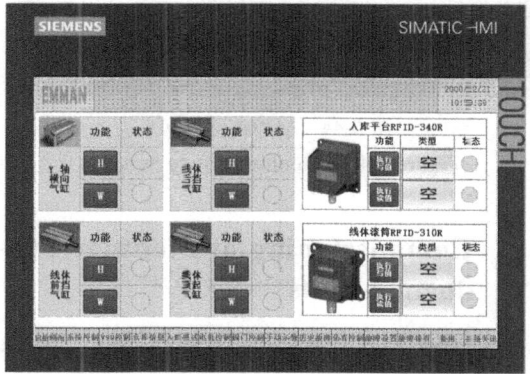

图 5-39　阀门控制页面

Figure 5-39　Valve Control Page

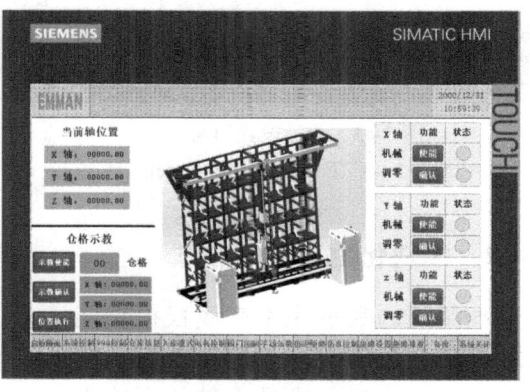

图 5-40　手动示教页面

Figure 5-40　Manual Teaching Page

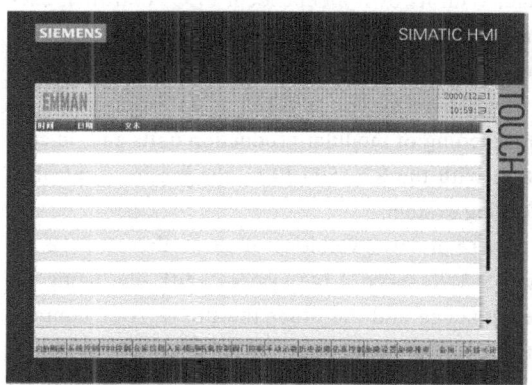

图 5-41　历史故障查看画面

Figure 5-41　Screen of History Fault Review

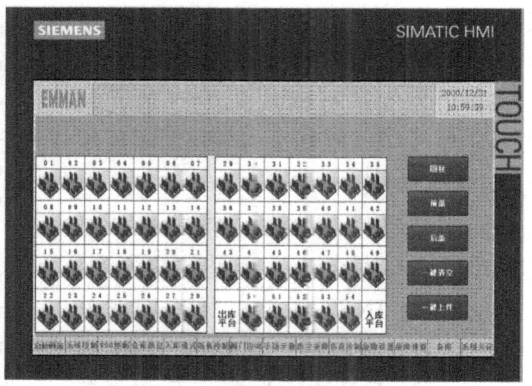

图 5-42　仿真控制上件画面

Figure 5-42　Screen of Simulated Control of Materials Loading

11) 进入故障设置画面，可对阻挡机构后端传感器、阻挡机构前端电磁阀等设备进行虚拟化故障设置，如图 5-43 所示。

12) 进入故障排除画面，可查看故障发生的具体原因，如图 5-44 所示。

11) Enter the fault setting screen, virtual fault settings can be performed on devices such as the sensor at the rear end and the solenoid valve at the front end of the blocking mechanism, as shown in Figure 5-43.

12) Enter the troubleshooting screen, and the specific cause of the malfunction can be seen, as shown in Figure 5-44.

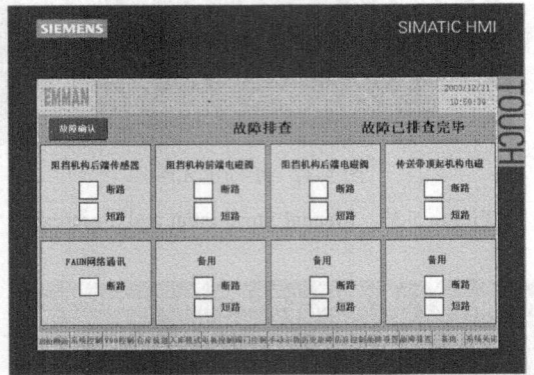

图 5-43　故障设置画面　　　　　　　　图 5-44　故障排除画面
Figure 5-43　Fault Setting Screen　　　　Figure 5-44　Troubleshooting Screen

5.4.3　模块化生产单元调试

本课程的数字化生产线根据功能需求将其划分为包含供料、分拣、搬运、加工、装配、成品仓储 6 个独立的单元模块，每个单元模块都具有各自独立的 PLC 控制器，各单元模块可完成其特定的实训项目。本次任务以基于 SINAMICS V90 伺服驱动系统的供料单元为例进行模块化生产单元调试，配合 S7-1500 PLC 完成供料系统的运动控制功能。

伺服驱动系统 SINAMICS V90 伺服驱动版本有两个，一个是脉冲序列，一个是 PROFINET 通信。脉冲控制的一个特点是具有脉冲位置控制。PROFINET 通信特点是适应性强，与大多数设备构建成工业以太网通信网络。供料单元采用 PROFINET 版本 SINAMICS V90 PN。

5.4.3 Commissioning of Modular Production Units

The digital production line of this course is divided into six independent unit modules based on functional requirements, including materials feeding, sorting, handling, machining, assembly, and finished product storage. With its own independent PLC controller, each unit module can complete its specific training programs. In this task, the feeding unit based on the SINAMICS V90 servo drive system is taken as an example for the commissioning of modular production units and, in conjunction with the S7-1500 PLC, the motion control of the feeding system.

There are two versions of servo drive of the SINAMICS V90 servo drive system, with one being the pulse sequence and the other the PROFINET communication. One characteristic of pulse control is that it has the control of pulse position, while the PROFINET communication is characterized by its strong adaptability and can be built into an industrial Ethernet communication network with most devices. The materials feeding unit adopts the PROFINET version of SINAMICS V90 PN.

(1) SINAMICS V90 组态

首先要检查所使用的 Protal 软件是否安装有 V90 HSP 支持包。若没有 V90 HSP 支持包，需要到西门子工业在线支持中心下载，打开 https://support.industry.siemens.com/ 到下载中心搜索 V90 HSP 支持包，如图 5-45 所示。

(1) SINAMICS V90 configuration

Firstly, check if the V90 HSP support package has been installed for the Protal software. If not, download it from the SIEMENS Industry Online Support by visiting https://support.industry.siemens.com, and searching for the V90 HSP support package in the download center, as shown in Figure 5-45.

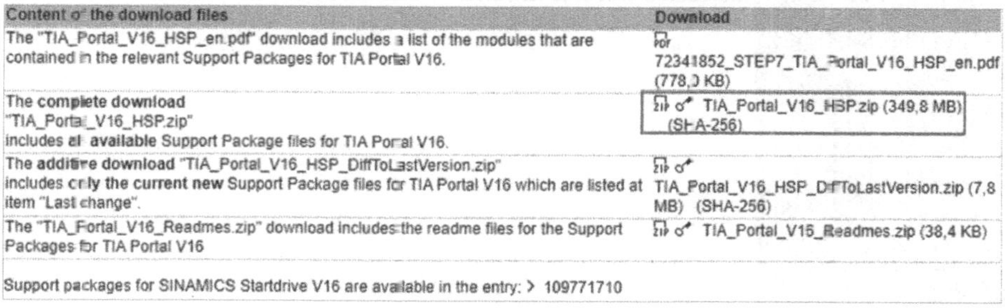

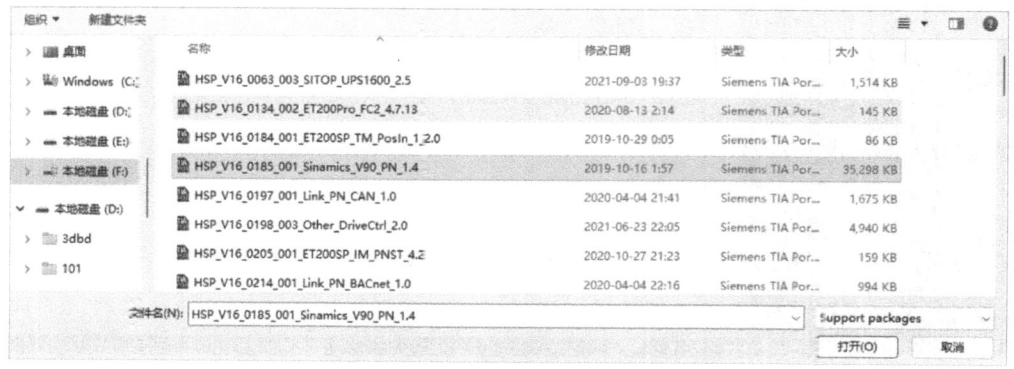

图 5-45　下载对应版本的 HSP 支持包

Figure 5-45　Downloading a corresponding version of the HSP support package

然后安装 HSP 支持包，打开 PORTAL V16 → 选项 → 支持包，如图 5-46 所示。

从文件系统添加，打开如下图 5-47 所示的文件安装即可。

Then install the HSP support package as follows: open PORTAL V16 → Options → Support Package, as shown in Figure 5-46.

Add from the file system, and open the file shown in Figure 5-47 to install.

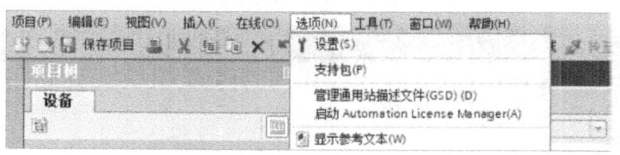

图 5-46　找到选项 → 支持包

Figure 5-46　Finding Options → Support Package

图 5-47　选择安装文件

Figure 5-47　Selecting the File to be Installed

(2) 硬件添加 SINAMICS V90-PN

装配模块单元采用 S7-1500 PLC 连接 SINAMICS V90 伺服系统，实现位置控制。通过在 Portal V16 工艺对象中添加定位轴组态 V90。

1) 创建新项目 (见图 5-48)。
2) 组态新设备 (见图 5-49)。

(2) Add SINAMICS V90-PN for hardware

The assembly module unit is connected to the SINAMICS V90 servo system using S7-1500 PLC to achieve position control. Configure the V90 by adding the positioning axis to the process object of Portal V16.

1) Create a new item (as shown in Figure 5-48).
2) Configure a new device (as shown in Figure 5-49).

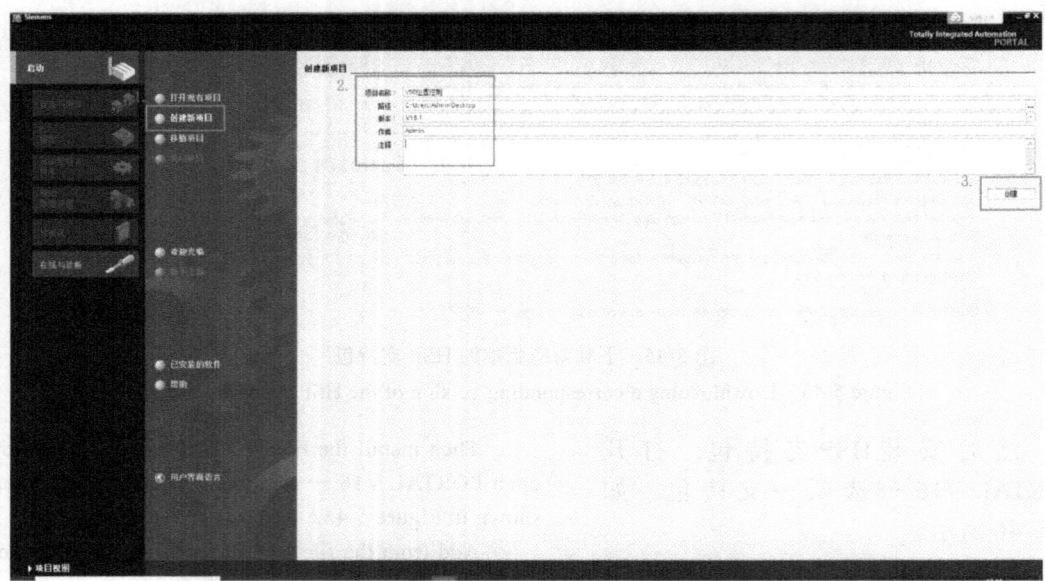

图 5-48　创建新项目
Figure 5-48　Create a New Item

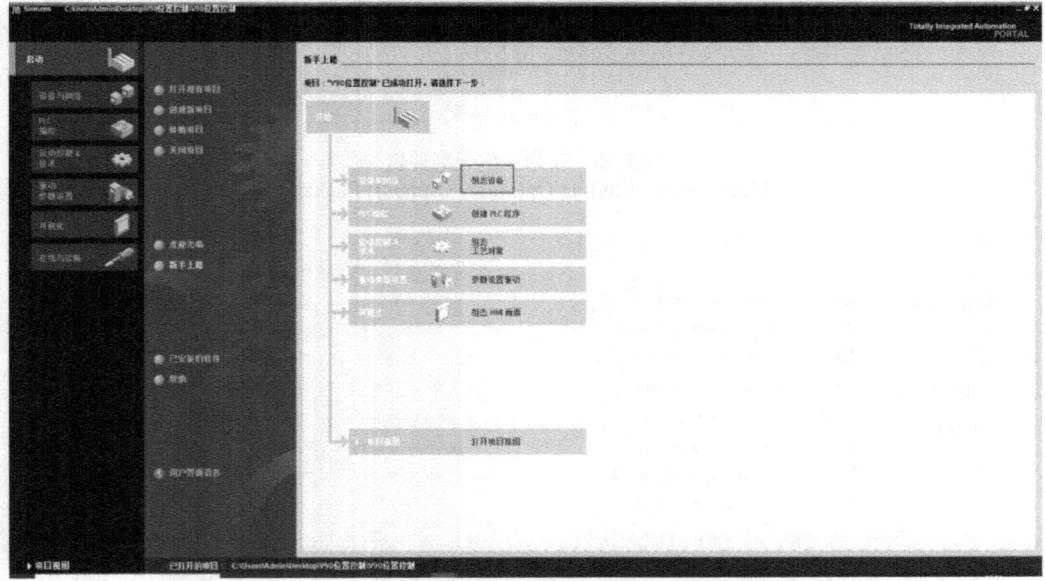

图 5-49　组态新设备
Figure 5-49　Configure a New Device

3) 添加 1516PLC(见图 5-50)。

4) 设置 1516PLC IP 地址(见图 5-51)。

3) Add the 1516PLC (as shown in Figure 5-50).

4) Set the IP address for 1516PLC (as shown in Figure 5-51).

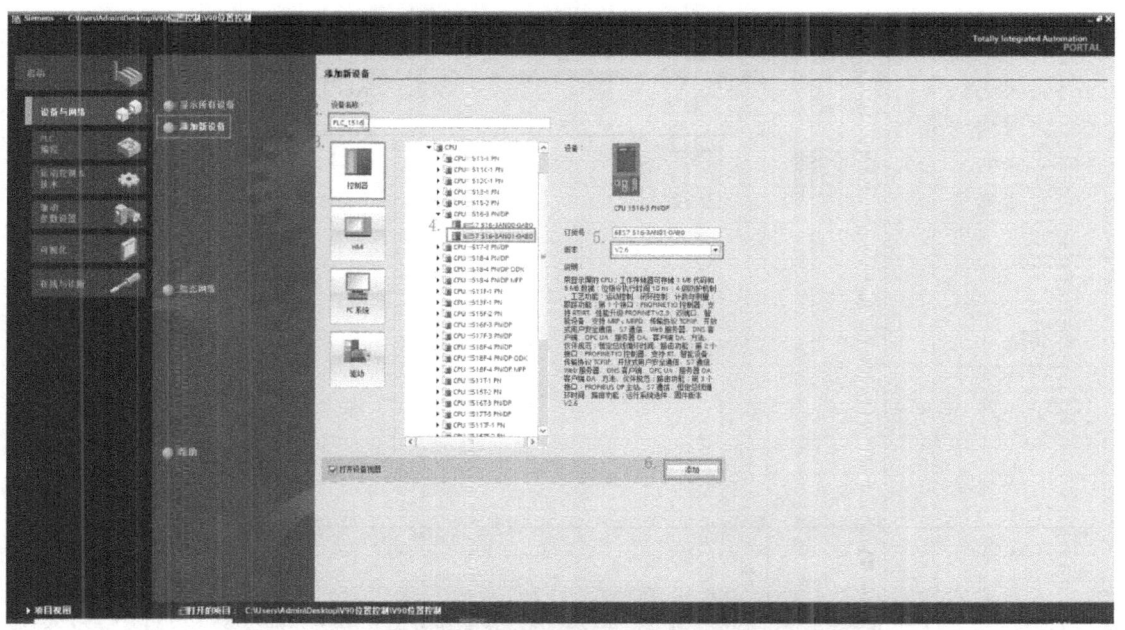

图 5-50　添加 1516PLC

Figure 5-50　Add the 1516PLC

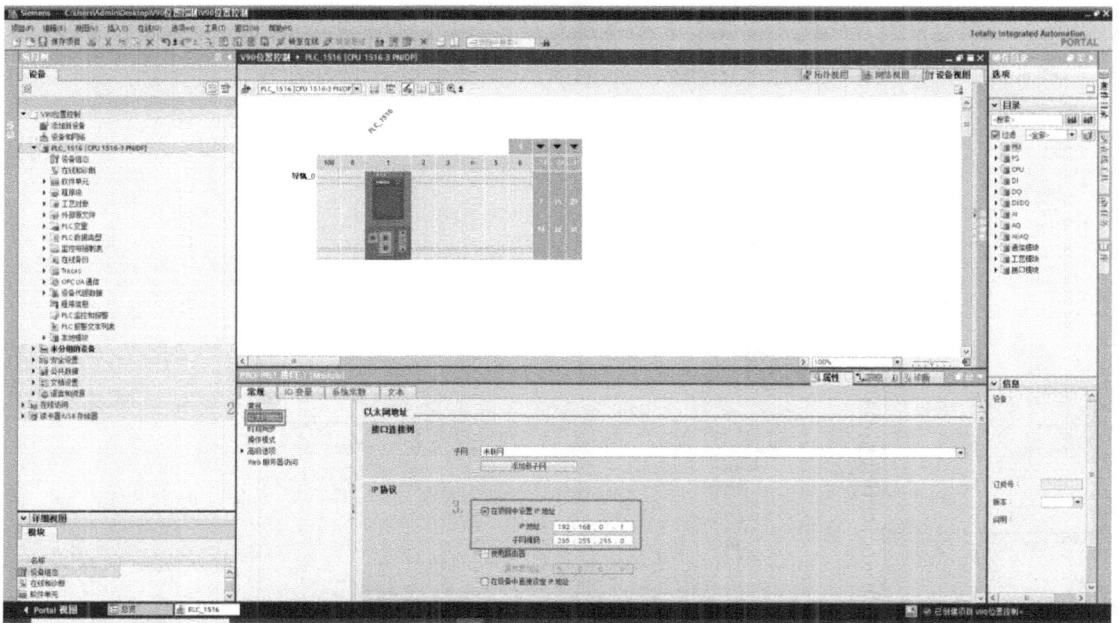

图 5-51　设置 1516PLC IP 地址

Figure 5-51　Set the IP Address for 1516PLC

5) 添加伺服驱动 (见图 5-52)。
6) 硬件属性设置伺服驱动 IP(见图 5-53)。

5) Add the servo drive (as shown in Figure 5-52).
6) Set the servo drive IP in the hardware attribute (as shown in Figure 5-53).

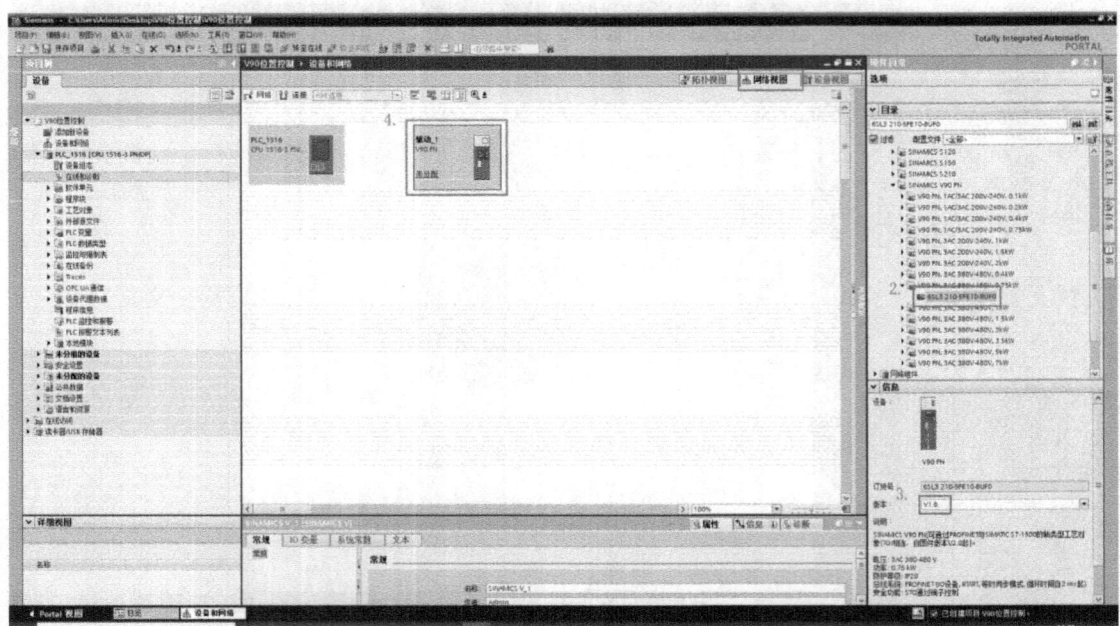

图 5-52　添加伺服驱动
Figure 5-52　Add the Servo Drive

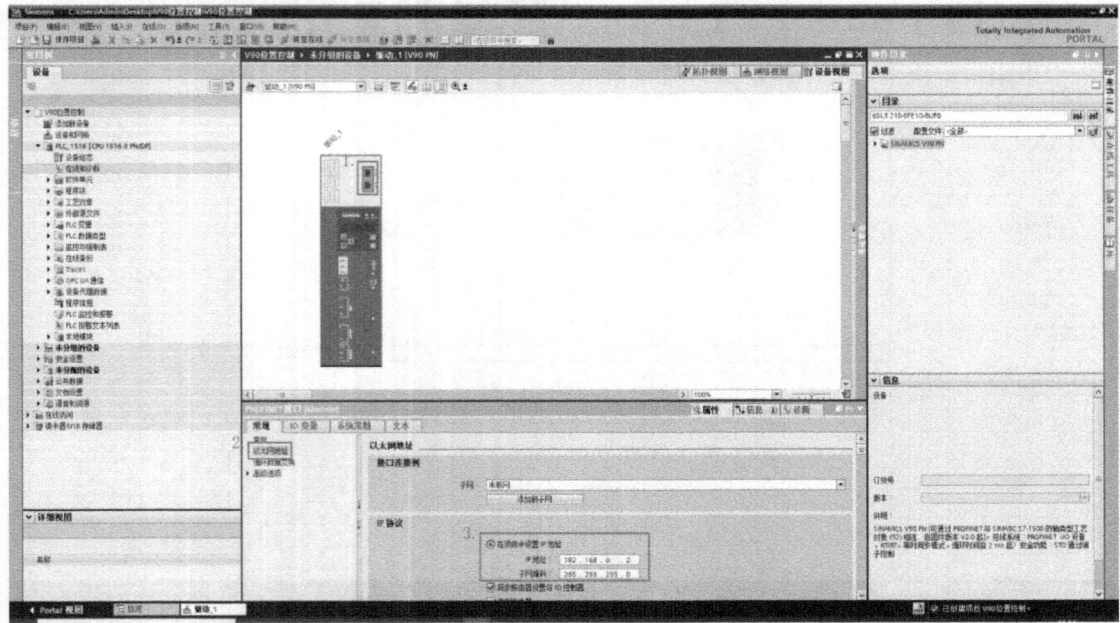

图 5-53　设置伺服驱动 IP
Figure 5-53　Set the Servo Drive IP

7) 使用鼠标拖拽设备网口连接，与 PLC 建立网络连接，如图 5-54 所示。

8) 使用鼠标拖拽设备网口连接，与 PLC 建立拓扑连接，如图 5-55 所示。

7) Establish a network connection with the PLC by dragging the device's network port with a mouse as shown in Figure 5-54.

8) Establish a topology connection with the PLC by dragging the device's network port with a mouse, as shown in Figure 5-55.

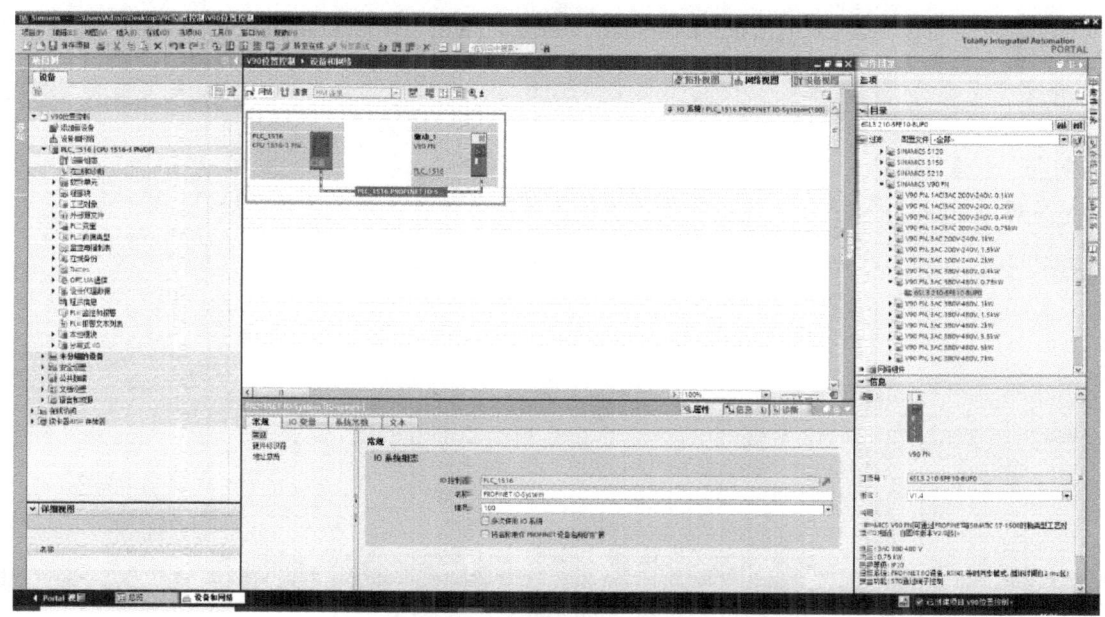

图 5-54　建立网络连接
Figure 5-54　Establish a Network Connection

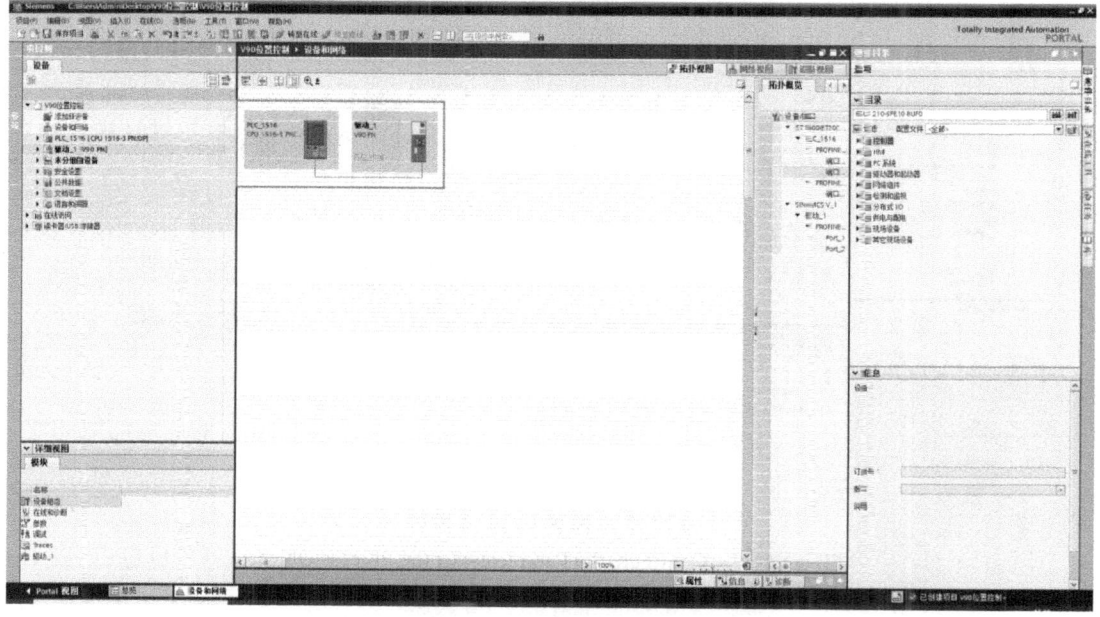

图 5-55　建立拓扑连接
Figure 5-55　Establish a Topology Connection

9) 设置伺服驱动参数(见图 5-56)。

10) 配置工艺对象参数(见图 5-57)。

9) Set the parameters of the servo drive (as shown in Figure 5-56).

10) Configure the parameters of the process object (as shown in Figure 5-57).

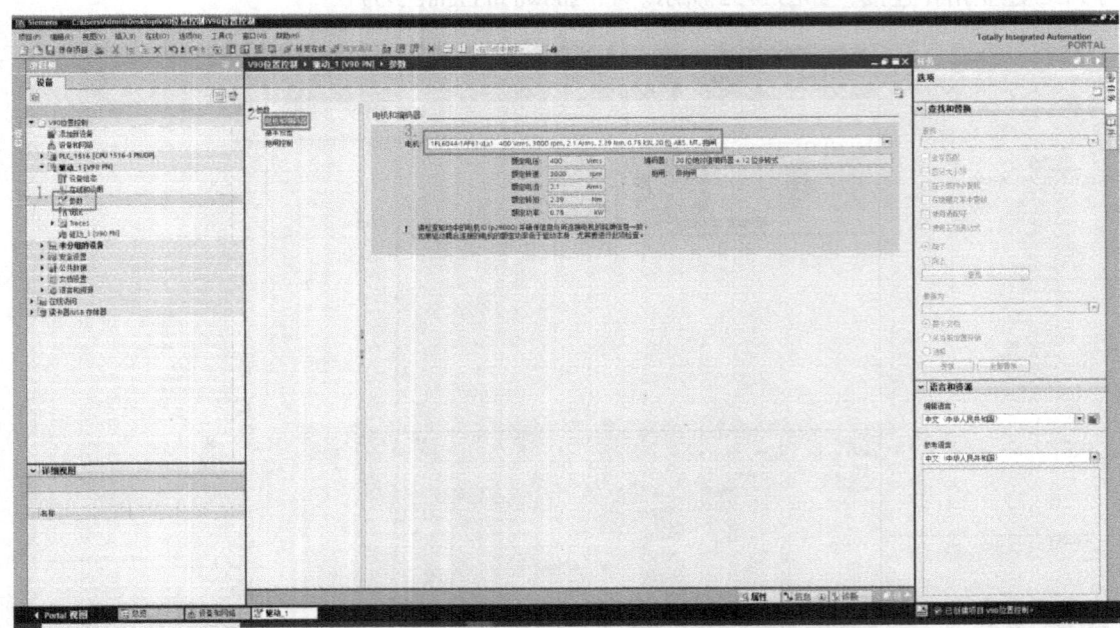

图 5-56 设置伺服驱动参数
Figure 5-56 Set the Parameters of the Servo Drive

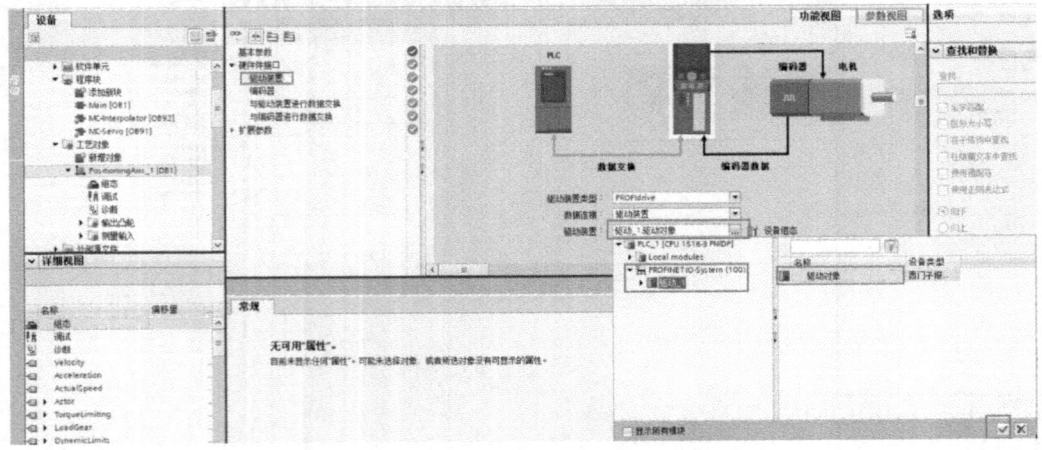

图 5-57 配置工艺对象参数
Figure 5-57 Configure the Parameters of the Process Object

项目 5　数字化生产线的构架及技术特点

11) 添加 V90 控制数据块操作变量（见图 5-58）。

12) 添加 V90 控制变量（见图 5-59）。

11) Add the operating variable of V90 control data block (as shown in Figure 5-58).

12) Add the V90 control variable (as shown in Figure 5-59).

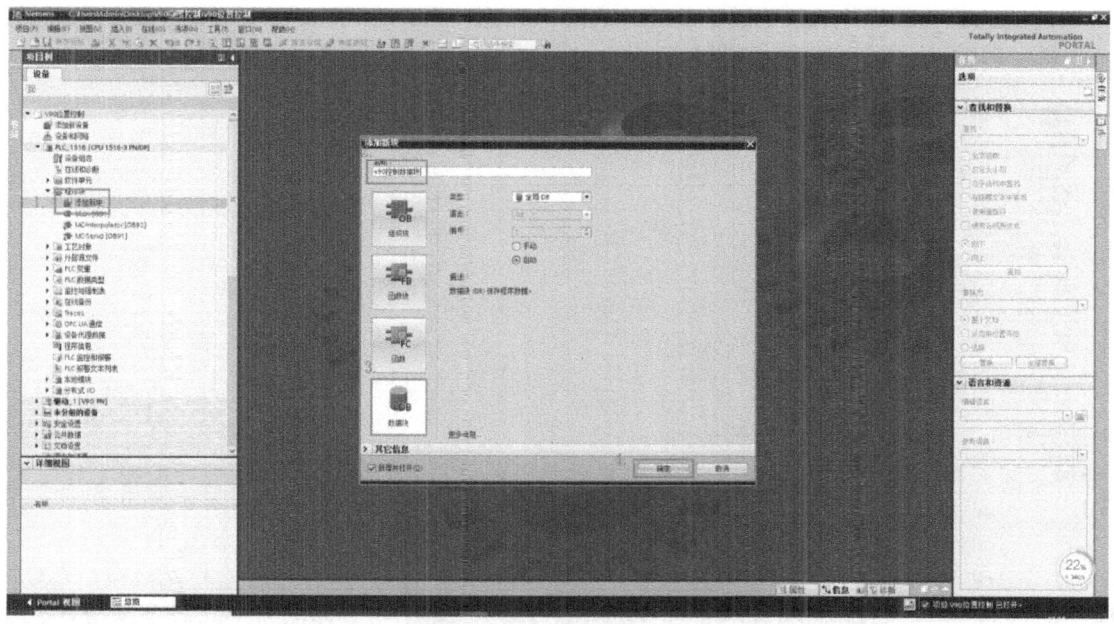

图 5-58　添加操作变量
Figure 5-58　Add the Operating Variable

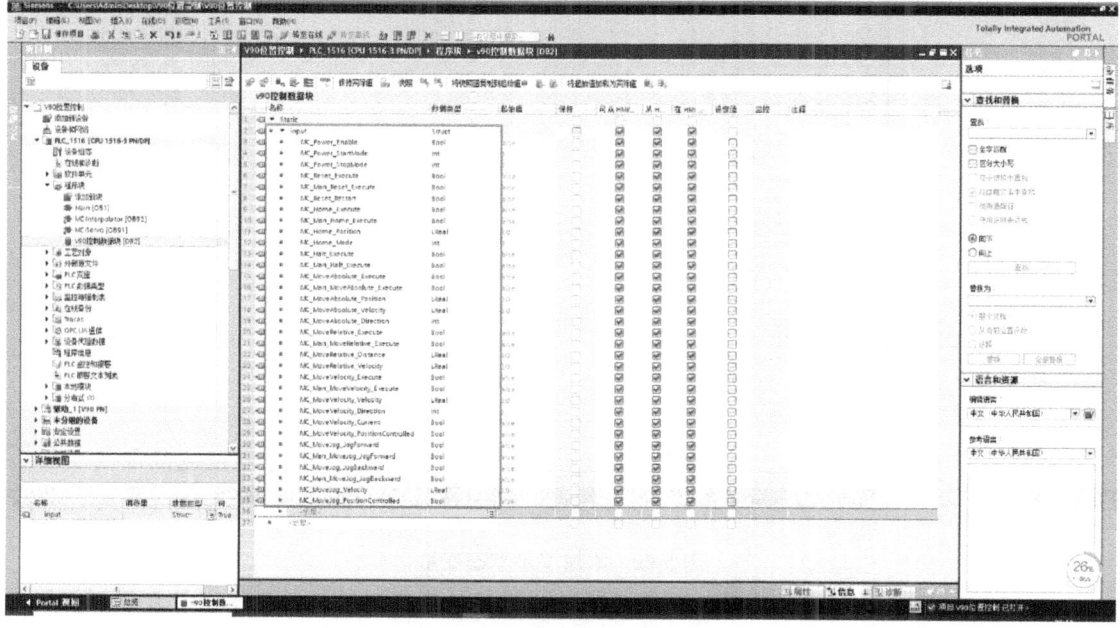

图 5-59　添加 V90 控制变量
Figure 5-59　Add the V90 Control Variable

13) 添加 V90 状态变量 (见图 5-60)。

14) 添加 V90 控制块操作指令 (见图 5-61)。

13) Add the V90 state variable (as shown in Figure 5-60).

14) Add the operating command of V90 control block (as shown in Figure 5-61).

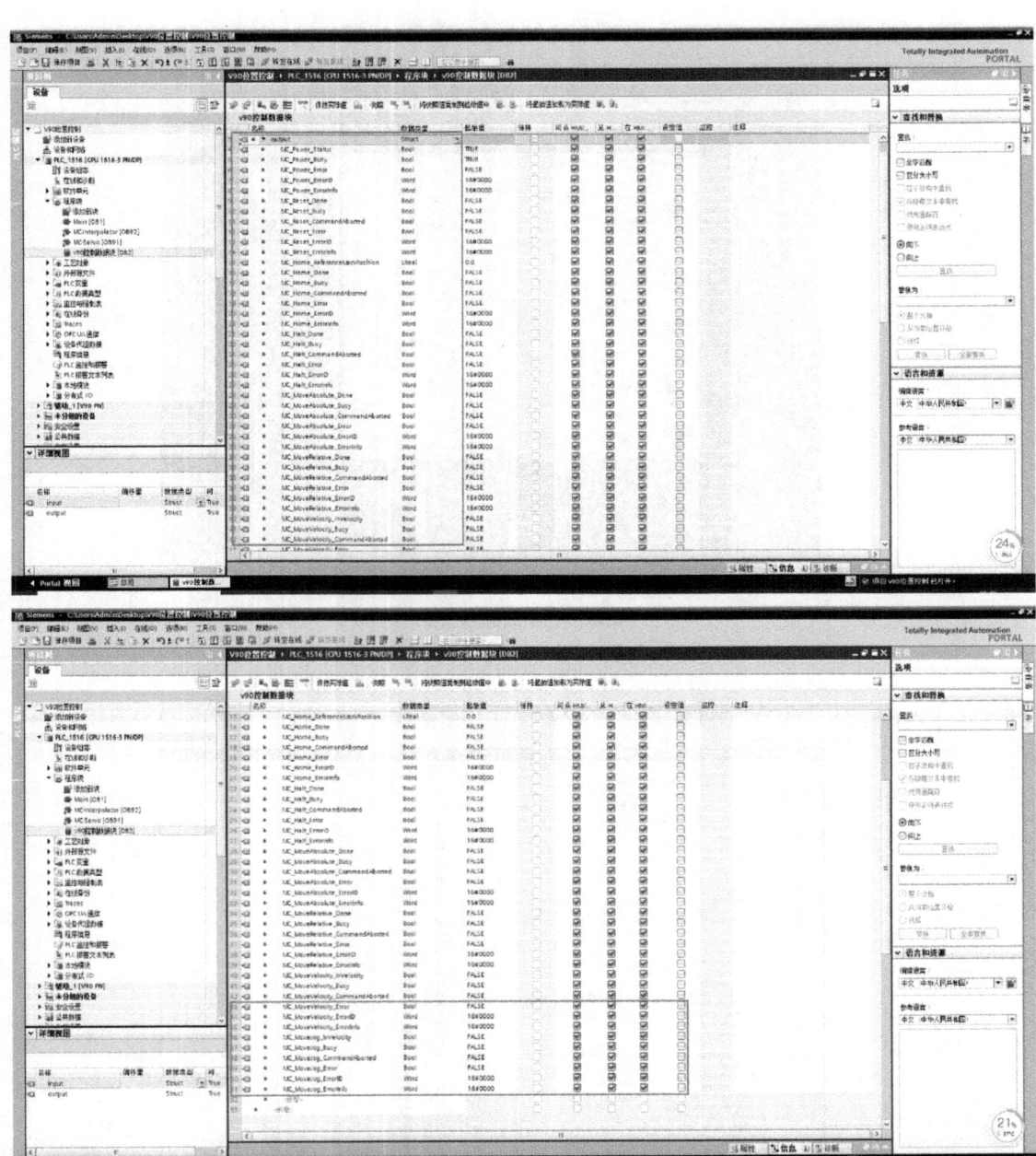

图 5-60　添加 V90 状态变量

Figure 5-60　Add the V90 State Variable

项目 5　数字化生产线的构架及技术特点

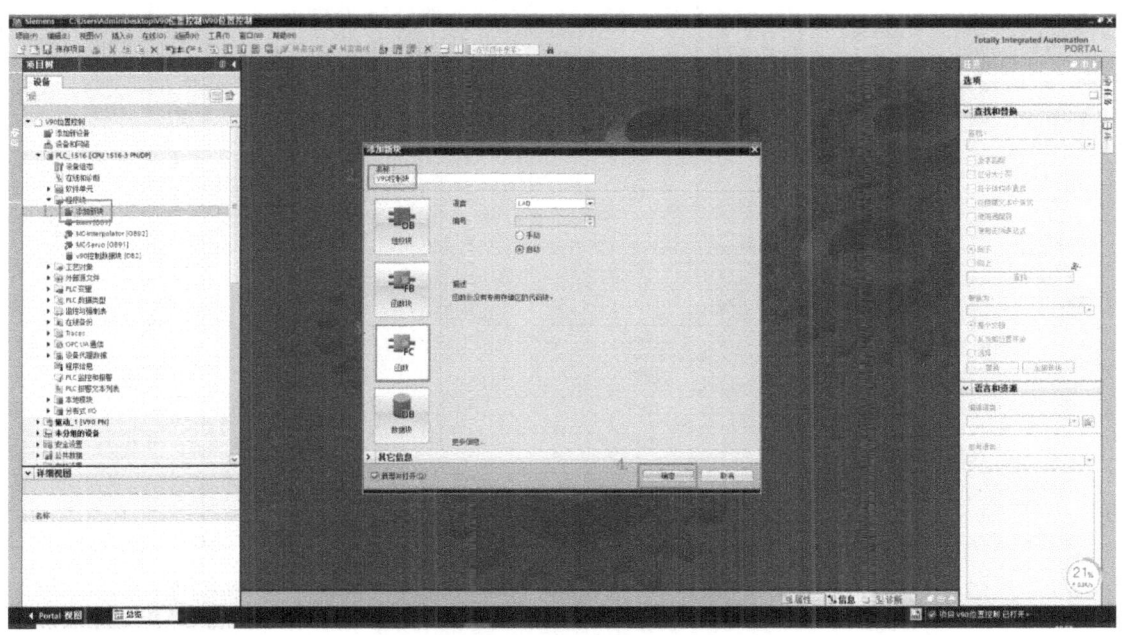

图 5-61　添加操作指令
Figure 5-61　Add the Operating Command

15) 添加轴使能指令 (见表 5-1 和图 5-62)。

15) Add the command of axis enable (as shown in Table 5-1 and Figure 5-62).

表 5-1　轴使能指令的参数
Table 5-1　Parameters of Axis Enable Command

参数 Parameter	数据类型 Data type	声明 Statement	说明 Remarks
Axis	TO_Axis	Input	工艺对象 Process object
Enable	Bool	Input	工艺对象：=1 启用，=0 禁用 Process object:=1 enable,=0 disable
Status	Bool	Ouput	工艺对象状态：=1 启用；=0 禁用 Process object status:=1 enable, =0 disable
Busy	Bool	Ouput	=1 作业正在处理中 =1 Operation is being processed
Error	Bool	Ouput	=1 出错 =1 Error
ErrorID	Word	Ouput	出错代号 Error code

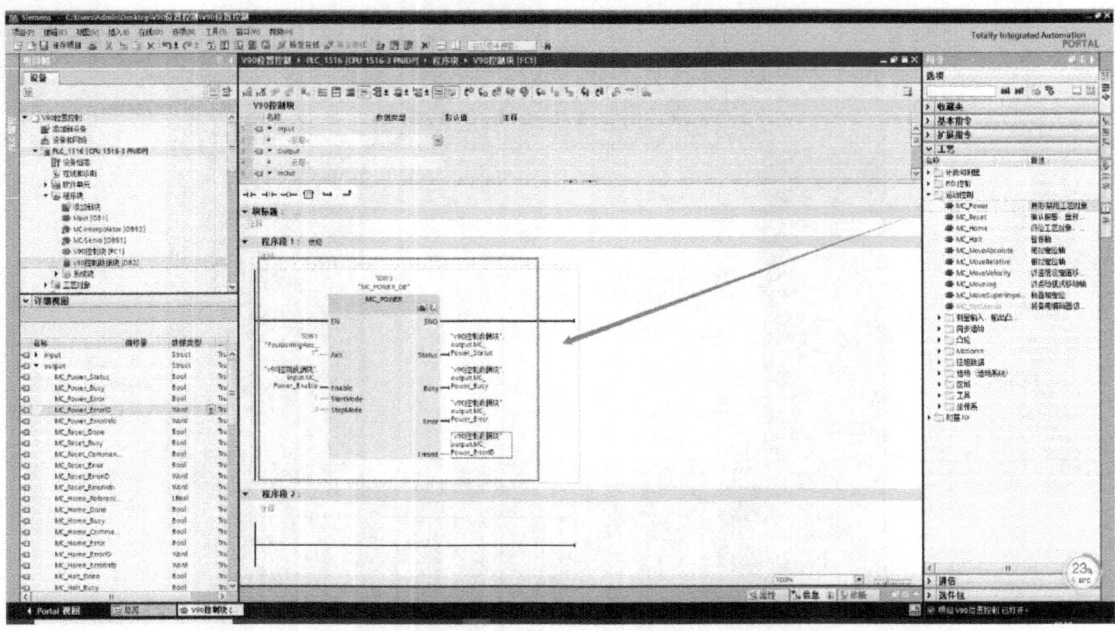

图 5-62 添加轴使能指令
Figure 5-62　Add the Command of Axis Enable

16) 添加轴复位指令 (见表 5-2 和图 5-63)。

17) 添加轴回原指令 (见表 5-3 和图 5-64)。

16) Add the command of axis reset (as shown in Table 5-2 and Figure 5-63).

17) Add the command of axis homing (as shown in Table 5-3 and Figure 5-64).

表 5-2　轴复位指令的参数
Table 5-2　Parameters of Axis Reset Command

参数 Parameter	数据类型 Data type	声明 Statement	说明 Remarks
Axis	TO_Axis	Input	工艺对象 Process object
Execute	Bool	Input	上升沿时启动作业 Start the operation at the rising edge
Busy	Bool	Ouput	=1 作业正在处理中 =1 Operation is being processed
Error	Bool	Ouput	=1 出错 =1 Error
ErrorID	Word	Ouput	出错代号 Error code

项目5 数字化生产线的构架及技术特点 319

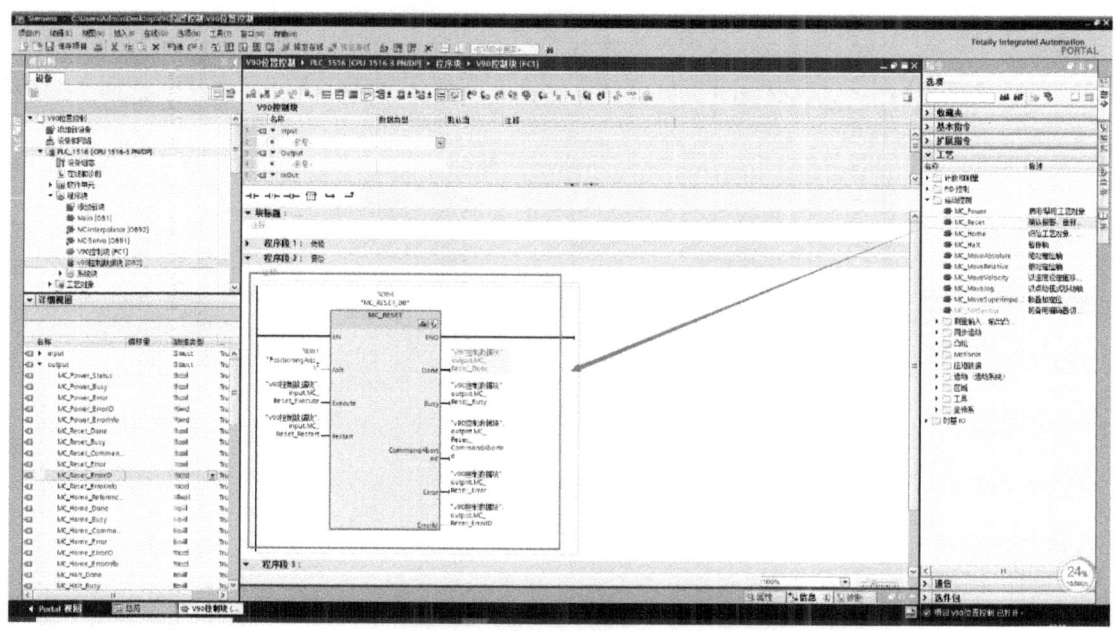

图 5-63 添加轴复位指令
Figure 5-63 Add the Command of Axis Reset

表 5-3 轴回原指令的参数
Table 5-3 Parameters of Axis Homing Command

参数 Parameter	数据类型 Data type	声明 Statement	说明 Remarks
Axis	TO_Axis	Input	工艺对象 Process object
Execute	Bool	Input	上升沿时启动作业 Start the operation at the rising edge
Mode	Int	Input	操作模式 Operating mode
Done	Bool	Ouput	作业已完成 Operating has been done
Busy	Bool	Ouput	=1 作业正在处理中 =1 Operation is being processed
Error	Bool	Ouput	=1 出错 =1 Error
ErrorID	Word	Ouput	出错代号 Error code

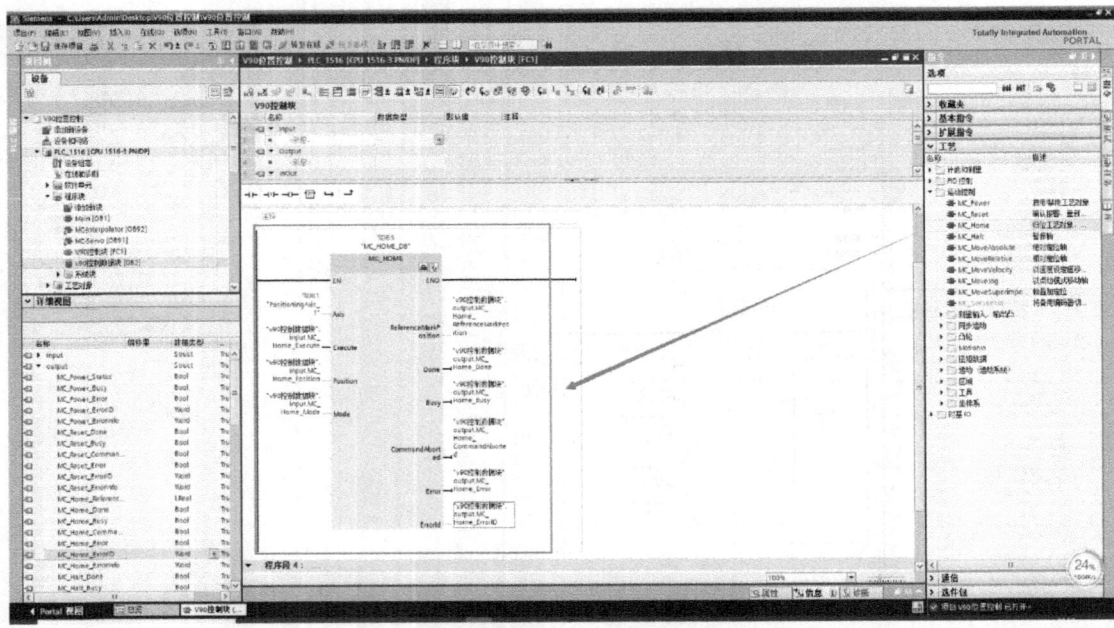

图 5-64 添加轴回原指令
Figure 5-64 Add the Command of Axis Homing

18) 添加轴暂停指令 (见表 5-4 和图 5-65)。

19) 添加轴绝对定位指令 (见表 5-5 和图 5-66)。

18) Add the command of axis pause (as shown in Table 5-4 and Figure 5-65).

19) Add the command of axis absolute positioning (as shown in Table 5-5 and Figure 5-66).

表 5-4 轴暂停指令的参数
Table 5-4 Parameters of Axis Pause Command

参数 Parameter	数据类型 Data type	声明 Statement	说明 Remarks
Axis	TO_Axis	Input	工艺对象 Process object
Execute	Bool	Input	上升沿时启动作业 Start the operation at the rising edge
Done	Bool	Ouput	作业已完成 Operation has been done
Busy	Bool	Ouput	=1 作业正在处理中 =1 Operation is being processed
Error	Bool	Ouput	=1 出错 =1 Error
ErrorID	Word	Ouput	出错代号 Error code

项目 5 数字化生产线的构架及技术特点

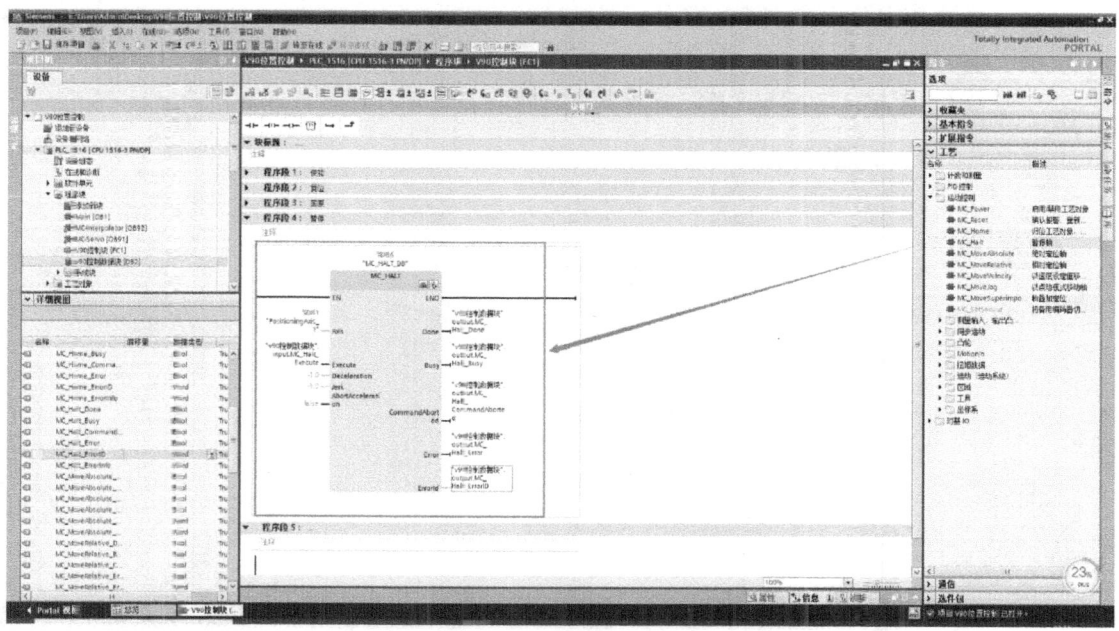

图 5-65　添加轴暂停指令
Figure 5-65　Add the Command of Axis Pause

表 5-5　轴绝对定位指令的参数
Table 5-5　Parameters of Axis Absolute Positioning Command

参数 Parameter	数据类型 Data type	声明 Statement	说明 Remarks
Axis	TO_Axis	Input	工艺对象 Process object
Execute	Bool	Input	上升沿时启动作业 Start the operation at the rising edge
Position	LReal	Input	绝对目标位置 Absolute target position
Velocity	LReal	Input	定位的速度设定值 Set value of the positioning speed
Direction	Int	Input	轴的运动方向 Direction of the axis motion
Done	Bool	Ouput	作业已完成 Operation has been done
Busy	Bool	Ouput	=1 作业正在处理中 =1 Operation is being processed
Error	Bool	Ouput	=1 出错 =1 Error
ErrorID	Word	Ouput	出错代号 Error code

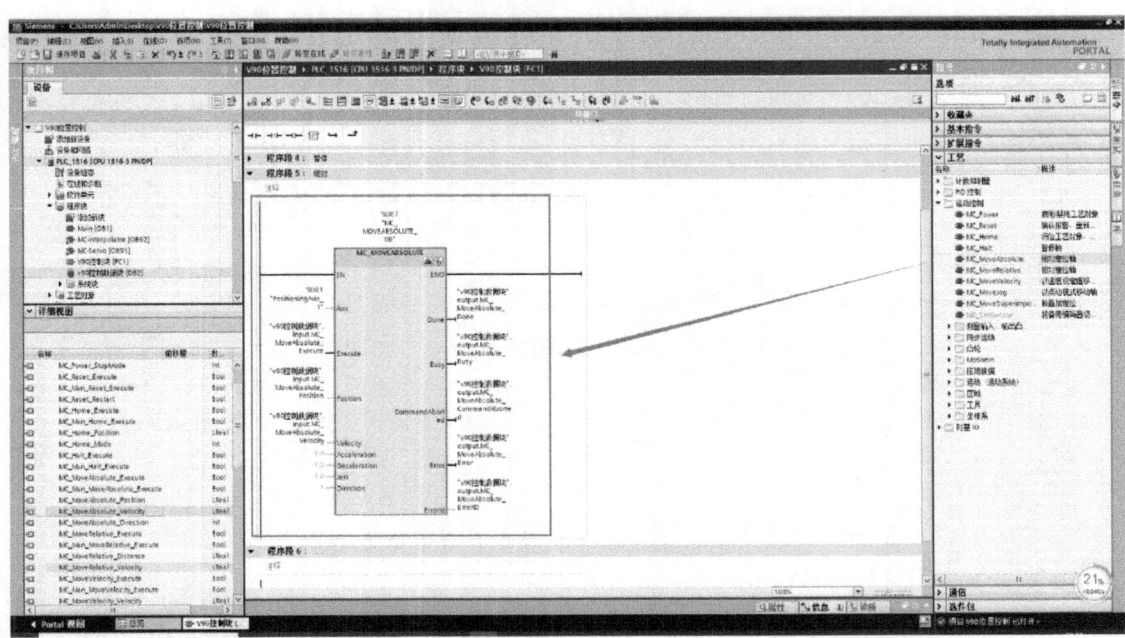

图 5-66 添加轴绝对定位指令
Figure 5-66　Add the Command of Axis Absolute Positioning

20) 添加轴相对定位指令（见表 5-6 和图 5-67）。

20) Add the command of axis relative positioning (as shown in Table 5-6 and Figure 5-67).

表 5-6　轴相对指令的参数
Table 5-6　Parameters of Axis Relative Positioning Command

参数 Parameter	数据类型 Data type	声明 Statement	说明 Remarks
Axis	TO_Axis	Input	工艺对象 Process object
Execute	Bool	Input	上升沿时启动作业 Start the operation at the rising edge
Velocity	LReal	Input	速度设定值 Set value of speed
Direction	Int	Input	定位距离 Positioning distance
Done	Bool	Ouput	作业已完成 Operation has been done
Busy	Bool	Ouput	=1 作业正在处理中 =1 Operation is being processed
Error	Bool	Ouput	=1 出错 =1 Error
ErrorID	Word	Ouput	出错代号 Error code

项目 5 数字化生产线的构架及技术特点

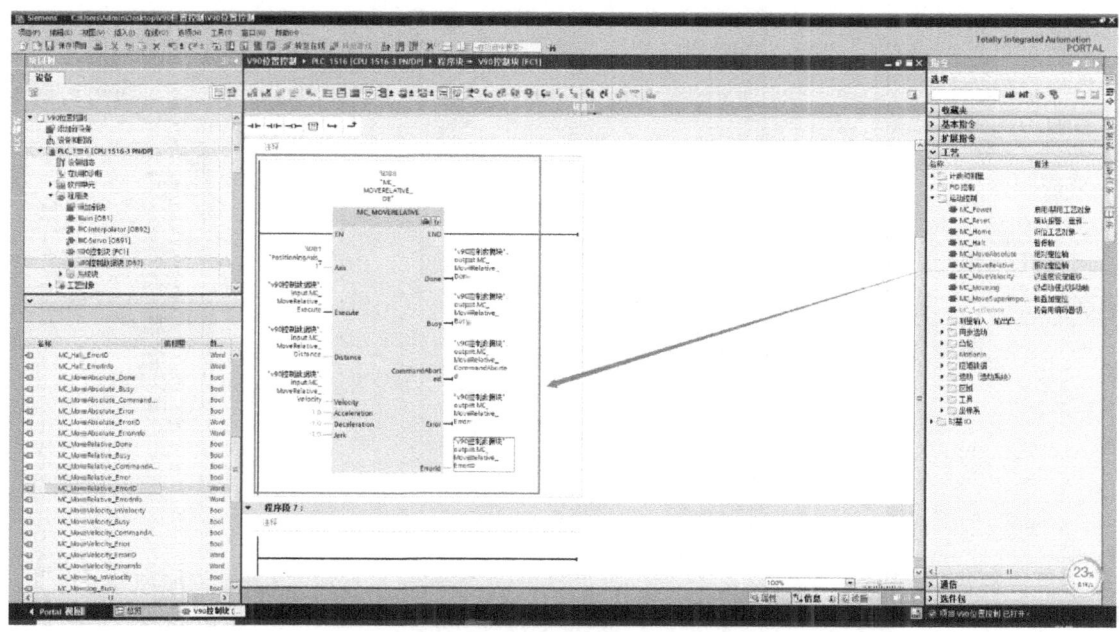

图 5-67 添加轴相对定位指令
Figure 5-67 Add the Command of Axis Relative Positioning

21) 添加轴点动指令 (见表 5-7 和图 5-68)。

21) Add the command of axis jogging (as shown in Table 5-7 and Figure 5-68).

表 5-7 轴点动指令的参数
Table 5-7 Parameters of Axis Jogging Command

参数 Parameter	数据类型 Data type	声明 Statement	说明 Remarks
Axis	TO_Axis	Input	工艺对象 Process object
JogForward	Bool	Input	正向移动 Forward motion
JogBackward	LReal	Input	反向移动 Backward motion
Velocity	Int	Input	速度设定值 Set value of speed
Done	Bool	Ouput	作业已完成 Operation has been done
Busy	Bool	Ouput	=1 作业正在处理中 =1 Operation is being processed
Error	Bool	Ouput	=1 出错 =1 Error
ErrorID	Word	Ouput	出错代号 Error code

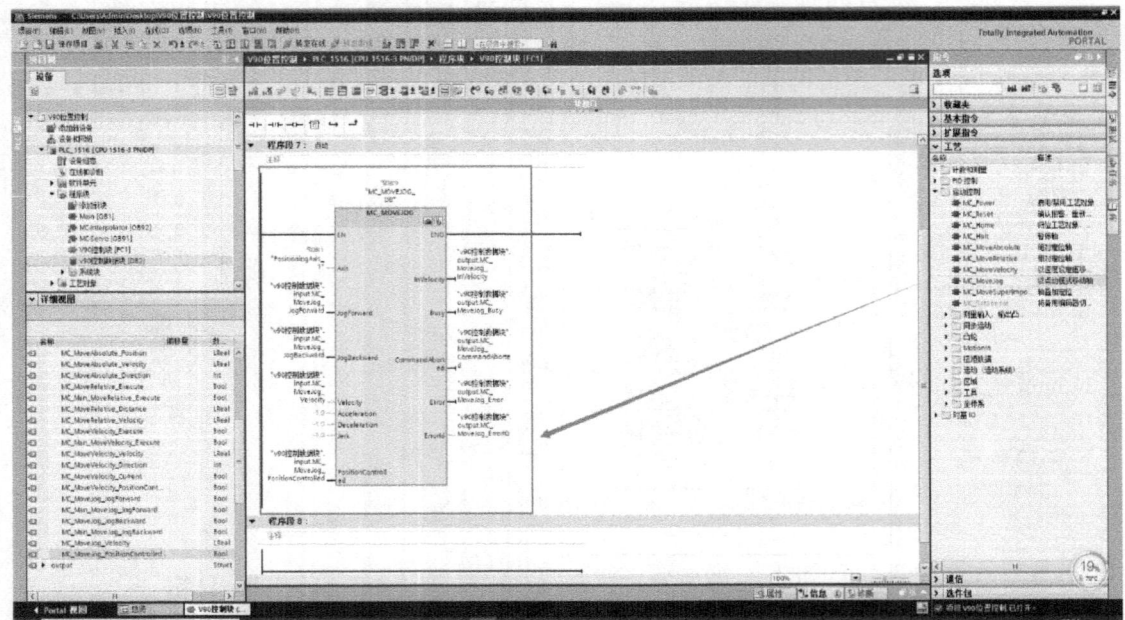

图 5-68 添加轴点动指令
Figure 5-68　Add the Command of Axis Jogging

【课后巩固】

[Consolidation after Class]

PROFINET 与 PROFIBUS 通信各有什么特点？

What are the characteristics of the PROFINET and PROFIBUS communication respectively?

任务 5.5　典型数字化生产线系统的运行和维护
Task 5.5　Operation and Maintenance of a Typical Digital Production Line System

【知识目标】

[Knowledge Objectives]

1. 了解 MES 概论。
2. 掌握典型数字化生产线系统的运行方法。
3. 掌握典型数字化生产线系统的维护方法。

1. To understand the general introduction to the MES.
2. To master the operation methods for typical digital production line systems.
3. To master the maintenance methods for typical digital production line systems.

【技能目标】

[Skill Objectives]

1. 能够对典型数字化生产线系统进行运行操作。
2. 能够对典型数字化生产线系统

1. To be able to operate typical digital production line systems.
2. To be able to maintain typical digital production

的问题进行维护。

【素质目标】

1. 具有分析与决策能力。
2. 具有发现问题、解决问题的能力。
3. 具有组织管理能力。

【任务情景】

MES 实现典型数字化生产线工艺流程教学任务。

【任务分析】

通过对典型数字化生产线系统进行操作与维护，完成学习任务。

【知识准备】

5.5.1 MES 概述

MES(Manufacturing Executions System) 又称为制造执行系统。它主要面向车间层级进行信息管理，MES 在下层工业控制执行和上层计划管理两者中间，如图 5-69 所示。与 ERP 系统进行对比，MES 具有独特的特点，且具有相互融合的功能。MES 是企业内部的执行层次，ERP 系统是企业内部的计划层次，ERP 系统与 MES 具有顺应性，不同点是作为企业的执行层次，MES 主要是优化管理生产过程，并适当加入基层工业控制系统，通过这种举措，构建企业生产信息化的集成框架。

在 MES 出现之前，车间生产管理依赖若干独立的单一功能软件，如车间作业计划系统、工序调度、工时管理、设备管理、库存控制、质量管理、数据采集等软件来完成。这些

line systems.

[Competence Objectives]

1. To have the ability to conduct analysis and decision-making.
2. To have the ability to discover and solve problems.
3. To have organizational and management abilities.

[Task Scenario]

Complete the task of teaching the technological process of typical digital production lines with the MES.

[Task Analysis]

Complete the learning task by operating and maintaining typical digital production line systems.

[Assumed Knowledge]

5.5.1 Overview of the MES

The MES (Manufacturing Executions System) is called Manufacturing Execution System in full. As it is focused on the information management at the layer of plant production, the MES is located between the lower industrial control execution and the upper planning management system, as shown in Figure 5-69. Compared with the ERP system, the MES has unique characteristics, which can be integrated themselves. For an enterprise, the MES is at the execution layer, while the ERP system is at the planning layer. The ERP system and MES are mutually adaptable, with the difference being that the MES which is at the execution layer of an enterprise mainly optimizes the management of production processes and appropriately incorporates grassroots industrial control systems, thus building an integrated framework diagram of information-based production for an enterprise.

Before the emergence of MES, the production activity management relied on several independent single functional software, such as the software for production activity control system, procedure scheduling, working hour management, equipment management, inventory

软件之间缺乏有效的集成与数据共享，难以达到车间生产过程的总体优化。

为了提高车间生产过程管理的自动化与智能化水平，必须对车间生产过程进行集成化管理，实现信息集成与共享，从而达到车间生产过程整体全局优化的目标。MES 是面向车间的生产过程管理与实时信息系统，它主要解决车间生产任务的执行问题，MES 的主要功能及概述见表 5-8。

control, quality management, and data collection. It was difficult to achieve overall optimization of the whole process of production, as the aforesaid software was lack of effective integration and data sharing.

In order to improve the automation and intelligence level of the management of production process, it is necessary to conduct integrated management of production process to have information integration and sharing, thus realizing the goal of overall optimization of the whole production process. The MES is focused on the management of production process and real-time information, and mainly used for solving the execution problems of production tasks. The main functions and their overview of the MES are shown in Table 5-8.

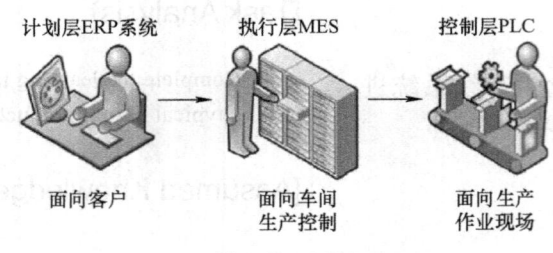

图 5-69　MES

Figure 5-69　The MES

表 5-8　MES 的主要功能及概述

Table 5-8　Main Functions and Their Overviews of the MES

序号 S/N	功能项目 Function	功能概述 Overview of function
1	资源分配与状态 Resource allocation and status	管理生产资源分配信息及状态 Manage the information of production resource allocation and status
2	操作/详细调度 Operating/detailed scheduling	生成操作计划，提供作业排序功能 Generate operating plans and provide the function of activity sorting
3	分派生产单位 Allocate production units	管理和控制生产单位的流程 Manage and control the flow of the production units
4	文档管理 Document management	管理、控制与生产单位相关的记录 Manage and control records related to production units
5	数据采集/获取 Data collection/acquisition	采集生产现场中各种必要的数据 Collect various necessary data from the production site
6	人力管理 Human Resources Management	提供最新的员工状态信息 Provide the information of latest employee status
7	质量管理 Quality assurance	记录、跟踪和分析产品及过程特性 Record, track, and analyze product and process characteristics

（续）

序号 S/N	功能项目 Function	功能概述 Overview of function
8	过程管理 Process management	监视生产，纠偏或提供决策支持 Monitor production, correct deviations or provide support for decision-making
9	维护管理 Maintenance management	跟踪和指导设备及工具的维护活动 Track and guide maintenance activities of equipment and tools
10	产品跟踪和谱系 Product tracking and spectrum	提供工件在任意时刻的位置及其状态信息 Provide the information of position and status of the workpiece at any time
11	性能分析 Performance analysis	提供最新的实际制造过程及对比结果报告 Provide the latest actual manufacturing process and comparison results report
12	物料管理 Materials management	管理物料的运动、缓冲与储存 Manage the movement, buffering, and storage of materials

5.5.2 系统生产计划实施

1）打开浏览器输入 MES 地址，直接打开界面，如图 5-70 所示。

5.5.2 Production Plan Implementation of the System

1) Open the browser and enter the MES address to open directly the interface, as shown in Figure 5-70.

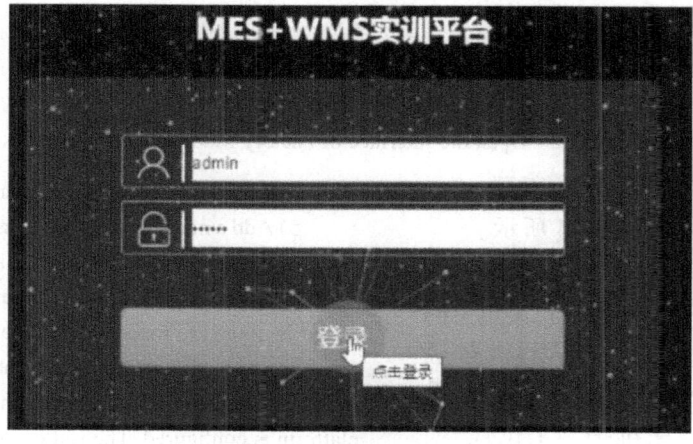

图 5-70　MES 界面
Figure 5-70　MES interface

2）账号登录后，单击系统管理，选择用户管理，打开界面如图 5-71 所示。

3）选择工厂建模及布局—车间，打开界面如图 5-72 所示。

2) Log in with the account, click the System Management and select the User Management to open the interface, as shown in Figure 5-71.

3) Select the Factory Modeling and Layout–Plant, and open the interface as follows, as shown in Figure 5-72.

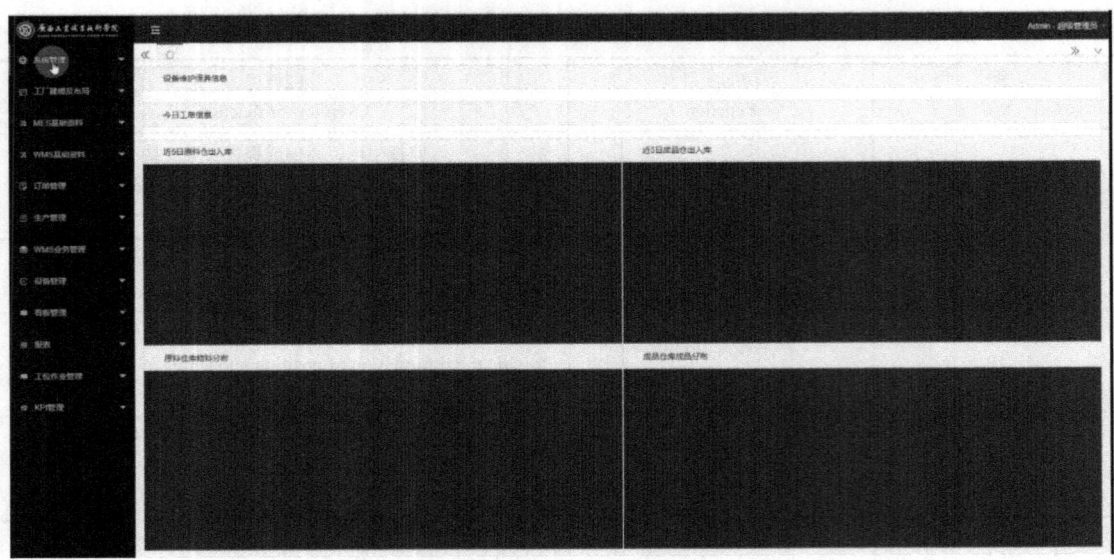

图 5-71　选择用户管理
Figure 5-71　Select the User Management

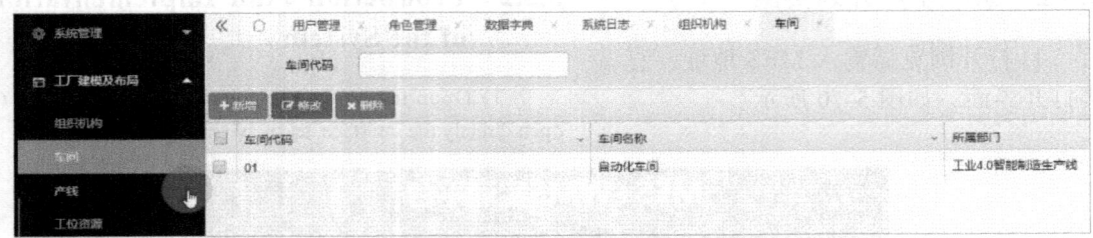

图 5-72　打开工厂建模及布局界面
Figure 5-72　Open the Interface of Factory Modeling and Layout

4) 新增车间，如图 5-73 所示。
5) 新增产线，如图 5-74 所示。
6) 新建工位资源，如图 5-75 所示。
　　每一个工作单元为一个工位（例如：斜轨机床、加工中心等），对应自动化配置代号暂且为空，等数据采集平台配置好后再来维护。工位代号、工位名称、所属产线为必填字段。

5.5.3　系统维护

1. 软件系统维护

1) 修改或删除车间，选中相应的车间单击修改或删除按钮即可完成对车间信息进行修改或删除，如图 5-76 所示。

4) Add a new plant, as shown in Figure 5-73.
5) Add a new production line, as shown in Figure 5-74.
6) Create new station resources, as shown in Figure 5-75.
Each working unit is considered as a station (such as the inclined rail machine tool, machining center) whose corresponding automation configuration code is temporarily blank and can be maintained when the data collection platform is configured. The station code, station name, and the production line to which the station belongs are mandatory fields that have to be filled in.

5.5.3　System Maintenance

1. Software system maintenance

1) To modify or delete a plant, select the corresponding plant and click the Modify or Delete button to complete the modification or deletion of the plant information, as shown in Figure 5-76.

项目 5　数字化生产线的构架及技术特点

图 5-73　新增车间
Figure 5-73　Add a New Plant

图 5-74　新增产线
Figure 5-74　Add a New Production Line

图 5-75 新建工位资源
Figure 5-75 Create New Station Resources

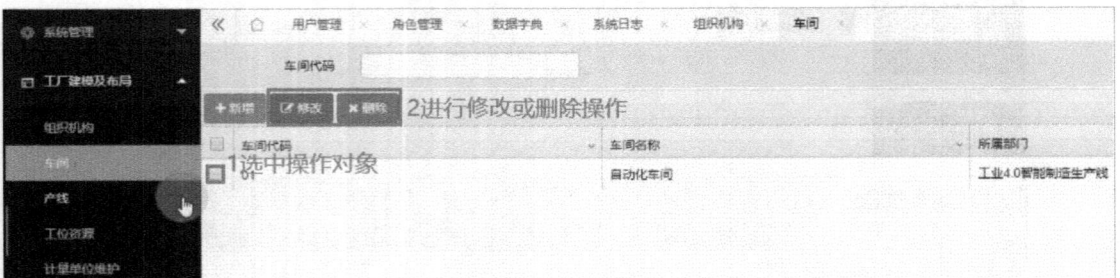

图 5-76 修改或删除车间
Figure 5-76 Modify or Delete a Plant

2) 修改或删除产线,选中相应的产线单击修改或删除按钮即可完成对产线信息进行修改或删除,如图 5-77 所示。

3) 修改或删除工位资源,选中相应的工位资源单击修改或删除按钮即可完成对工位资源信息进行修改或删除,如图 5-78 所示。

2) To modify or delete a production line, select the corresponding production line and click the Modify or Delete button to complete the modification or deletion of the production line information, as shown in Figure 5-77.

3) To modify or delete station resource, select the corresponding station resource and click the Modify or Delete button to complete the modification or deletion of the station resource information, as shown in Figure 5-78.

项目 5　数字化生产线的构架及技术特点 　331

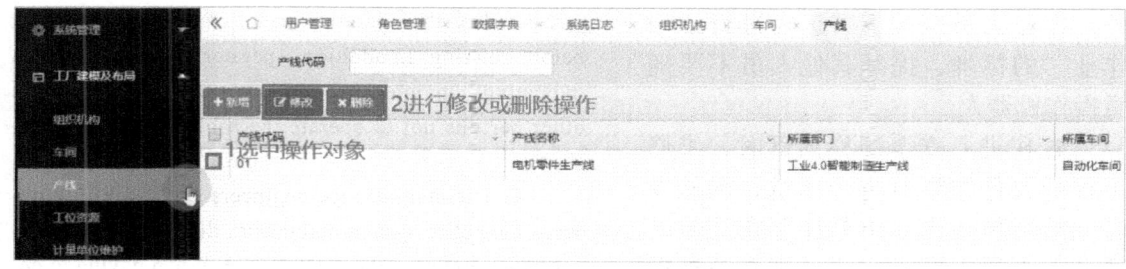

图 5-77　修改或删除产线
Figure 5-77　Modify or Delete a Production Line

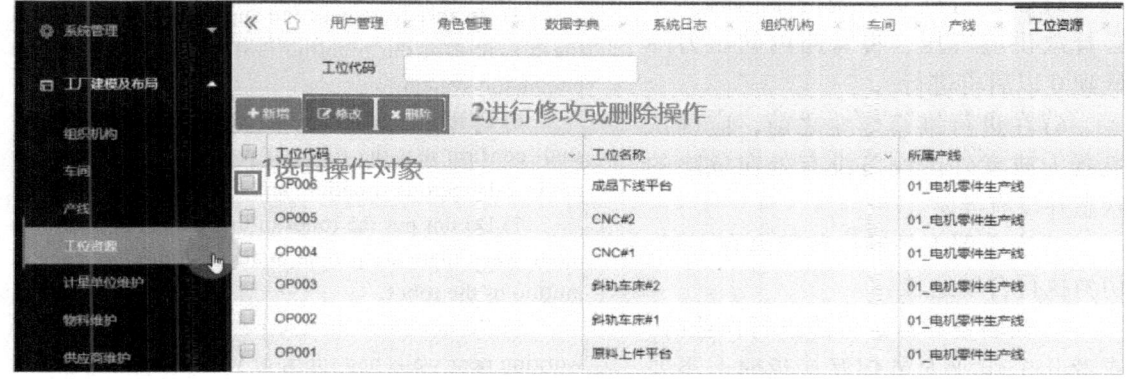

图 5-78　修改或删除工位资源
Figure 5-78　Modify or Delete Station Resource

2. 硬件维护与故障维修

以 FANUC 机器人维护和维修为例，为了确保维修工程师的安全，应充分注意下列事项。

1) 在机器人运转过程中切勿进入机器人的动作范围内。

2) 应尽可能在断开机器人和系统电源的状态下进行维护和维修。当接通电源时，有的作业有触电的危险。此外，应根据需要上好锁，以使其他人员不能接通电源。即使是在维修工程师由于迫不得已而需要接通电源后再进行作业的情形下，也应尽量按下急停按钮后再进行作业。

3) 作业人员在通电中因迫不得已的情况而需要进入机器人的动作范围内时，应在按下操作箱/操作面板或者示教操作盘的急停按钮后再入内。此

2. Hardware maintenance and fault repair

The maintenance and repair of FANUC robots is taken as an example to demonstrate the precautions that shall be fully considered as follows in order to ensure the safety of the maintenance engineers.

1) Do not enter the area within the range of motion of the robot during its operation.

2) The work of maintenance and repair shall be performed with the robot and system disconnected to the power supply whenever as possible, as there is a risk of electric shock for maintenance engineers who perform certain work with the power on. In addition, corresponding device or facilities shall be locked as required to prevent other personnel from hooking up the power. Even in situations where the maintenance engineers have to work with the power on, it is advisable to press the emergency stop button whenever possible before proceeding with the work.

3) When the power is on and the operators have to enter the area within the action range of the robot, they shall press the emergency stop button on the operating

外，作业人员应挂上"正在进行维修作业"的标牌，提醒其他人员不要随意操作机器人。

4) 在进入安全栅栏内部时，维修工程师要仔细察看整个系统，确认没有危险后再入内。如果在存在危险的情形下不得不进入栅栏，则必须把握系统的状态，同时要十分小心谨慎地入内。

5) 在进行气动系统的维修时，务必释放供应气压，将管路内的压力降低到 0 以后再进行。

6) 在进行维修作业之前，应确认机器人或者外围设备没有处在危险的状态并没有异常。

7) 当机器人的动作范围内有人时，切勿执行自动运转。

8) 在墙壁和器具旁边进行作业时，或者几个作业人员相互接近时，不要堵住其他作业人员的逃生通道。

9) 当机器人上备有工具时，以及除了机器人外还有传送带等可动器具时，应充分注意这些装置的运动。

10) 作业时应在操作箱/操作面板的旁边配置一名熟悉机器人系统且能够察觉危险的人员，使其处在任何候都可以按下急停按钮的状态。

11) 需要更换部件时，请向生产厂商洽询。在客户独自的判断下进行作业，有可能导致意想不到的事故，致使机器人损坏，或作业人员受伤。

12) 在检修控制装置内部时，如要触摸到单元、印制电路板等，为了预防触电，务必先断开控制装置的主断路器的电源，而后再进行作业。2 台机柜的情况下，请断开其各自的断路器的电源。

13) 在更换部件或重新组装时，避免异物的黏附或者异物的混入。

box/panel or the operating panel of the teach pendant before entering. In addition, operators shall hang a sign indicating "maintenance work in progress" to remind other persons not to operate the robot without permission.

4) The maintenance engineers shall carefully observe the entire system and ensure that there is no danger before entering the safety fence. When having to enter the fence in a dangerous situation, they must ensure that the state of the system is under control and be very careful when entering.

5) Be sure to release the supply air pressure until the pressure in the pipeline is 0 before repairing the pneumatic system.

6) Before conducting the maintenance work, they shall confirm that the robot or peripheral equipment is not in a dangerous situation, and not abnormal.

7) Do not put the robot in the automatic operation mode when there are people in areas within the range of motion of the robot.

8) Do not block the escape routes of others when working near walls and tools, or when several operators are approaching each other.

9) When there are tools on the robot, or movable devices such as conveyor belts in addition to the robot, full attention shall be paid to the movement of these devices.

10) During the performance of maintenance work, a person familiar with the robot system and able to detect danger shall be assigned next to the operation box/panel and in a state that he/she can press the emergency stop button at any time.

11) Please consult manufacturer when component component replacement is needed, as the performance of corresponding work under the sole judgment of the customer may lead to unexpected accidents, resulting in damage to the robot or injury to the operator.

12) When repairing the interior of the control device, and touching of the unit or printed circuit board, etc., is required, first disconnect the power supply of the main circuit breaker of the control device in order to prevent electric shock, then proceed with the operation. In the case of 2 cabinets, please disconnect both of the power supplies of their respective circuit breakers.

13) When replacing or reassembling components, avoid the adhesion or mixing of foreign objects.

14) 更换部件务必使用生产厂商指定的部件。若使用指定部件以外的部件，则有可能导致机器人的错误操作和破损。特别是熔丝如果使用额定值不同者，不仅会导致控制装置内部的部件损坏，而且还可能引发火灾，因此，切勿使用此类熔丝。

15) 维修作业结束后重新启动机器人系统时，应事先充分确认机器人动作范围内是否有人，机器人和外围设备是否有异常。

16) 在拆卸电动机和制动器时，应采取以吊车等来吊运措施后再拆除，以避免手臂等落下来。

17) 注意不要因为洒落在地面的润滑脂而滑倒。应尽快擦掉洒落在地面上的润滑脂，排除可能发生的危险。

3. 机器人故障

1) 三合一夹具部分故障及排除方法见表5-9。

14) For component replacement, always use the components specified by manufacturer when replacing the old ones. If components other than the specified ones are used, it may lead to incorrect operation by and damage to the robot. In particular, do not use fuses with different ratings other than the rated value, as the use of such fuses will not only cause damage to the internal components of the control device, but also a fire.

15) When restarting the robot system after the maintenance operation is completed, it shall be confirmed in advance that there are no people at areas within the action range of the robot, and that no abnormalities exist in the robot and the peripheral equipment.

16) When disassembling the motor and brake, lifting and handling devices such as cranes shall be taken and available before dismantling to prevent the arms, etc. from falling off.

17) Be careful of the slippery floor on which the lubricating grease has spilled. The lubricating grease spilled on the floor shall be wiped off as soon as possible to eliminate potential hazards.

3. The Robot Faults

1) For part of the faults of the three-in-one fixture, and the failed closing of the gripper, are shown in Table 5-9.

表5-9 三合一夹具部分故障及排除方法
Table 5-9 Part of the Faults and Troubleshooting Methods of the Three-in-one Fixture

序号 S/N	可能引起的原因 Possible cause	排除方法 Troubleshooting method
1	气阀没有打开 The air valve is not opened	打开气阀 Open the air valve
2	油水过滤器调压阀压力调节过小，或者油水过滤器损坏 The pressure of the pressure control valve of the oil-water filter is too low, or the filter is damaged	调高压力或者更换油水过滤器 Increase the pressure or replace the oil-water filter
3	抓手气缸进出气管漏气或者打折 Air leakage or folding happens to the inlet or outlet air tubes of the gripper cylinder	检查气管并更换或者理顺气管打折处 Check the air tubes, replace them or straighten their folding parts
4	抓手气缸损坏 The gripper cylinder is damaged	更换损坏的抓手气缸 Replace the damaged gripper cylinder
5	导轨处有东西阻挡滑块运动 The slider is blocked at the guide rail	清理导轨并上油 Clean and oil the guide rail
6	抓手处滑块损坏 The slider at the gripper is damaged	更换滑块 Replace the slider

2) 输送线及自动控制系统部分故障及排除方法见表 5-10。

2) Part of the faults of the conveying line and automatic control system are shown in Table 5-10.

表 5-10 输送线及自动控制系统部分故障及排除方法
Table 5-10 Part of the Faults and Troubleshooting Methods of the Conveying Line and Automatic Control System

序号 S/N	故障 Fault	引起的原因 Cause	排除方法 Troubleshooting method
1	上料台无法上升 The loading table cannot rise	(1) 上料台的光电检测开关没有检测到托盘 (1) The photoelectric detection switch of the loading table does not detect the tray (2) 升降机到底部卡死 (2) The elevator is stuck at the bottom	(1) 调节光电检测开关位置或者推动托盘 (1) Adjust the position of the photoelectric detection switch or push the tray (2) 先将此升降机打到手动操作模式，再将此电动机的变频器速度调低，再按提升电动机按钮 (2) First put the elevator into the manual operating mode, then lower the speed of the frequency converter of this motor, and press the button of the lifting motor
2	输送机运行中出现托盘卡住的情况 The tray is stuck when the conveyor is running	检查托盘和滚筒两侧挡圈的距离，挡圈是否跑偏导致 Check the distance between the retaining rings on both sides of the tray and drum, and check if the tray is stuck due the the deviation of the retaining rings	调整滚筒挡圈位置 Adjust the position of the retaining ring of the drum
3	机器人动作与预设不一致 The robot does not act as preset	程序号或者程序行未按要求调整 The program number or program line has not been adjusted as required	按要求调整程序号或者程序行 Adjust the program number or program line as required
4	顶托盘气缸无法收回 The top tray cylinder cannot be retracted	气缸上的磁感应开关故障 The fault of the magnetic switch on the cylinder	检查气缸上的磁感应开关检测是否正常 Check if the magnetic switch on the cylinder has normal detection
5	控制柜电源指示灯不亮 The power indicator of the control cabinet is not on	电源总开关是否处于开启状态 Check if the switch of the main power is on	开启开关 Turn on the switch
6	输送线无法启动 The conveying line cannot be started	(1) 调速变频器是否还在归零的状态 (1) If the speed control frequency converter is still in the return-to-zero state (2) 电箱及生产线两旁的急停按钮是否被按下 (2) If the emergency stop buttons on both sides of the electric box and production line are pressed	(1) 恢复变频器归零状态 (1) Restore the return-to-zero state of the frequency converter (2) 松开急停按钮 (2) Release the emergency stop buttons

【课后巩固】

1. 铝壳电机关键构件数字化生产线系统维护的基本原则是什么？

2. 铝壳电机关键构件数字化生产线系统运行的基本步骤是什么？

[Consolidation after Class]

1. What are the basic principles for the maintenance of the digital production line for a key component of a motor with an aluminum shell?

2. What are the basic steps for the operation of the digital production line for a key component of a motor with an aluminum shell?

参考文献

References

[1] 陈友动，谭珠珠，等．工业机器人集成与应用[M]．北京：机械工业出版社，2021．

[2] 谭立新，张宏立．工业机器人系统集成[M]．北京：北京理工大学出版社，2021．

[3] 林燕文，魏志丽．工业机器人系统集成与应用[M]．北京：机械工业出版社，2018．

[4] 周书兴．工业机器人工作站系统与应用[M]．北京：机械工业出版社，2020．

[5] 祝春来，宋春胜，等．工业机器人集成应用技术[M]．哈尔滨：哈尔滨工程大学出版社，2021．

[1] Chen Youdong, Tan Zhuzhu, et al. Integration and Application of Industrial Robots [M]. Beijing: China Machine Press, 2021.

[2] Tan Lixin, Zhang Hongli. System Integration of Industrial Robots [M].Beijing: Beijing Institute of Technology Press, 2021.

[3] Lin Yanwen, Wei Zhili. System Integration and Application of Industrial Robots [M]. Beijing: China Machine Press, 2018.

[4] Zhou Shuxing. Workstation System and Application of Industrial Robots[M]. Beijing: China Machine Press, 2020.

[5] Zhu Chunlai, Song Chunsheng, et al. Integrated Application Technology of Industrial Robots [M]. Harbin: Harbin Engineering University Press, 2021.

本书配套数字资源的获取与使用

本教材配套数字资源已上线超星学习通数字教材,师生可通过学习通获取本书配套的 PPT 课件、微课视频、在线测验等。

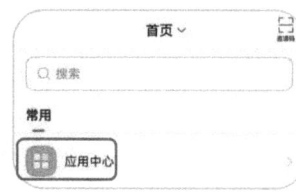

 教师端

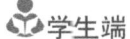 学生端